SOLUTIONS MANUAL

To accompany

TRANSPORT PHENOMENA IN

MATERIALS PROCESSING*

E.J. Poirier and D.R. Poirier

*D.R. Poirier and G.H. Geiger, *Transport Phenomena in Materials Processing*, The Minerals, Metals, and Materials Society, Warrendale, Pennsylvania, 1994

CONTENTS

1.1 Compute the steady-state momentum flux (N m^{-2}) in a lubricating oil, viscosity equal to 2×10^{-2} N s m^{-2}, that is contained between a stationary plate and one that is moving with a velocity of 61.0 cm s^{-1}. The distance between the plates is 2 mm. Next, show the direction of the momentum flux and the shear stress with respect to the x-y axis system in the diagram below.

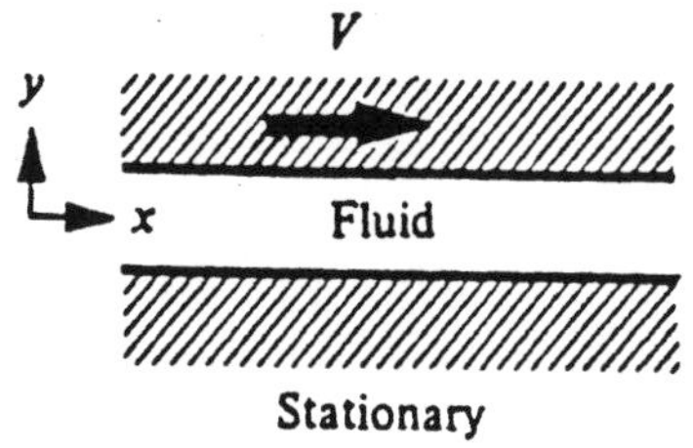

$$\text{Momentum Flux} = \frac{F}{A} = -\eta \frac{\bar{v}}{y}$$

$$\frac{F}{A} = \frac{-2\times10^{-2}\,N\,s}{m^2}\left|\frac{61.0\,cm}{s}\right|\frac{}{2\,mm}\left|\frac{10\,mm}{cm}\right. = -6.1\,N\,s^{-1}$$

The momentum flux is downward, given by the minus sign; from high momentum to low momentum.

The shear stress is numerically equal to the momentum flux, but it is visualized as parallel to the plates' surfaces.

$$\tau_{yx} = -6.1\ N\ s^{-1}$$

1.2 Near the surface of a flat plate, water has a velocity profile given by

$$v_x = 3y - y^3$$

with y in mm, v_x in cm s^{-1}, and $0 \le y \le 1$ mm. The density and the kinematic viscosity of water are 10^3 kg m^{-3} and 7×10^{-7} m^2 s^{-1}, respectively.

a) What is the shear stress at x_1 on the plate?

b) What is the momentum flux at $y = 0.8$ mm and $x = x_1$ in the y-direction?

c) Is there momentum flux in the x-direction at $y = 0.8$ mm and $x = x_1$ in the x-direction? If so, evaluate.

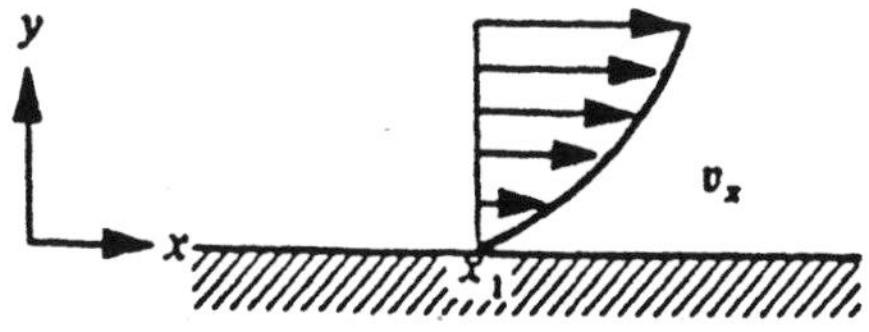

a. $\eta = \nu\rho = \frac{7\times10^{-7}\,m^2}{s}\Big|\frac{10^3\,kg}{m^3} = 7\times10^{-4}\,N\,s\,m^{-2}$

$\tau_{yx} = -\eta\frac{dv_x}{dy}, \quad \frac{dv_x}{dy} = 3 - 3y^2;\ cm\,s^{-1}\,mm^{-1}$

$@\ y = 0, \quad \frac{dv_x}{dy} = \frac{3\,cm}{s\,mm}\Big|\frac{10\,mm}{cm} = 30\,s^{-1}$

$\tau_{yx} = -\frac{7\times10^{-4}\,N\,s}{m^2}\Big|\frac{30}{s} = -2.1\times10^{-2}\,N\,m^{-2}$

b. $@\ y = 0.8\,mm, \quad \frac{dv_x}{dy} = \left[3 - 3(0.8)^2\right]\frac{cm}{s\,mm}\Big|\frac{10\,mm}{cm} = 10.8\,s^{-1}$

$\tau_{yx} = -\frac{7\times10^{-4}\,N\,s}{m^2}\Big|\frac{10.8}{s} = -7.56\times10^{-3}\,N\,m^{-2}$

c. Yes. Mom. Trans. $= \rho v_x v_x = \frac{10^3\,kg}{m^3}\left[(3)(0.8) - (0.8)^3\right]^2\frac{cm^2}{s^2}\left[\frac{m}{100\,cm}\right]^2$

$= 3.565\times10^{-1}\,\frac{kg\,m^2}{m^3\,s^2}\Big|\frac{1\,N\,s^2}{1\,kg\,m} = 3.565\times10^{-1}\,N\,m^{-2}$

1.3 In fluidized bed processes, operating temperatures can vary from process to process or within the beds itself. Therefore, it is necessary to know how these temperature variations affect the properties, such as viscosity of the fluidizing gas in the reactor. Taking air as the fluidizing gas, estimate its viscosity at 313 K and 1073 K when considered as a one component gas with the parameters given in Table 1.1. Repeat the estimates, only now consider the air to be 79% N_2, 21% O_2. Compare these results to the experimental values of 0.019 cP and 0.0438 cP at 313 K and 1073 K, respectively.

a. $\eta = 2.67\times10^{-5}\frac{(MT)^{1/2}}{\sigma^2\Omega_\eta}$ (Poise)

Table 1.1 $M = 28.97, \ \sigma = 3.711, \ \frac{\epsilon}{K_B} = 78.6$

$$\frac{313\ K_B}{\varepsilon} = 3.98, \quad \frac{1073\ K_B}{\varepsilon} = 13.65$$

$$\Omega_\eta(313\text{K}) = 0.9721, \quad \Omega_\eta(1073\text{K}) = 0.7946$$

$$\eta(313\text{ K}) = 2.67 \times 10^{-5} \frac{\left[(28.97)(313)\right]^{1/2}}{(3.711)^2(0.9721)} = 1.90 \times 10^{-4}\ \text{Poise} = 0.0190\ \text{cP}$$

Compares very well with experimental value.

$$\eta(1073\text{K}) = 2.67 \times 10^{-5} \frac{\left[(28.97)(1073)\right]^{1/2}}{(3.711)^2(0.7946)} = 4.30 \times 10^{-4}\ \text{Poise} = 0.0430\ \text{cP}$$

Compares well with experimental value.

b. $$\eta_{mix} = \sum_{i=1}^{2}\left[\frac{x_i \eta_i}{\sum_{j=1}^{2} x_j \Phi_{ij}}\right]$$

$$\Phi_{ij} = \frac{1}{(8)^{1/2}}\left[1 + \frac{M_i}{M_j}\right]^{-1/2}\left[1 + \left(\frac{\eta_i}{\eta_j}\right)\left(\frac{M_j}{M_i}\right)^{1/4}\right]$$

x_{O_2}	x_{N_2}	M_{O_2}	M_{N_2}	σ_{O_2}	σ_{N_2}	$\frac{\varepsilon}{K_B}\Big\|_{O_2}$	$\frac{\varepsilon}{K_B}\Big\|_{N_2}$
0.21	0.79	32.0	28.013	3.467	3.798	106.7	71.4

T	$\frac{T K_B}{\varepsilon}\Big\|_{O_2}$	$\frac{T K_B}{\varepsilon}\Big\|_{N_2}$	$\Omega_{\eta_{O_2}}$	$\Omega_{\eta_{N_2}}$	η_{O_2} Poise	η_{N_2} Poise
313 K	2.93	4.38	1.080	0.956	2.058×10^{-4}	1.813×10^{-4}
1073 K	10.06	15.03	0.8237	0.7835	4.997×10^{-4}	4.096×10^{-4}

$$\eta_{mix} = \frac{x_1 \eta_1}{x_1 + x_2 \Phi_{12}} + \frac{x_2 \eta_2}{x_1 \Phi_{21} + x_2}$$

$$\Phi_{12} = \frac{1}{(8)^{1/2}}\left[1 + \frac{M_1}{M_2}\right]^{-1/2}\left[1 + \left(\frac{\eta_1}{\eta_2}\right)^{1/2}\left(\frac{M_2}{M_1}\right)^{1/4}\right]^2$$

$$\Phi_{21} = \frac{1}{(8)^{1/2}}\left[1 + \frac{M_2}{M_1}\right]^{-1/2}\left[1 + \left(\frac{\eta_2}{\eta_1}\right)^{1/2}\left(\frac{M_1}{M_2}\right)^{1/4}\right]^2$$

Let $1 \Rightarrow O_2$ and $2 \Rightarrow N_2$

Typical calculation at $T = 313K$

$$\Phi_{12} = \frac{1}{(8)^{1/2}}\left[1 + \frac{32.0}{28.013}\right]^{-1/2}\left[1 + \left(\frac{2.058 \times 10^{-4}}{1.813 \times 10^{-4}}\right)^{1/2}\left(\frac{28.013}{32.0}\right)^{1/4}\right]^2 = 0.996$$

$T = 313K$, $\Phi_{12} = 0.996$, $\Phi_{21} = 1.002$

$T = 1073K$, $\Phi_{12} = 1.033$, $\Phi_{21} = 0.968$

$$\eta_{mix}(313K) = \frac{(0.21)(2.058 \times 10^{-4})}{0.21 + (0.79)(.996)} + \frac{(0.79)(1.813 \times 10^{-4})}{(0.79) + (0.21)(1.002)} = 1.866 \times 10^{-4}\ P$$

$= 0.019\ cP$ - Compares very well with experimental value.

$$\eta_{mix}(1073K) = \frac{(0.21)(4.997 \times 10^{-4})}{(0.21) + (0.79)(1.033)} + \frac{(0.79)(4.096 \times 10^{-4})}{(0.79) + (0.21)(0.968)} = 4.28 \times 10^{-4}\ P$$

$= 0.0428\ cP$ - Compares well with experimental value.

Notice that $\Phi_{12} \simeq \Phi_{21} \simeq 1$. When this holds, then we can use a simple law of mixtures.

$$\eta_{mix} \cong x_1 \eta_1 + x_2 \eta_2$$

$$\eta_{MIX}(313K) \simeq (0.21)(2.658\times10^{-4}) + (0.79)(1.813\times10^{-4}) = 1.86\times10^{-4}P = 0.0186\,cP$$

$$\eta_{MIX}(1073K) \simeq (0.21)(4.997\times10^{-4}) + (0.79)(4.096\times10^{-4}) = 4.29\times10^{-4}P = 0.0429\,cP$$

1.4 Consider that the binary gas A-B is such that at a given temperature $\eta_A = \eta_B$. Plot the ratio of η_{mix}/η_A versus x_B if the ratio of the molecular weights of the two species, M_A/M_B, is: a) 100; b) 10; c) 1.

a. $\frac{M_A}{M_B} = 100$, $x_A = 1 - x_B$, $\eta_A = \eta_B$

$$\eta_{MIX} = \frac{x_A \eta_A}{x_A + x_B \Phi_{AB}} + \frac{x_B \eta_B}{x_A \Phi_{BA} + x_B}$$

$$\frac{\eta_{mix}}{\eta_A} = \frac{(1-x_B)}{(1-x_B) + x_B \Phi_{AB}} + \frac{x_B}{x_B + (1-x_B)\Phi_{BA}}$$

$$\Phi_{AB} = \frac{1}{(8)^{1/2}}\left[1 + \frac{M_A}{M_B}\right]^{-1/2}\left[1 + \left(\frac{M_B}{M_A}\right)^{1/4}\right]^2$$

$$\Phi_{BA} = \frac{1}{(8)^{1/2}}\left[1 + \frac{M_B}{M_A}\right]^{-1/2}\left[1 + \left(\frac{M_A}{M_B}\right)^{1/4}\right]^2$$

$\frac{M_A}{M_B}$	Φ_{AB}	Φ_{BA}
100	0.06095	6.0948
10	0.26020	2.6020
1	1.00000	1.0000

Plots of $\left(\frac{\eta_{MIX}}{\eta_A}\right)$ as a function of x_B for varying $\left(\frac{M_A}{M_B}\right)$ ratios.

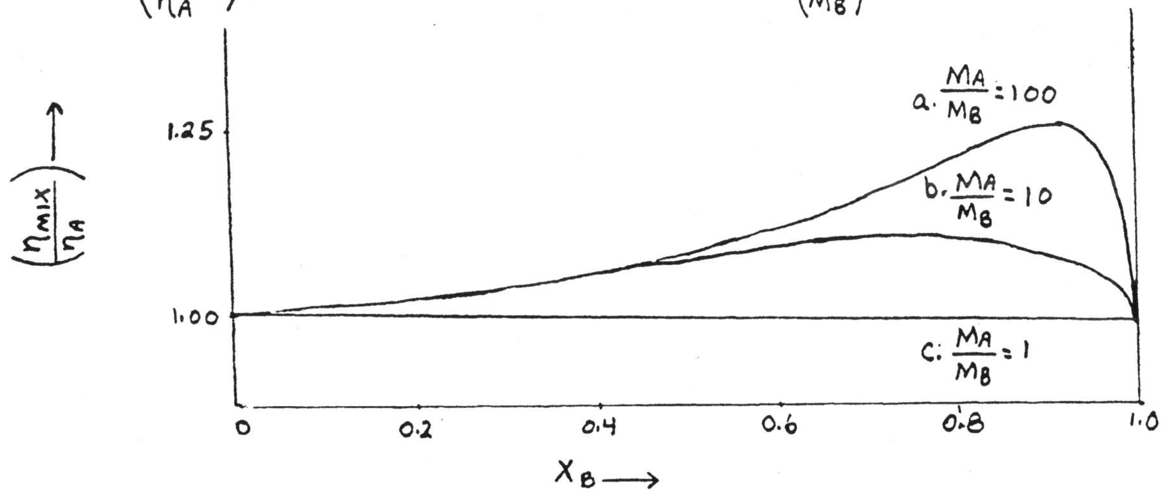

1.5 A quick estimate of the viscosity of a binary gas A-B is

$$\eta_{mix} = x_A \eta_A + x_B \eta_B.$$

What are the maximum errors in the viscosities estimated by the above equation when compared to the results calculated for a), b) and c) in problem 1.4? Briefly discuss the errors.

using $\eta_{MIX} = x_A \eta_A + x_B \eta_B$ in problem 1.4,

a. when $\frac{M_A}{M_B} = 100$, maximum error of $\frac{\eta_{MIX}}{\eta_A} = 25\%$,

b. when $\frac{M_A}{M_B} = 10$, maximum error of $\frac{\eta_{MIX}}{\eta_A} = 10\%$,

c. when $\frac{M_A}{M_B} = 1$, no error

These calculations show that the law of mixtures holds exactly when $M_A = M_B$. But even when $\frac{M_A}{M_B} = 100$, the maximum error is only 25% and when $\frac{M_A}{M_B} = 10$, the maximum error is only 10%. Notice that by adding the lighter gas (B) to the heavier gas (A), it changes the viscosity very little. The effect of adding the heavier gas (A) to the lighter gas (B), however, changes the viscosity more drastically.

1.6 At 920 K the viscosity of methane (CH_4) is 2.6×10^{-5} N s m^{-2}, and the viscosity of nitrogen is 3.8×10^{-5} N s m^{-2}. Plot the viscosity of methane-nitrogen mixtures versus mole fraction methane at 920 K.

$T = 920\text{ K} \quad \eta_{CH_4} = 2.6 \times 10^{-5}\text{ N s m}^{-2}$

$\eta_{N_2} = 3.8 \times 10^{-5}\text{ N s m}^{-2}$

Let $1 = CH_4$, $2 = N_2$

$M_{CH_4} = 16.04$, $M_{N_2} = 28.013$

$$\Phi_{12} = \frac{1}{(8)^{1/2}}\left[1 + \frac{16.04}{28.013}\right]^{-1/2}\left[1 + \left[\frac{2.6\times10^{-5}}{3.8\times10^{-5}}\right]^{1/2}\left[\frac{28.013}{16.04}\right]^{1/4}\right]^2 = 1.073$$

$$\Phi_{21} = \frac{1}{(8)^{1/2}}\left[1 + \frac{28.013}{16.04}\right]^{-1/2}\left[1 + \left[\frac{3.8\times10^{-5}}{2.6\times10^{-5}}\right]^{1/2}\left[\frac{16.04}{28.013}\right]^{1/4}\right]^2 = 0.897$$

$$\eta_{MIX} = \frac{x_1(2.6\times10^{-5})}{x_1 + (1-x_1)(1.073)} + \frac{(1-x_1)(3.8\times10^{-5})}{x_1(0.897) + (1-x_1)}$$

x_1	$\eta_{MIX} \times 10^{-5}$ N s m^{-2}
0.0	3.800
0.2	3.595
0.4	3.374
0.6	3.135
0.8	2.878
1.0	2.600

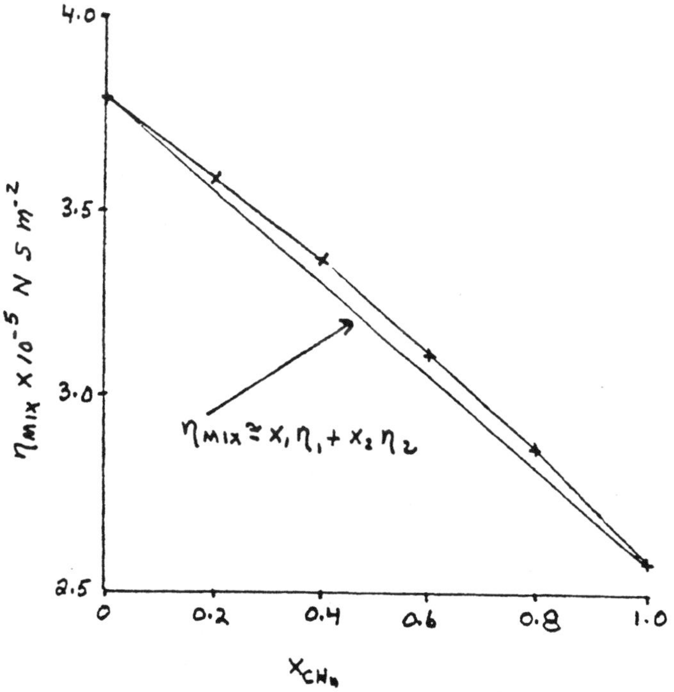

Notice how closely the simple mixture rule approximates the result.

1.7 Estimate the viscosity of liquid beryllium at 1575 K. The following data are available: atomic weight, 9.01 g mol^{-1}; melting point, 1550 K; density at 293 K, 1850 kg m^{-3}; crystal structure, hcp; atomic radius, 0.114 nm.

$$\frac{\epsilon_p T_m}{K_B} = (5.20)(1550) = 8060\ \text{K}$$

$$T^* = \frac{1575}{8060} = 0.1954, \quad \frac{1}{T^*} = 5.118$$

From Fig. 1.10 $\eta^*(V^*)^2 \approx 3.6$

$$V^* = \frac{1}{n\delta^3} = \frac{1}{\left[\frac{6.023\times10^{23}}{9.01}\right](1.85)(1.14\times10^{-8})^3} = 0.546$$

$$\eta^* = \frac{3.6}{(0.546)^2} = 12.08$$

$$\eta = \frac{\eta^*(MRT)^{1/2}}{\delta^2 N_0}$$

$$(MRT)^{1/2} = \left[\frac{9.01\ \text{g}}{\text{mol}}\Big|\frac{8.3144\ \text{J}}{\text{mol K}}\Big|1575\ \text{K}\right]^{1/2} = 343.5\ (\text{Jg})^{1/2}\text{mol}^{-1}$$

$$\eta = 12.08\Big|\frac{343.5\ \text{J}^{1/2}\text{g}^{1/2}}{\text{mol}}\Big|\frac{}{(0.114\times10^{-9})^2\text{m}^2}\Big|\frac{\text{mol}}{6.023\times10^{23}}$$

$$= \frac{0.5301\ \text{J}^{1/2}\text{g}^{1/2}}{\text{m}^2}\Big|\frac{1\ \text{kg}^{1/2}}{10^{3/2}\text{g}^{1/2}}\Big|\frac{1\ \text{kg}^{1/2}\text{m}}{1\ \text{J}^{1/2}\text{s}} = 1.68\times10^{-2}\ \text{kg m}^{-1}\ \text{s}^{-1}\ (16.8\ \text{cP})$$

1.8 Chromium melts at approximately 2148 K. Estimate its viscosity at 2273 K given the following data: atomic weight, 52 g mol^{-1}; density, 7100 kg m^{-3}; interatomic distance, 0.272 nm.

$$\frac{\epsilon T_m}{K_B} = (5.20)(2148) = 1.11696 \times 10^4 \text{ K}$$

$$T^* = \frac{2273}{1.11696 \times 10^4} = 0.2035, \quad \frac{1}{T^*} = 4.91$$

From Fig. 1.10 $\eta^* (V^*)^2 \approx 3.4$

$$V^* = \frac{1}{n\delta^3} = \frac{1}{\left[\frac{6.023 \times 10^{23}}{52}\right](7.1)(2.72 \times 10^{-8})^3} = 0.604$$

$$\eta^* = \frac{3.4}{(0.604)^2} = 9.32$$

$$\eta = \frac{\eta^* (MRT)^{1/2}}{\delta^2 N_0}$$

$$(MRT)^{1/2} = \left[\frac{52\text{ g}}{\text{mol}} \middle| \frac{8.3144 \text{ J}}{\text{mol K}} \middle| \frac{2273 \text{ K}}{}\right]^{1/2} = 991 \, (\text{J g})^{1/2} \text{ mol}^{-1/2}$$

$$\eta = 9.32 \left| \frac{991 \text{ J}^{1/2} \text{g}^{1/2}}{} \right| \frac{}{(0.272 \times 10^{-9})^2 \text{ m}^2} \left| \frac{\text{mol}}{6.023 \times 10^{23}} \right.$$

$$= \frac{0.2073 \text{ J}^{1/2} \text{g}^{1/2}}{\text{m}^2} \left| \frac{1 \text{ kg}^{1/2}}{10^{3/2} \text{g}^{1/2}} \right| \frac{1 \text{ kg}^{1/2} \text{m}}{1 \text{ J}^{1/2} \text{s}} = 6.56 \times 10^{-3} \text{ kg m}^{-1} \text{s}^{-1} \, (6.56 \text{ cP})$$

1.9 At 1273 K a melt of Cu-40% Zn has a viscosity of 5 cP and at 1223 K this same alloy has a viscosity of 6 cP. Using this information, estimate the viscosity of the alloy at 1373 K.

$$\eta = A \exp\left[\frac{\Delta G^{\ddagger}}{RT}\right]$$

$$\frac{\eta_1}{\eta_2} = \exp\left[\frac{\Delta G^{\ddagger}}{R}\left(\frac{1}{T_1} - \frac{1}{T_2}\right)\right]$$

$$\frac{\Delta G^{\ddagger}}{R} = \frac{T_1 T_2}{T_1 - T_2} \ln\left(\frac{\eta_1}{\eta_2}\right)$$

$$\frac{\Delta G^{\ddagger}}{R} = \frac{(1273)(1223)}{(1273-1223)} \ln\frac{6}{5} = 5677\ K$$

$$\eta\,(1273\ K) = 5\ cP$$

$$\eta\,(1373\ K) = 5 \exp\left[5677\left[\frac{1}{1373} - \frac{1}{1273}\right]\right] = 3.61\ cP$$

1.10 Many metal processing operations, such as steelmaking, employ a slag as a means by which impurities are removed from the processing operation. Thus, it becomes necessary to characterize certain properties of the slag, such as viscosity, at temperatures it will encounter in the processing operation. Using Bills' method, estimate the viscosity of a 50 wt.% SiO_2, 30 wt.% CaO, 20 wt.% Al_2O_3 slag at 1773 K.

Converting to mol fraction

Oxide	Wt. %	M. W.	Moles	Mole Fraction
CaO	30	56.08	0.53495	0.34218
SiO_2	50	60.08	0.83222	0.53235
Al_2O_3	20	101.96	0.19615	0.12547

Fig. 1.15 Alumina has a silica equivalence X_a, which depends on the Al_2O_3/CaO ratio and on the total Al_2O_3 content.

$$\frac{X_{Al_2O_3}}{X_{CaO}} = \frac{0.12547}{0.34218} = 0.367$$

First locate $X_{Al_2O_3}$ on the vertical axis (~0.125), then locate the corresponding curve at the ratio equal to 0.367. The silica equivalence is read from the bottom axis as $X_a = 0.150$.

$$X_{SiO_2} + X_a = 0.5325 + 0.150 = 0.682$$

From Fig. 1.16 the silica equivalence of 0.682 is found on the bottom axis and carried up to the intersection with the 1500°C (1773 K) isotherm.

The corresponding $\log_{10}\eta$ is read from the vertical axis.

$\log_{10}\eta = 1.6$, $\eta = 39.8$ Poise

1.11 Assume that the viscosity of a glass varies with temperature according to Eq. (1.18). At 1700 K it has a viscosity of 20 N s m^{-2}; at 1500 K it is 100 N s m^{-2}. What is the viscosity at 1450 K?

$$\eta = A \exp\left[\frac{\Delta G^{\ddagger}}{RT}\right]$$

$$\frac{\Delta G^{\ddagger}}{R} = \frac{T_1 T_2}{T_1 - T_2} \ln\left(\frac{\eta_1}{\eta_2}\right) = \frac{(1700)(1500)}{(1700-1500)} \ln\left(\frac{100}{200}\right) = 2.052 \times 10^4 \text{ K}$$

$$\eta(1500\text{ K}) = 100 \text{ N s m}^{-2}$$

$$\eta(1450\text{K}) = 100 \exp\left[2.052 \times 10^4 \left[\frac{1}{1450} - \frac{1}{1500}\right]\right]$$

$$\eta(1450\text{K}) = 100 \exp(0.472) = 160.3 \text{ N s m}^{-2}$$

1.12 When oxides such as MgO and CaO are added to molten silica, the activation energy of viscosity is reduced from 135 kcal/mol for pure SiO_2 to approximately 39 kcal/mol for 0.5 mole fraction of SiO_2. We can observe even more dramatic effects when oxides such as Li_2O and Na_2O are added to silica. For example, for 0.5 mol fraction SiO_2, the activation energy is only 23 kcal/mol. Explain with the aid of any appropriate sketches, why this phenomenon occurs.

In both cases O:Si = 3:1 so the original SiO_2 network is broken down to some extent. Therefore the answer must be due to differences between the free ions, i.e., Ca^{+2} when CaO is added (as in Fig. 1.12) and Li^+ when Li_2O is added. Divalent Ca^{+2} ions attract two vertices of neighboring tetrahedra (see Fig. 1.12) to a localized region and thus hinder motion of the large silicate ions. On the other hand, monovalent Li^{+1} ions only associate with one vertex and do not hinder motion.

1.13 The glass transition temperature for the soda-lime-silicate glass shown below is approximately 720 K. Test the applicability of Eq. (1.18). Does it apply? Explain why or why not.

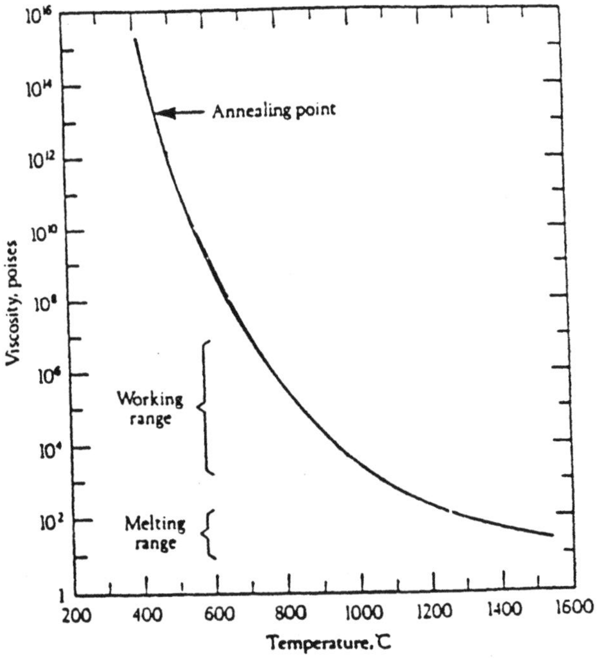

$$\eta = A \exp\left[\frac{\Delta G^{\ddagger}}{RT}\right]$$

$$\ln \eta = \ln A + \frac{\Delta G^{\ddagger}}{RT}$$

$$\ln \eta = C_1 + \frac{C_2}{T} \quad \text{where } C_1 = \ln A, \; C_2 = \frac{\Delta G^{\ddagger}}{RT}$$

The following points are plotted:

η (P)	ln η	T (°C)	T^{-1} (K^{-1})
40	3.7	1500	5.6×10^{-4}
10^2	4.6	1250	6.6×10^{-4}
10^3	6.9	1070	7.4×10^{-4}
10^4	9.2	940	8.2×10^{-4}
10^6	13.8	770	9.6×10^{-4}

η (P)	ln η	T (°C)	T^{-1} (K^{-1})
10^8	18.4	650	11×10^{-4}
10^{10}	23.0	560	12×10^{-4}
10^{12}	27.6	485	13×10^{-4}
10^{14}	32.2	430	14×10^{-4}
10^{16}	37.0	400	15×10^{-4}

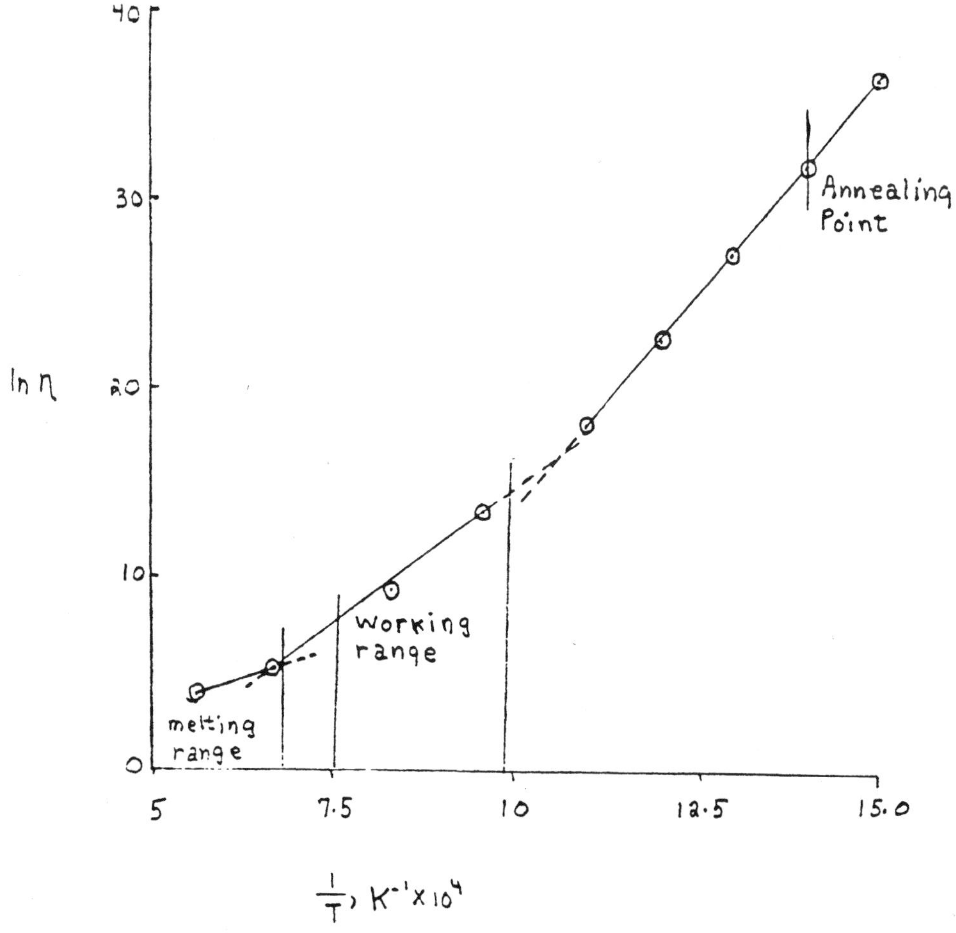

A linear plot over the entire temperature range is not obtained. Hence, Eq. (1.18) applies only to limited ranges, at best. The breaks in the curve correspond to structural changes in the glass.

1.14 Equation (1.30) differs in form from Eq. (1.18). Based on Eq. (1.30) and Table 1.6, calculate the viscosity of 0.5 Al_2O_3-0.5 SiO_2 (mol fraction) in the temperature range of 2100 to 2500 K. Plot your results as ln η vs. T^{-1}. Does Eq. (1.18) adequately represent your results?

Eq. 1.30 $\eta = AT \exp\left[\frac{1000B}{T}\right]$

$\ln\eta = \ln A + \ln T + \frac{1000B}{T}$

From Table 1.6 – For mol fraction $Al_2O_3 = 0.5$, $-\ln A = 18.54$, $B = 25.85$

$\ln\eta = -18.54 + \ln T + \frac{2.585\times10^4}{T}$

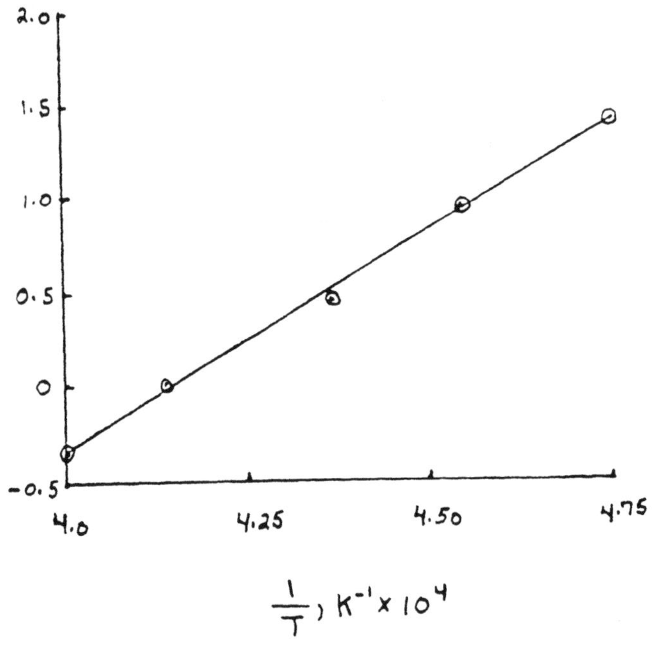

Since a plot of ln η versus $\frac{1}{T}$ is linear, then Eq. 1.18 would adequately represent the results.

1.15 Estimate the viscosities of LiCl and LiBr at 1000 K. Which is greater? Can you explain your results on the basis of ionic bonding?

$$\eta = AT \exp\left[\frac{1000\,B}{T}\right]$$

Table 1.7

	$-\ln A$	B
LiCl	15.91	4.52
LiBr	15.14	3.65

$\ln \eta\,(\text{LiCl}) = -15.91 + \ln(1000) + 4.52 = -4.48$

$\eta\,(\text{LiCl}) \approx 0.0113$ Poise

$\ln \eta\,(\text{LiBr}) = -15.14 + \ln(1000) + 3.65 = -4.58$

$\eta\,(\text{LiBr}) \approx 0.0103$ Poise

η(LiCl) is greater than η(LiBr). The electronegativity difference between Li and Cl is slightly greater than that between Li and Br. Hence, to displace the anions relative to the cations (and vice-versa) by shearing requires a greater stress in LiCl than in LiBr.

1.16 Estimate the percent change in viscosity of polymer melts on changing the shear strain rate from 10^{-4} s^{-1} to 10^{-1} s^{-1}.

As an example, select the polymer to be polystyrene. Pg. 33 gives a forming temperature in the range: $T \sim 1.5 T_g = 560 K$; $T \sim 1.3 T_g = 485 K$.

$\therefore 485 \leq T \leq 560 K$.

Use $N = 1000$ and $N = 10\,000$ for the degrees of polymerization.

Now use Eq. (1.35), with $k(T)$ from Table 1.8.

Now use Eq. (1.34). The following computer program gives the percent change in viscosity for different shear stains for the range of forming temperatures and the two different degrees of polymerization.

```
10  PRINT " for N select from 1000 to 10000"
20  PRINT " for T select from 485 to 560 K "
30 'Problem 1.16    set up for polystyrene
40  INPUT N : INPUT T              'N is degree of polymerization, T is temperatur'
50  LPRINT "*******************************************************************

60  LPRINT " Temperature is";T;"K      Degree of polymerization is";N
70  LPRINT "     "
80  LPRINT "   Rate, 1/s      Viscosity, N s/m2     Percent change"
90  LPRINT "   *********      ******************    **************"
100      KT = 2.7E+16/T^6 - 9.51        'KT is k(T) in Table 1.8
110      NETA00 = 3.4*LOG(N)/2.303 + KT   'Eq (1.35)
120      NETA0 = 10^NETA00               'viscosity at zero strain rate
130 'constants for Eq (1.34)
140  A1 = 6.12*10^(-2.645) : A2 = 2.85*10^(-3.645) : ALPHA = .355
150  RATE = 10^(-4)
160     RATIO = 1 + A1*(RATE*NETA0)^ALPHA + A2*(RATE*NETA0)^(2*ALPHA) 'Eq (1.34)
170 '   NETA = NETA0*RATIO              'viscosity at T,strain rate
180     IF RATE > .00011 THEN 200
190     NETA1 = NETA                'viscosity at 10^-4 strain rate
200     PCT = ( (NETA - NETA1)/NETA1 )*100    'percent change in viscosity, com-
              pared to that at strain rate of 10^-4
210     IF RATE > .11 THEN 240
220     LPRINT USING"    #.####              ##.##^^^^              ######.##";RATE,NETA
,PCT
230     RATE = 10*RATE : GOTO 160
240 END
```

****

Temperature is 485 K Degree of polymerization is 1000

Rate, 1/s	Viscosity, N s/m2	Percent change
0.0001	5.82E+02	0.00
0.0010	5.86E+02	0.67
0.0100	5.95E+02	2.29
0.1000	6.20E+02	6.46

Temperature is 560 K Degree of polymerization is 1000

Rate, 1/s	Viscosity, N s/m2	Percent change
0.0001	3.67E+01	0.00
0.0010	3.68E+01	0.24
0.0100	3.70E+01	0.81
0.1000	3.75E+01	2.16

Temperature is 485 K Degree of polymerization is 10000

Rate, 1/s	Viscosity, N s/m2	Percent change
0.0001	1.60E+06	0.00
0.0010	1.88E+06	17.58
0.0100	2.90E+06	81.11
0.1000	7.16E+06	346.60

Temperature is 560 K Degree of polymerization is 10000

Rate, 1/s	Viscosity, N s/m2	Percent change
0.0001	9.49E+04	0.00
0.0010	9.96E+04	4.97
0.0100	1.14E+05	19.79
0.1000	1.63E+05	71.64

The percentage change in viscosity increases with shear stain rate and decreases with increasing temperature but is dramatically changed by the degree of polymerization.

1.17 Estimate the viscosity of poly(ε-caprolactam) at 526 K (see Table 1.8). Compare its viscosity to that of a branched form (octachain). On the basis of the structure of polymers, explain the effect of branching on the viscosity.

Table 1.8 Poly(ε-Caprolactam)-Linear $K = -8.0$, $N = 340$

Poly(ε-Caprolactam)-Octachain $K = -8.7$, $N = 550$

$$\log \eta_0 = 3.4 \log(N) + K(T)$$

$$\log \eta_0 (\text{linear}) = 3.4 \log 340 - 8 = 0.6071$$

$$\eta_0 (\text{linear}) = 4.05 \text{ N s m}^{-2}$$

$$\log \eta_0 (\text{octachain}) = 3.4 \log 550 - 8.7 = 0.616$$

$$\eta_0 (\text{octachain}) = 4.13 \text{ N s m}^{-2}$$

Effect of branching is to increase the viscosity.

1.18 Refer to Example 1.6. Estimate the viscosities at the glass transition temperature (T_g) and 10 K above T_g. Comment on the sensitivity of the viscosity with temperature near the glass transition temperature.

$$\log \eta_0 = 3.4 \log N + K(T)$$

Using the data of example 1.6

$$\log \eta = 3.4 \log (9608) + \frac{2.7\times 10^{16}}{T^6} - 9.51$$

$T_g = 373\ K$

$$\log \eta = 3.4 \log (9608) + \frac{2.7\times 10^{16}}{373^6} - 9.51 = 13.54 + 10 - 9.51 = 14.03$$

$$\eta(373\ K) \approx 10^{14}$$

$T_g + 10K = 383K$

$$\log \eta = 3.4 \log (9608) + \frac{2.7\times 10^{16}}{383^6} - 9.51 = 13.54 + 70.50 - 9.51 = 74.53$$

$$\eta(383\ K) = 10^{74}$$

Viscosity changes drastically near the transition temperature.

2.1 Refer to the results of Example 2.1. The viscosity of the glass is 1 N s m^{-2}, and the viscosity of the metal is 3×10^{-3} N s m^{-2}. The densities of the glass and the metal are 3.2 kg m^{-3} and 7.0 kg m^{-3}, respectively. For $\beta = \pi/8$ and $\delta_1 = 1$ mm and $\delta_2 = 2$ mm, calculate the maximum velocities and average velocities of the glass and the metal.

$$\nu_g = \frac{1}{3.2} = 0.313 \text{ m}^2\text{s}^{-1}; \quad \nu_m = \frac{3\times10^{-3}}{7} = 4.286\times10^{-4} \text{ m}^2\text{ s}^{-1}$$

$$C_2 = \frac{(3.2-7.0)(1\times10^{-3})(9.8)}{3\times10^{-3}} \cos\left(\frac{\pi}{8}\right) = -11.468 \text{ s}^{-1}$$

$$C_3 = \frac{(2\times10^{-3})^2(9.8)}{(2)(4.286\times10^{-4})} \cos\left(\frac{\pi}{8}\right) + (-11.468)(2\times10^{-3}) = 1.931\times10^{-2} \text{ m s}^{-1}$$

$$C_1 = \left[\frac{(2\times10^{-3})^2 - (1\times10^{-3})^2}{4.286\times10^{-4}} + \frac{(1\times10^{-3})^2}{0.313}\right] \frac{9.8}{2} \cos\left(\frac{\pi}{8}\right) + (-11.468)(1\times10^{-3}) = 2.024\times10^{-2} \text{ m s}^{-1}$$

Glass: Max. vel. is at $x=0$: $v_z = -\frac{x^2 g \cos\beta}{2\nu_g} + C_1 \quad \therefore v_z = C_1 = 2.024\times10^{-2} \text{ m s}^{-1}$

Min. vel. is at $x = 1\times10^{-3}$ m: $v_z = -\frac{(1\times10^{-3})^2(9.8)}{(2)(0.313)}\cos\left(\frac{\pi}{8}\right) + 2.024\times10^{-2}$

$v_z = 2.022\times10^{-2} \text{ m s}^{-1}$

Avg. vel. $\bar{v}_z = \frac{1}{\delta}\int_0^\delta v_z\, dx = \frac{1}{\delta}\int_0^\delta \left[-\frac{x^2 g\cos\beta}{2\nu_g} + C_1\right] dx = C_1 - \frac{\delta^2 g \cos\beta}{6\nu_g}$

$$= 2.023\times10^{-2} - \frac{(0.001)^2(9.8)}{(6)(0.312)}\cos\left(\frac{\pi}{8}\right) = 2.023 \text{ m s}^{-1}$$

Metal: Max. vel. is at $x = 1$ mm where the glass is at its min. vel.

$v_z = 2.022\times10^{-2} \text{ m s}^{-1}$

Min. vel. $= 0$

Avg. Vel. $\bar{v}_z = \frac{1}{\delta_2-\delta_1}\int_{\delta_1}^{\delta_2} v_z\, dx = \frac{1}{\delta_2-\delta_1}\int_{\delta_1}^{\delta_2}\left[-\frac{x^2 g\cos\beta}{6\nu_m} - C_2 x + C_3\right] dx$

$$= C_3 - \frac{C_2(\delta_2+\delta_1)}{2} - \frac{g\cos\beta}{6\nu_m}\left[\frac{\delta_2^3-\delta_1^3}{\delta_2-\delta_1}\right]$$

$$= 1.9313\times10^{-2} - \frac{(-11.46)(0.002+0.001)}{2} - \frac{(9.8)\cos\left(\frac{\pi}{8}\right)}{(6)(4.286\times10^{-4})}\left[\frac{(0.002)^3-(0.001)^3}{(0.002)-(0.001)}\right]$$

$= 1.1863\times10^{-2} \text{ m s}^{-1}$

```
10  'Problem 2.1
20  PI = 3.1416  : G = 9.8                    'G is gravitational acceleration
30  '
40  'Input properties    G for glass     M for metal
50  VISCG = 1: VISCM = .003: RHOG = 3.2: RHOM = 7  'viscosities and densities
60  BETA = PI/8 ; DEL1 = .001 : DEL2 = .002       'film thicknesses
70  NETAM = VISCM/RHOM : NETAG = VISCG/RHOG       'kinematic viscosities
80  '
90  'Evaluate constants
100  C2 = (RHOG - RHOM)*DEL1*G*COS(BETA)/VISCM
110  C3 = DEL2*DEL2*G*COS(BETA)/(2*NETAM) + C2*DEL2
120  C1 = ( (DEL2^2 - DEL1^2)/NETAM + DEL1^2/NETAG ) * G*COS(BETA)/2
         + C2*(DEL2-DEL1)
130 '
140 'Calculate velocities in the glass
150  VMAXG = C1                                  'max. velocity at X = 0
160  VMING = C1 - G*DEL1*DEL1*COS(BETA)/(2*NETAG)  'min. velocity at X = DEL1
170  VAVGG = C1 - G*DEL1*DEL1*COS(BETA)/(6*NETAG)  'avg. velocity
180 '
190 'Calculate velocities in the metal
200  VMAXM =  - G*DEL1*DEL1*COS(BETA)/(2*NETAM) - C2*DEL1 + C3   'max. velocity
                                                             at X = DEL1
210  VMINM =  0                                   'min. velocity at X = DEL2
220  VAVGM = C3 - C2*(DEL2+DEL1)/2
         - (G*COS(BETA)/(6*NETAM))*(DEL2^3 - DEL1^3)/(DEL2 - DEL1) 'avg. vel.

230 '
240 LPRINT "The maximum, minimum, and average velocities in the glass are"
250 LPRINT VMAXG;"m/s  ";VMING;"m/s  ";VAVGG;"m/s  " : LPRINT
260 LPRINT "The maximum, minimum, and average velocities in the metal are"
270 LPRINT VMAXM;"m/s  ";VMINM;"m/s  ";VAVGM;"m/s  " : LPRINT
280 END
```

```
The maximum, minimum, and average velocities in the glass are
 2.023513E-02 m/s   2.022064E-02 m/s   .0202303 m/s

The maximum, minimum, and average velocities in the metal are
 2.022064E-02 m/s   0 m/s   1.187082E-02 m/s
```

2.2 A continuous sheet of metal is cold-rolled by passing vertically between rolls. Before entering the rolls, the sheet passes through a tank of lubricating oil equipped with a squeegee device that coats both sides of the sheet uniformly as it exits. The amount of oil that is carried through can be controlled by adjusting the squeegee device. Prepare a control chart that can be used to determine the thickness of oil (in mm) on the plate just before it enters the roll as a function of the mass rate of oil (in kg per hour). Values of interest for the thickness of the oil film range from 0-0.6 mm. *Data:* Oil density, 962 kg m^{-3}; oil viscosity, 4.1×10^{-3} N s m^{-2}; width of sheet, 1.5 m; velocity of sheet, 0.3 m s^{-1}.

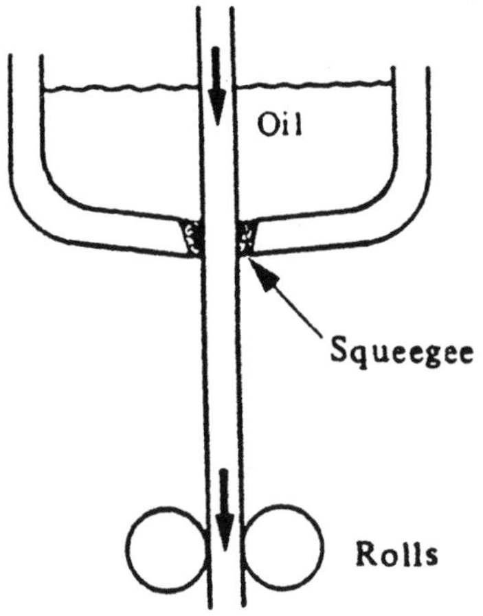

$v_z = V_m$ at $x = \delta$; $\beta = 0$

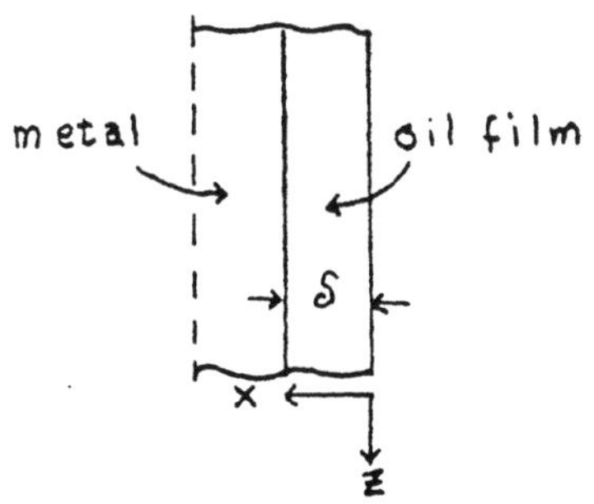

Vol. flow rate (two sides): $Q = 2\bar{v}_z W \delta$

$$\therefore Q = 2\left(\frac{\rho g \delta^2}{3\eta} + V_m\right) W\delta \quad \text{Eq. (2.15)}$$

Mass flow – $\dot{W} = \rho Q$

$$\dot{W} = \frac{(2)}{(3)}\left(\frac{962\ \text{kg}}{\text{m}^3}\right)^2\left(\frac{9.8\ \text{m}}{\text{s}^2}\right)\left(\frac{1}{4.1\times10^{-3}}\frac{\text{m}^2}{\text{N s}}\right)\left(\frac{1\ \text{N s}^2}{\text{kg m}}\right)\delta^3 + (2)\left(\frac{0.3\ \text{m}}{\text{s}}\right)(1.5\ \text{m})\left(\frac{962\ \text{kg}}{\text{m}^3}\right)\delta$$

$$\dot{W} = 2.212\times10^9\frac{\text{kg}}{\text{s m}^3}\delta^3 + 8.658\times10^2\frac{\text{kg}}{\text{s m}}\delta = 7.963\times10^3\frac{\text{kg}}{\text{hr. mm}^3}\delta^3 + 3.117\times10^3\frac{\text{kg}}{\text{hr. mm}}\delta$$

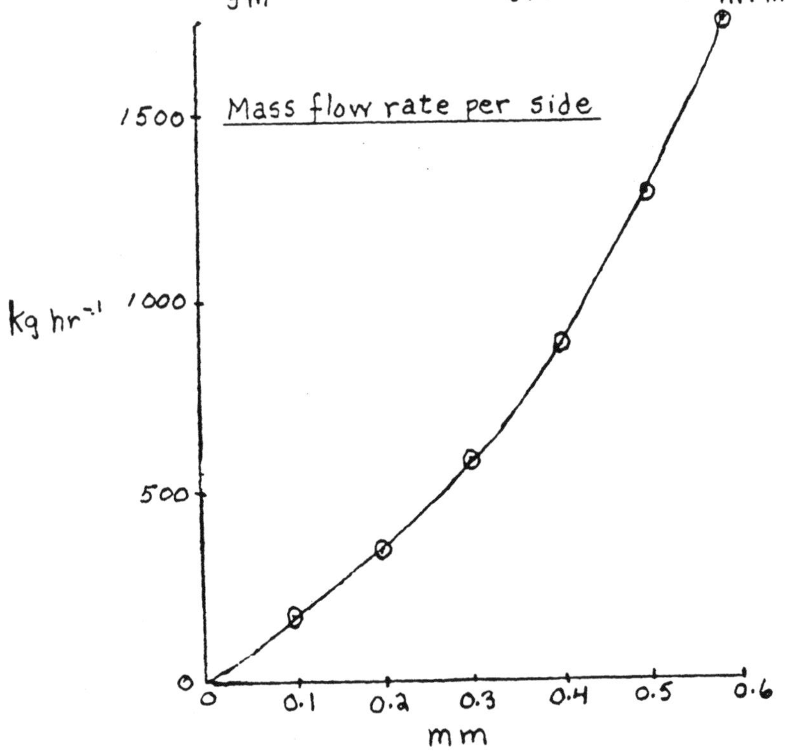

2.3 A Newtonian liquid flows simultaneously through two parallel and vertical channels of different geometries. Channel "A" is circular with a radius R, and "B" is a slit of thickness 2δ and width W; $2\delta \ll W$. Assume fully developed flow in both channels and derive an equation which gives the ratio of the volume flow rate through A to that through B.

Channel "A"

$$Q_A = \left(\frac{P_0 - P_L}{L} + \rho g\right)\frac{\pi R^4}{8\eta}$$

Channel "B"

$$Q_B = \left(\frac{P_0 - P_L}{L} + \rho g\right)\frac{2 W \delta^3}{3\eta}$$

$$\therefore \frac{Q_A}{Q_B} = \frac{3\pi R^4}{(8)(2) W \delta^3} = \frac{3\pi R^4}{16 W \delta^3}$$

2.4 Develop expressions for the flow of a fluid between vertical parallel plates. The plates are separated by a distance of 2δ. Consider fully developed flow and determine

a) the velocity distribution,

b) the volume flow rate.

Compare your expressions with Eqs. (2.20) and (2.23).

a. Eq. (2.20) with flow in the direction of gravity.

Vel. Distribution: $v_x = \left(\frac{\rho g}{2\eta} + \frac{P_0 - P_L}{2\eta L}\right)(\delta^2 - y^2)$

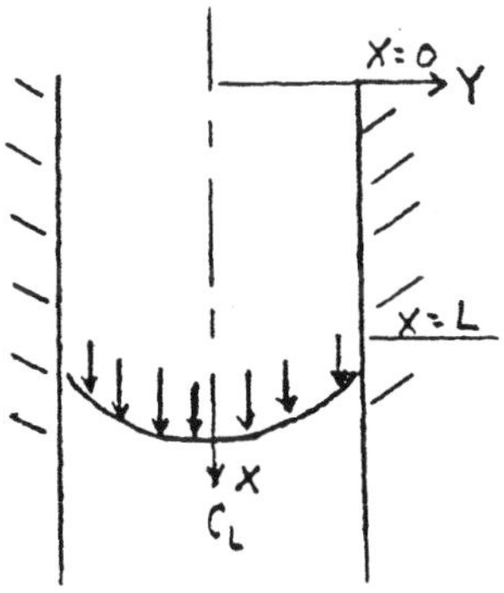

b. Eq. (2.23) with flow in the direction of gravity.

Vol. Flow Rate: $Q = \frac{2}{3}\frac{\delta^3 W}{\eta}\left(\rho g + \frac{P_0 - P_L}{L}\right)$

A comparison with Eqs. (2.20) and (2.23) shows the effect of gravity on the velocity distribution and the volume flow rate.

2.5 Repeat Problem 2.4 but now orient the plates at an angle β to the direction of gravity and obtain expressions for

a) the velocity distribution,
b) the volume flow rate.

Compare your expressions with the results of Problem 2.4 and Eqs. (2.20) and (2.23).

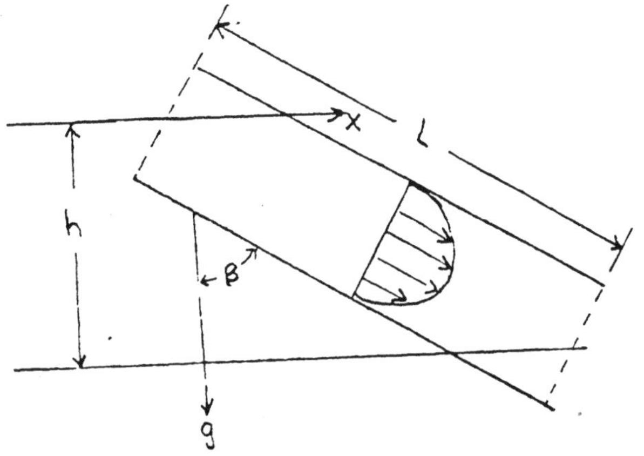

$$v_x = \frac{1}{2\eta}\left(\delta^2 - y^2\right)\left[\frac{P_0 - P_L}{L} + \rho g \cos\beta\right]$$

$$Q = \frac{2}{3}\frac{W\delta^3}{\eta}\left[\frac{P_0 - P_L}{L} + \rho g \cos\beta\right]$$

In this case the force due to gravity is not ρg but rather $\rho g \cos\beta$. The term in the bracket as it appears here is more general in the sense that the expressions in the brackets of problem 2.4 are special cases. The term in the brackets may also be viewed differently. For example the volume flow rate may be written:

$$Q = \frac{2}{3}\frac{W\delta^3}{\eta L}\left[\left(P_0 - P_L\right) + \rho g h\right]$$

The brackets can now be viewed as a "total pressure drop" in that the $\rho g h$, itself, can be viewed as a pressure, e.g., the pressure in a fluid at a point is given by the height of a fluid above that point.

2.6 A liquid is flowing through a vertical tube 0.3 m long and 2.5 mm in I.D. The density of the liquid is 1260 kg m^{-3} and the mass flow rate is 3.8×10^{-5} kg s^{-1}.

a) What is the viscosity in N s m^{-2}?

b) Check on the validity of your results.

Assume that $P_0 - P_L = 0$: $Q = \frac{\rho g \pi R^4}{8\eta} = \frac{\dot{W}}{\rho}$

$$\eta = \frac{\rho^2 g \pi R^4}{8\dot{W}} = \left(\frac{1260\ kg}{m^3}\right)^2 \frac{9.8\ m}{s^2} \frac{\pi}{8} \left(\frac{0.0025\ m}{2}\right)^4 \frac{1}{3.8\times10^{-5}\ kg} \frac{s}{} \frac{N\ s^2}{kg\ m} = 0.3925\ N\ s\ m^{-2}$$

check for laminar flow; $Re = \frac{4\rho Q}{\pi D \eta} = \frac{4\rho}{\pi D \eta}\left(\frac{\dot{W}}{\rho}\right) = \frac{4\dot{W}}{\pi D \eta}$

$$Re = \frac{4}{\pi} \frac{3.8\times10^{-5}\ kg}{s} \frac{1}{0.0025\ m} \frac{1}{0.3925\ N\ s} \frac{m^2}{} \frac{N\ s^2}{kg\ m} = 4.931\times10^{-2}$$

$Re < 2100$, flow is laminar.

2.7 Water (viscosity 10^{-3} N s m^{-2}) flows parallel to a flat horizontal surface. The velocity profile at $x = x_1$ is given by

$$v_x = 6 \sin\left[\frac{\pi}{2}\right] y$$

with v_x in m s^{-1} and y as distance from surface in mm.

a) Find the shear stress at the wall at x_1. Express results in N m^{-2}.

b) Farther downstream, at $x = x_2$, the velocity profile is given by

$$v_x = 4 \sin\left[\frac{\pi}{2}\right] y$$

Is the flow "fully developed"? Explain.

c) Is there a y-component to flow (i.e., v_y)? Explain with the aid of the continuity equation.

a. $\left.\frac{\partial v_x}{\partial Y}\right|_{x=x_1} = 3\pi \cos\left(\frac{\pi}{2}\right)Y$; $\tau_{Yx} = -\eta \frac{\partial v_x}{\partial Y}$

$\left.\frac{\partial v_x}{\partial Y}\right|_{Y=0} = \frac{3\pi \text{ m}}{\text{s mm}} \cdot \frac{1 \times 10^3 \text{ mm}}{\text{m}} = 3 \times 10^3 \pi \text{ s}^{-1}$; $\tau_{Yx} = \frac{-10^{-3} \text{ N s}}{\text{m}^2} \cdot \frac{3 \times 10^3 \pi}{\text{s}} = -3\pi \text{ N m}^{-2}$

b. $v_x = f(x)$ not fully developed.

c. $\frac{\partial v_x}{\partial x} + \frac{\partial v_Y}{\partial Y} + \frac{\partial v_z}{\partial z} = 0$; Assume ρ = const. and $\frac{\partial v_z}{\partial z} = 0$

$\frac{\partial v_x}{\partial x} = -\frac{\partial v_Y}{\partial Y}$; $\partial v_Y = -\left(\frac{\partial v_x}{\partial x}\right)\partial Y$; $v_Y = -\left(\frac{\partial v_x}{\partial x}\right)Y + c_1$

There is a Y component to flow

2.8 For a polymeric melt that follows a power law for shear stress versus shear strain rate, derive an equation for the velocity profile and volume flow rate for flow between parallel plates.

a. Eq.(2.19) $\tau_{YX} = \left(\frac{P_0 - P_L}{L}\right) Y$ for flow between parallel plates.

Power law $\tau_{YX} = -\eta_0 \left[\frac{dv_x}{dY}\right]^n$

$$dv_x = \left[\frac{P_L - P_0}{\eta_0 L}\right]^{\frac{1}{n}} Y^{\frac{1}{n}} \, dY$$

$$v_x = -\frac{n}{n+1}\left[\frac{P_0 - P_L}{\eta_0 L}\right]^{\frac{1}{n}} Y^{\frac{n+1}{n}} + C_1$$

Boundary Condition: at $Y = \delta$, $v_x = 0$

$$\therefore \; C_1 = \frac{n}{n+1}\left[\frac{P_0 - P_L}{\eta_0 L}\right]^{\frac{1}{n}} \delta^{\frac{n+1}{n}}$$

$$v_x = \left(\frac{n}{n+1}\right)\left[\frac{P_0 - P_L}{\eta_0 L}\right]^{\frac{1}{n}}\left[\delta^{\frac{n+1}{n}} - Y^{\frac{n+1}{n}}\right]$$

$$V_x^{MAX} = \left(\frac{n}{n+1}\right)\left[\frac{P_0 - P_L}{\eta_0 L}\right]^{\frac{1}{n}} \delta^{\frac{n+1}{n}}$$

$$v_x = V_x^{max} - \left(\frac{n}{n+1}\right)\left[\frac{P_0 - P_L}{\eta_0 L}\right]^{\frac{1}{n}} Y^{\frac{n+1}{n}} = V_x^{max} - \frac{V_x^{max}}{\delta^{\frac{n+1}{n}}} Y^{\frac{n+1}{n}} = V_x^{max}\left[1 - \left(\frac{Y}{\delta}\right)^{\frac{n+1}{n}}\right]$$

b. $$\bar{v}_x = \frac{1}{\delta}\int_0^{\delta} v_x \, dY = \frac{1}{\delta} V_x^{max}\left[Y\right]_0^{\delta} - \left(\frac{1}{\delta}\right)\left(\frac{1}{\delta^{\frac{n+1}{n}}}\right) V_x^{max}\left[\frac{Y^{\frac{2n+1}{n}}}{\frac{2n+1}{n}}\right]_0^{\delta} = V_x^{max} - V_x^{max}\left(\frac{n}{2n+1}\right)$$

$$\bar{v}_x = V_x^{max}\left(\frac{n+1}{2n+1}\right)$$

$$Q = \bar{v}_x (2\delta W) = V_x^{max}\left(\frac{n+1}{2n+1}\right)(2\delta W) = 2\left(\frac{n}{2n+1}\right)\left[\frac{P_0 - P_L}{\eta_0 L}\right]^{\frac{1}{n}} W \delta^{\frac{2n+1}{n}}$$

2.9 The power law polymer of Problem 2.8 has constants $\eta_0 = 1.2 \times 10^4$ N s m^{-2} and $n = 0.35$. It is injected through a gate into a thin cavity, which has a thickness of 2 mm, a width of 10 mm and a length of 20 mm. If the injection rate is constant at 200 mm^3 s^{-1}, estimate the time to fill the cavity and the injection pressure at the gate.

Cavity Volume $= (2\times10^{-3}\text{m})(1\times10^{-2}\text{m})(2\times10^{-2}\text{m}) = 4\times10^{-7}\text{m}^3$; $Q = 2\times10^{-7}\text{m}^3\text{s}^{-1}$

Time to fill the cavity $= \dfrac{4\times10^{-7}\text{m}^3}{} \Big| \dfrac{\text{s}}{2\times10^{-7}\text{m}^3} = 2\text{ s}$

$$Q = 2\left(\frac{n}{2n+1}\right)\left[\frac{P_0 - P_L}{\eta_0 L}\right]^{\frac{1}{n}} W\,\delta^{\frac{2n+1}{n}}$$

$$P_0 - P_L = \left[\frac{Q}{2}\left(\frac{2n+1}{n}\right)\frac{1}{W}\right]^n \frac{1}{\delta^{2n+1}}\,\eta_0 L$$

$$P_0 - P_L = \left[\left(\frac{2\times10^{-7}}{2}\right)\left(\frac{0.7+1}{0.35}\right)\left(\frac{1}{0.01}\right)\right]^{0.35} \frac{(1.2\times10^4)(2\times10^{-2})}{(1\times10^{-3})^{1.70}} = 9.342\times10^5\text{ N m}^{-2}$$

Assume that $P_L = 1\text{ atm} = 1.014\times10^5\text{ N m}^{-2}$. Then $P_0 = 10.356\times10^5\text{ N m}^{-2}$

2.10 A wire is cooled after a heat treating operation by being pulled through the center of an open-ended, oil-filled tube which is immersed in a tank. In a region in the tube where end effects are negligible, obtain an expression for the velocity profile assuming steady state and all physical properties constant.

Tube inner radius: R
Wire radius: KR
Wire velocity: U

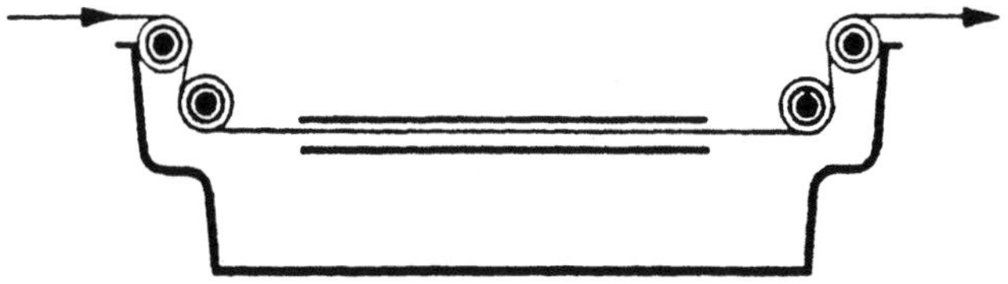

Assume constant properties and fully developed flow.

z component of momentum: Eq. (F) in Table 2.3; $0 = -\dfrac{\partial P}{\partial z} + \eta\left[\dfrac{1}{r}\dfrac{\partial}{\partial r}\left(r\dfrac{\partial v_z}{\partial r}\right)\right] + \rho g_z$

$P_0 = P_i$ so $\dfrac{\partial P}{\partial z} = 0$; horizontal, so $\rho g_z \to 0$

$$\therefore \frac{d}{dr}\left(r\frac{dv_z}{dr}\right) = 0$$

$$r\frac{dv_z}{dr} = C_1;\; dv_z = C_1\frac{dr}{r};\; v_z = C_1 \ln r + C_2$$

$$\text{at } r = R;\; v_z = 0 = C_1 \ln R + C_2;\; C_2 = -C_1 \ln R$$

$$\text{at } r = KR;\; v_z = u = C_1(\ln KR - \ln R) = C_1 \ln K;\; C_1 = \frac{u}{\ln K}$$

$$\therefore\; v_z = \left(\frac{u}{\ln K}\right)\left[\ln r - \ln R\right] = \left(\frac{u}{\ln K}\right)\ln\left(\frac{r}{R}\right)$$

2.11 Starting with the *x*-component of the momentum equation (Eq. (2.52)), develop the *x*-component for the Navier-Stokes' equation (constant ρ and η, (Eq. 2.63)).

1. Eq. (2.52)
$$\frac{d}{dt}\rho v_x = -\left(\frac{d}{dx}\rho v_x v_x + \frac{d}{dY}\rho v_Y v_x + \frac{d}{dz}\rho v_z v_x\right)$$
$$-\left(\frac{d}{dx}\tau_{xx} + \frac{d}{dY}\tau_{Yx} + \frac{d}{dz}\tau_{zx}\right) - \frac{dP}{dx} + \rho g_x$$

2. Expand the convective momentum terms:

$$\frac{d}{dx}(\rho v_x) v_x = \rho v_x \frac{dv_x}{dx} + v_x \frac{d}{dx}\rho v_x$$

$$\frac{d}{dY}(\rho v_x) v_Y = \rho v_x \frac{dv_Y}{dY} + v_Y \frac{d}{dY}\rho v_x$$

$$\frac{d}{dz}(\rho v_x) v_z = \rho v_x \frac{dv_z}{dz} + v_z \frac{d}{dz}\rho v_z$$

Combine 2 with 1 with ρ = const.

$$\rho\frac{dv_x}{dt} = -\rho\left[\underbrace{v_x\left(\frac{dv_x}{dx} + \frac{dv_Y}{dY} + \frac{dv_z}{dz}\right)}_{=0 \text{ by continuity}} + v_x\frac{dv_x}{dx} + v_Y\frac{dv_x}{dY} + v_z\frac{dv_x}{dz}\right] - \left[\cdots\cdots\right.$$

From Eq. (2.56), (2.59), (2.61) with $\tau_{xx} = -\eta\left[\frac{dv_x}{dx} + \frac{dv_x}{dx}\right]$ for Eq. (2.56)

$$\frac{\partial}{\partial x}\tau_{xx} = -\eta\frac{\partial^2 v_x}{\partial x^2} - \eta\frac{\partial}{\partial x}\left(\frac{\partial v_x}{\partial x}\right)$$

$$\frac{\partial}{\partial Y}\tau_{Yx} = -\eta\frac{\partial^2 v_x}{\partial Y^2} - \eta\frac{\partial}{\partial x}\left(\frac{\partial v_Y}{\partial Y}\right)$$

$$\frac{\partial}{\partial z}\tau_{zx} = -\eta\frac{\partial^2 v_x}{\partial z^2} - \eta\frac{\partial}{\partial x}\left(\frac{\partial v_z}{\partial z}\right)$$

$$\frac{\partial}{\partial x}\tau_{xx} + \frac{\partial}{\partial Y}\tau_{Yx} + \frac{\partial}{\partial z}\tau_{zx} = -\eta\left[\frac{\partial^2 v_x}{\partial x^2} + \frac{\partial^2 v_x}{\partial Y^2} + \frac{\partial^2 v_x}{\partial z^2}\right] - \eta\frac{\partial}{\partial x}\underbrace{\left[\frac{\partial v_x}{\partial x} + \frac{\partial v_Y}{\partial Y} + \frac{\partial v_z}{\partial z}\right]}_{=0 \text{ by continuity}}$$

$$\therefore \rho\left[\frac{\partial v_x}{\partial t} + v_x\frac{\partial v_x}{\partial x} + v_Y\frac{\partial v_x}{\partial Y} + v_z\frac{\partial v_x}{\partial z}\right] = -\frac{\partial P}{\partial x} + \eta\left[\frac{\partial^2 v_x}{\partial x^2} + \frac{\partial^2 v_x}{\partial Y^2} + \frac{\partial^2 v_x}{\partial z^2}\right] + \rho g_x$$

2.12 Air at 289 K flows over a flat plate with a velocity of 9.75 m s^{-1}. Assume laminar flow and a) calculate the boundary-layer thickness 50 mm from the leading edge; b) calculate the rate of growth of the boundary layer at that point; i.e., what is $d\delta/dx$ at that point? *Properties of air at 289 K*: density: 1.22 kg m^{-3}; viscosity: 1.78×10^{-5} N s m^{-2}.

a. Eq. (2.100) $\delta = 5.0\left(\frac{\nu x}{V_\infty}\right)^{1/2}$: $\nu = \frac{\eta}{\rho} = \frac{1.78\times10^{-5}}{1.22} = 1.459\times10^{-5}\ m^2 s^{-1}$

$$\delta = 5.0\left[\frac{1.459\times10^{-5}\ m^2}{s^{-1}}\left|\frac{0.05\ m}{}\right|\frac{s}{9.75\ m}\right]^{1/2} = 1.368\times10^{-3}\ m = 1.368\ mm$$

$$b.\ \frac{d\delta}{dx} = \frac{5}{2}\left(\frac{\nu}{V_\infty}\right)^{1/2}\frac{1}{x^{1/2}} = \frac{5}{2}\left[\frac{1.459\times10^{-5}\ m^2}{s}\left|\frac{s}{9.75\ m}\right.\right]^{1/2}\frac{1}{(0.05)^{1/2}\ m^{1/2}} = 1.368\times10^{-2}$$

$\frac{d\delta}{dx} = 1.368\times10^{-2}$ at $x = 50$ mm

2.13 A fluid flows upward through a vertical cylindrical annulus of length L. Assume that the flow is fully developed. The inner radius of the annulus is κR, and the outer radius is R. a) Write the momentum equation in terms of velocity. b) Solve for the velocity profile. c) Solve for the maximum velocity.

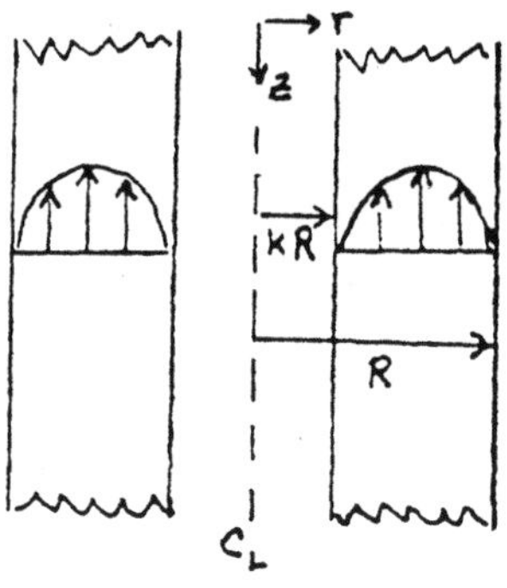

a. Fully developed flow, $v_z = f(r)$

Table 2.3 Eq. (F): $0 = -\frac{dP}{dz} + \eta\left[\frac{1}{r}\frac{d}{dr}\left(r\frac{dv_z}{dr}\right)\right] + \rho g_z;\ g_z = g$

$$\therefore \frac{d}{dr}\left(r\frac{dv_z}{dr}\right) = \frac{r}{\eta}\left(\frac{dP}{dz} - \rho g\right)$$

b. $r\frac{dv_z}{dr} = \left(\frac{dP}{dz} - \rho g\right)\frac{r^2}{2\eta} + C_1 :\ v_z = \frac{1}{4\eta}\left(\frac{dP}{dz} - \rho g\right)r^2 + C_1 \ln r + C_2$

B.C. at $r = \kappa R,\ v_z = 0 :\ r = R,\ v_z = 0$

$$\therefore C_1 = \frac{1}{4\eta}\left(\frac{dP}{dz} - \rho g\right)\frac{(1-\kappa^2)R^2}{\ln \kappa};\ C_2 = -\frac{1}{4\eta}\left(\frac{dP}{dz} - \rho g\right)R^2\left[1 + (1-\kappa^2)\frac{\ln R}{\ln \kappa}\right]$$

$$v_z = \frac{1}{4\eta}\left(\frac{dP}{dz} - \rho g\right)\left[r^2 - R^2\left[1 - (1-\kappa^2)\frac{\ln\left(\frac{r}{R}\right)}{\ln \kappa}\right]\right]$$

c. $v_z = V_z^{max.}$ when $\frac{dv_z}{dr} = 0$

$$\frac{dv_z}{dr} = 0 = \left(\frac{dP}{dz} - \rho g\right)\frac{r}{2\eta} + \frac{1}{4r\eta}\left(\frac{dP}{dz} - \rho g\right)\frac{(1-\kappa^2)R^2}{\ln \kappa}$$

$$\frac{r}{2\eta} = \left(\frac{1}{4r\eta}\right)\frac{(\kappa^2-1)}{\ln \kappa}R^2;\ r^2 = \left(\frac{\kappa^2-1}{2\ln \kappa}\right)R^2$$

$$\therefore v_z = V_z^{max.} \text{ where } r = \left(\frac{\kappa^2-1}{2\ln \kappa}\right)^{1/2}R$$

2.14 In steelmaking, deoxidation of the melt is accomplished by the addition of aluminum, which combines with the free oxygen to form alumina, Al_2O_3. It is then hoped that most of these alumina particles will float up to the slag layer for easy removal from the process, because their presence in steel can be detrimental to mechanical properties. Determine the size of the smallest alumina particles that will reach the slag layer from the bottom of the steel two minutes after the steel is deoxidized. It may be assumed that the alumina particles are spherical in nature. For the purpose of estimating the steel's viscosity use the data for Fe-0.5 wt pct C in Fig. 1.11. *Data*: Temperature of steel melt: 1873 K; Steel melt depth: 1.5 m; Density of steel: 7600 kg m^{-3}; Density of alumina: 3320 kg m^{-3}.

Force balance: $F_s = F_g + F_K$

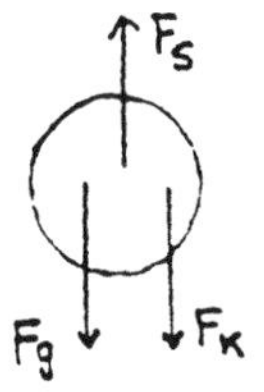

$$\frac{4}{3}\pi R^3 \rho_{(l)} g = \frac{4}{3}\pi R^3 \rho_{(s)} g + 6\pi\eta R V_t$$

$$\frac{4}{3}\pi R^3 g\left(\rho_{(l)} - \rho_{(s)}\right) = 6\pi\eta R V_t$$

$$R = \left[\frac{9\,\eta\,V_t}{2\,g\,(\rho_{(s)} - \rho_{(l)})}\right]^{1/2}$$

$$V_t = \frac{1.5\text{ m}}{2\text{ min.}}\left|\frac{\text{min.}}{60\text{ s}}\right. = 1.25\times10^{-2}\text{ m s}^{-1}, \quad \rho_{(s)} - \rho_{(l)} = 7600 - 3320 = 4280\text{ kg m}^{-3}$$

From Fig 1.11 $\eta \simeq 6.1\text{ cP} = 6.1\times10^{-3}\text{ N s m}^{-2}$

$$R = \left[\frac{9}{2}\left|\frac{6.1\times10^{-3}\text{ N s}}{\text{m}^2}\right|\frac{1.25\times10^{-2}\text{ m}}{\text{s}}\left|\frac{\text{s}^2}{9.8\text{ m}}\right|\frac{\text{m}^3}{4.280\times10^3\text{ kg}}\left|\frac{\text{kg m}}{\text{N s}^2}\right.\right]^{1/2} = 9.045\times10^{-5}\text{ m}$$

$R = 9.045\times10^{-5}\text{ m} = 90.45\ \mu\text{m}$

2.15

a) Consider a very large flat plate bounding a liquid that extends to $y = +\infty$. Initially, the liquid and the plate are at rest; then suddenly the plate is set into motion with velocity V_0 as shown in the figure below. Write (1) the pertinent differential equation in terms of velocity, for constant properties, that applies from the instant the plate moves, and (2) the appropriate boundary and initial conditions. The solution to these equations will be discussed in Chapter 9.

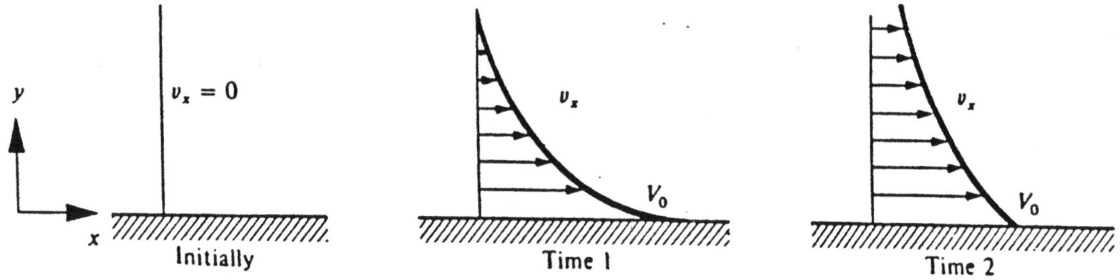

b) A liquid flows upward through a long vertical conduit with a square cross section. With the aid of a clearly labeled sketch, write (1) a pertinent differential equation that describes the flow for constant properties, and (2) the appropriate boundary conditions. Consider only that portion of the conduit where flow is fully developed and be sure that your sketch and equations correspond to one another.

a. conservation of momentum: Table 2.2 eq. D

$$\rho\left(\frac{\partial v_x}{\partial t} + v_x\frac{\partial v_x}{\partial x} + v_y\frac{\partial v_x}{\partial y} + v_z\frac{\partial v_x}{\partial z}\right) = -\frac{\partial P}{\partial x} + \eta\left(\frac{\partial^2 v_x}{\partial x^2} + \frac{\partial^2 v_x}{\partial y^2} + \frac{\partial^2 v_x}{\partial z^2}\right) + \rho g_x$$

$$\frac{\partial v_x}{\partial x} = 0, \; v_y = 0, \; v_z = 0, \; g_x = 0, \; \frac{\partial^2 v_x}{\partial x^2} = 0, \; \frac{\partial^2 v_x}{\partial z^2} = 0, \; \frac{\partial P}{\partial x} = 0$$

$$\therefore \; \rho\frac{\partial v_x}{\partial t} = \eta\frac{\partial^2 v_x}{\partial y^2} \quad \text{or} \quad \frac{\partial v_x}{\partial t} = \nu\frac{\partial^2 v_x}{\partial y^2}$$

Initial Condition: $v_x(y,0) = 0$

Boundary Conditions: $v_x(0,t) = V_0$

$v_x(\infty,t) = 0$

b.

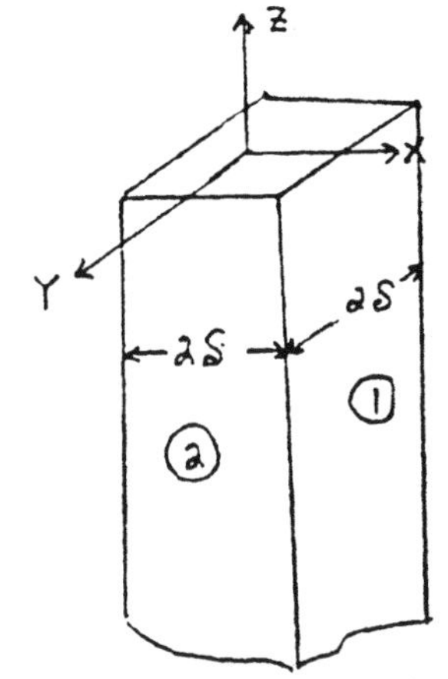

view ①

view ②

Conservation of momentum: Table 2.2 eq. F

$$\rho\left(\frac{\partial v_z}{\partial t} + v_x \frac{\partial v_x}{\partial x} + v_Y \frac{\partial v_z}{\partial Y} + v_z \frac{\partial v_z}{\partial z}\right) = -\frac{\partial P}{\partial z} + \eta\left(\frac{\partial^2 v_z}{\partial x^2} + \frac{\partial^2 v_z}{\partial Y^2} + \frac{\partial^2 v_z}{\partial z^2}\right) + \rho g_z$$

With fully developed flow,

$$\frac{\partial v_z}{\partial t} = 0, \quad v_x = 0, \quad v_Y = 0, \quad \frac{\partial v_z}{\partial z} = 0, \quad \frac{\partial^2 v_z}{\partial z^2} = 0$$

(1) $\therefore \quad \dfrac{\partial P}{\partial z} - \rho g_z = \eta\left(\dfrac{\partial^2 v_z}{\partial x^2} + \dfrac{\partial^2 v_z}{\partial Y^2}\right)$

(2) Boundary Conditions: $v_z(\delta, Y) = 0, \quad v_z(x, \delta) = 0$

$\tau_{xz}(0, Y) = 0, \quad \tau_{Yz}(x, 0) = 0$

2.16 Molten aluminum is degassed by gently bubbling a 75%N_2-25%Cl_2 gas through the melt. The gas passes through a graphite tube at a volumetric flow rate of 6.6×10^{-5} m^3 s^{-1}. Calculate the pressure that should be maintained at the tube entrance if the pressure over the bath is 1.014×10^5 N m^{-2} (1 atm). *Data*: Tube dimensions: $L = 0.9$ m; inside diameter = 2 mm. Temperature of aluminum melt is 973 K; density of aluminum is 2500 kg m^{-3}.

$$Q = \left[\frac{P_0 - P_L}{L} + \rho_{MIX}\, g\right]\frac{\pi R^4}{8\eta_{MIX}}$$

$$P_0 = P_L + L\left[\frac{8\eta Q}{\pi R^4} - \rho_{MIX}\, g\right]$$

$$P_L = 1.014\times10^5\ N m^{-2} + \rho g L = 1.014\times10^5 \frac{N}{m^2} + \frac{2500\ kg}{m^3}\left|\frac{9.8\ m}{s^2}\right|\frac{0.9\ m}{}\left|\frac{N s^2}{kg\ m}\right.$$

$$= 1.235\times10^5\ N m^{-2}$$

$$\eta_{MIX} = X_{N_2}\eta_{N_2} + X_{Cl_2}\eta_{Cl_2} = 0.75\,\eta_{N_2} + 0.25\,\eta_{Cl_2}$$

at $T = 973$ K

$\eta_{N_2} = 3.75\times10^{-5}\ N s m^{-2}$ Fig. 1.7

For Cl_2 – $M = 70.91$, $\sigma = 4.217$, $\frac{\varepsilon}{k_B} = 316.0$, $\frac{k_B T (973K)}{\varepsilon} = 3.08$, $\Omega_\eta = 1.064$

$$\eta_{Cl_2} = \frac{2.67\times10^{-5}\left[(70.91)(973)\right]^{1/2}}{(4.217)^2(1.064)} = 37.1\times10^{-5}\ P = 3.71\times10^{-5}\ N s m^{-2}$$

$$\eta_{MIX} = (0.75)(3.75\times10^{-5}) + (0.25)(3.71\times10^{-5}) = 3.74\times10^{-5}\ N s m^{-2}$$

at 298 K – $\rho_{N_2} = 1.185$ kg m^{-3}, $\rho_{Cl_2} = 2.956$ kg m^{-3}

$$\rho_{MIX} = (0.75)(1.185) + (0.25)(2.956) = 1.628\ kg\ m^{-3}$$

at 973 K – $\rho_{MIX} = 1.628\left(\frac{298}{973}\right) = 0.449$ kg m^{-3}

$$P_0 = 1.235\times10^5\frac{N}{m^2} + 0.9\,m\left[\frac{8}{\pi}\left|\frac{3.74\times10^{-5}\ N s}{m^2}\right|\frac{6.6\times10^{-5}\ m^3}{s}\left|\frac{1}{(1\times10^{-3})^4 m^4}\right| - \frac{0.449\ kg}{m^3}\left|\frac{9.8\ m}{s^2}\right.\right]$$

$$P_0 = 1.291\times10^5\ N m^{-2}$$

2.17 Glass flows through a small orifice by gravity to form a fiber. The free-falling fiber does not have a uniform diameter; furthermore as it falls through the air it cools so that its viscosity changes. a) Write the momentum equation for this situation. b) Write appropriate boundary conditions.

$V_\theta = 0$ but because the fiber diameter is changing, then both V_r and V_z exist. Assume steady state; but flow is not fully developed. The continuity equation, with $\rho \simeq$ const., is:

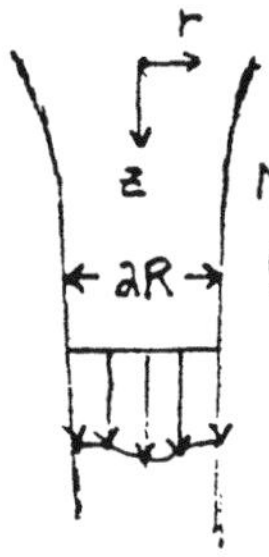

Notice: $R = R(z)$

$$\frac{1}{r}\frac{\partial}{\partial r}(r v_r) + \frac{\partial v_z}{\partial z} = 0$$

a. Table 2.3 (with $g_z = g$)

$$(A) \quad \rho\left(v_r \frac{\partial v_r}{\partial r} + v_z \frac{\partial v_z}{\partial z}\right) = -\left(\frac{1}{r}\frac{\partial}{\partial r}(r\,\tau_{rr}) + \frac{\partial \tau_{rz}}{\partial z}\right)$$

$$(C) \quad \rho\left(v_r \frac{\partial v_z}{\partial r} + v_z \frac{\partial v_z}{\partial z}\right) = -\left(\frac{1}{r}\frac{\partial}{\partial r}(r\,\tau_{rz}) + \frac{\partial \tau_{zz}}{\partial z}\right) + \rho g$$

Eqs. (A) and (c) can be simplified further by recognizing terms that can be ignored. Start with Eq. (A) and refer to Table 2.6. We expect $\frac{\partial V_r}{\partial r}$ to be small and certainly $v_r \ll v_z$

$$\therefore \; v_r \frac{\partial v_r}{\partial r} \simeq 0 \quad \text{and} \quad \frac{\partial}{\partial r}(r\,\tau_{rr}) \simeq 0$$

$$(A) \quad \rho\left(v_z \frac{\partial v_r}{\partial z}\right) = -\frac{\partial \tau_{rz}}{\partial z}$$

Now proceed with Eq. (c): $\frac{1}{r}\frac{\partial}{\partial r}(r\,\tau_{rz}) \gg \frac{\partial \tau_{zz}}{\partial z}$

$$(c) \quad \rho\left(v_r \frac{\partial v_z}{\partial r} + v_r \frac{\partial v_z}{\partial z}\right) = -\left(\frac{1}{r}\frac{\partial}{\partial r}(r\,\tau_{rz})\right) + \rho g$$

b. B.C. $v_r(r,0) = 0$; $v_z(r,0) = V_z$ (uniform); $\tau_{rz}(R) = 0$; $\frac{\partial v_r}{\partial r}(0,z) = 0$; $\frac{\partial v_z}{\partial r}(0,z) = 0$

2.18 A liquid flows upward through a tube, overflows, and then flows downward as a film on the outside.

a) Develop the pertinent momentum balance that applies to the falling film, for steady-state laminar flow, neglecting end effects.

b) Develop an expression for the velocity distribution.

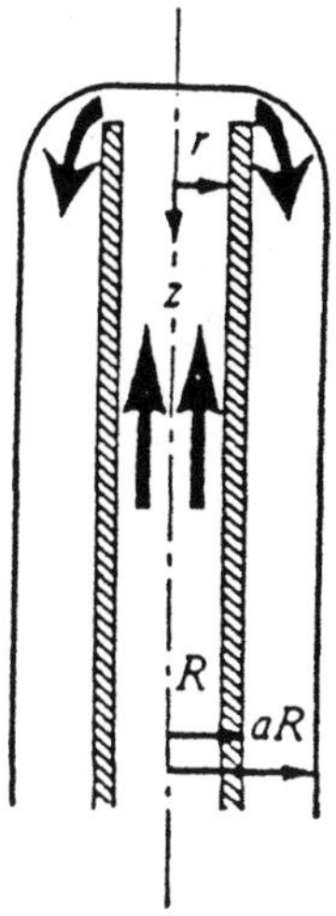

a. Table 2.3, Eq. F with $g_z = -g$ gives

$$\frac{d}{dr}\left(r\frac{dv_z}{dr}\right) = \frac{\rho g}{\eta} r \qquad \text{or} \qquad \frac{dv_z}{dr} = \frac{\rho g}{2\eta} r + \frac{c_1}{r}$$

b. at $r = aR$, $\frac{dv_z}{dr} = 0$

$$c_1 = -\frac{\rho g a^2 R^2}{2\eta}$$

$$\frac{dv_z}{dr} = \frac{\rho g r}{2\eta}\left[1 - a^2\frac{R^2}{r^2}\right]$$

$$v_z = \frac{\rho g}{4\eta} r^2 + c_1 \ln r + c_2$$

at $r = R$, $v_z = 0$

$$c_2 = -\frac{\rho g}{4\eta} R^2 - c_1 \ln R$$

So that

$$v_z = \frac{\rho g}{4\eta}(r^2 - R^2) + c_1 \ln\frac{r}{R}$$

or

$$v_z = \frac{\rho g R^2}{4\eta}\left[\left(\frac{r^2}{R^2} - 1\right) - 2a^2 \ln\frac{r}{R}\right]$$

3.1 Water at 300 K is flowing through a brass tube that is 30.0 m long and 13 mm in diameter (inner). The water is moving through the tube at a rate of 3.2×10^{-3} m^3 s^{-1}. The density of water is 1000 kg m^{-3}, and its viscosity is 8.55×10^{-4} N s m^{-2}. Calculate the pressure drop in Pa that accompanies this flow.

$$Re = \frac{4}{\pi}\left|\frac{3.2\times10^{-3}\,m^3}{s}\right|\frac{1\times10^3\,kg}{m^3}\left|\frac{1}{1.3\times10^{-2}\,m}\right|\frac{m^2}{8.55\times10^{-4}\,N\,s} = 3.666\times10^5 \;\therefore \text{Turbulent}$$

For brass $\frac{\varepsilon}{D} = \frac{0.0015}{13} = 1.2\times10^{-4}$; From Fig 3.2, $f = 0.0041$

From Eq. (3.17) $\Delta P = 4f\left(\frac{L}{D}\right)\left(\frac{1}{2}\right)\rho\bar{V}^2 = \frac{4 \mid 4.1\times10^{-3} \mid 30\,m \mid 1 \mid 1\times10^3\,kg \mid 4 \mid 3.2\times10^{-3}\,m^3}{\mid \mid 1.3\times10^{-2}\,m \mid 2 \mid m^3 \mid \pi \mid s \mid (1.3\times10^{-2})^2\,m^2}$

$\Delta P = 4.56\times10^5$ Pa

3.2 Evaluate the pressure drop in a horizontal 30 m length of galvanized rectangular duct (30.0 mm x 75.0 mm) for the following conditions:

a) An average air flow velocity of 0.46 m s^{-1} at 300 K and atmospheric pressure.
b) An average air flow velocity of 4.6 m s^{-1} at 300 K and atmospheric pressure. The density and viscosity are 1.16 kg m^{-3} and 1.85×10^{-5} N s m^{-2}.

$$De = \frac{2(30)(75)}{(30+75)} = 42.86\,mm = 4.286\times10^{-2}\,m$$

a. $Re = \frac{De\bar{V}\rho}{\eta} = \frac{4.286\times10^{-2}\,m \mid 0.46\,m \mid 1.16\,kg \mid m^2}{\mid s \mid m^3 \mid 1.85\times10^{-5}\,N\,s} = 1.236\times10^3$

$\therefore$ Laminar flow-apply Fig. 3.3: $\frac{z_1}{z_2} = \frac{30}{75} = 0.4 \;\therefore \phi = 0.98$

$$f = \frac{16}{\phi Re} = \frac{16}{(0.98)(1.236\times10^3)} = 1.32\times10^{-2}$$

$$P_0 - P_L = \frac{2fL\rho\bar{V}^2}{De} = \frac{2 \mid 1.32\times10^{-2} \mid 30\,m \mid 1.16\,kg \mid 0.46^2\,m^2}{\mid \mid \mid m^3 \mid s^2 \mid 4.286\times10^{-2}\,m} = 4.54\,Pa$$

b. $Re = 1.236\times10^3\left(\frac{4.6}{0.46}\right) = 12\,360 \;\therefore$ Turbulent flow – apply Fig. 3.2 or Eq. (3.20)

$$f = 0.0791\,Re^{-1/4} = (0.0791)(1.236\times10^4)^{-1/4} = 7.50\times10^{-3}$$

$$P_0 - P_L = \frac{2 \mid 7.50\times10^{-3} \mid 30\,m \mid 1.16\,kg \mid 4.6^2\,m^2}{\mid \mid \mid m^3 \mid s^2 \mid 4.286\times10^{-2}\,m} = 258\,Pa$$

3.3 For flow in tubes (smooth wall) the friction factor is given by Eq. (3.20) for $2.1 \times 10^3 < Re < 10^5$. What is the percent change in the pressure drop if the tube diameter is doubled for the same volume flow rate, same fluid and the same tube length? Assume that Eq. (3.20) for the friction factor applies.

Eq. (3.20) $f = 0.0791\, Re^{-\frac{1}{4}}$

Eq. (3.17) $f = \frac{1}{4}\frac{D}{L}\left[\frac{\Delta P}{\frac{1}{2}\rho \bar{V}^2}\right]$

Compare two pressure drops: ΔP_1 and ΔP_2

$$\frac{\Delta P_2}{\Delta P_1} = \frac{4 f_2 \left(\frac{L}{D_2}\right)\frac{1}{2}\rho \bar{V}_2^2}{4 f_1 \left(\frac{L}{D_1}\right)\frac{1}{2}\rho \bar{V}_1^2} = \frac{f_2\, D_1\, \bar{V}_2^2}{f_1\, D_2\, \bar{V}_1^2}$$

For $D_2 = 2D_1$ and $\bar{V}_2 D_2^2 = \bar{V}_1 D_1^2$ - same volume flow rate

$$\frac{D_1}{D_2} = \frac{1}{2} \text{ and } \left(\frac{\bar{V}_2}{\bar{V}_1}\right)^2 = \left(\frac{D_1}{D_2}\right)^4 = \frac{1}{16}$$

$$\therefore \frac{\Delta P_2}{\Delta P_1} = \left(\frac{f_2}{f_1}\right)\left(\frac{1}{2}\right)\left(\frac{1}{16}\right) = \frac{1}{32}\left(\frac{f_2}{f_1}\right)$$

$$\text{or } \frac{\Delta P_2}{\Delta P_1} = \frac{1}{32}\left(\frac{Re_1}{Re_2}\right)^{1/4} = \frac{1}{32}\left(\frac{D_1 \bar{V}_1}{D_2 \bar{V}_2}\right)^{\frac{1}{4}} = \frac{1}{32}\left(\frac{1}{2}\right)^{\frac{1}{4}}\left(4\right)^{\frac{1}{4}} = 3.716 \times 10^{-2}$$

$$\text{Pct. change} = \frac{\Delta P_1 - \Delta P_2}{\Delta P_1}(100) = \left(1 - \frac{\Delta P_2}{\Delta P_1}\right)(100)$$

Pct. change = 96.28 % decrease

3.4 Show that for flow through a slit with a spacing that is much less than the width, Eq. (3.22) gives the friction factor for laminar flow.

Flow through a slit is approximated as flow between parallel plates. Equation (2.22) applies.

$$\bar{v} = \frac{\delta^2 (P_0 - P_L)}{3\eta L}, \quad \delta = \text{semithickness}$$

$$D_e = \frac{4A}{P_W} = \frac{(4)(2\delta W)}{2W} = 4\delta \quad (W \gg \delta)$$

$$\therefore \bar{v} = \frac{D_e^2 (P_0 - P_L)}{48 \eta L}$$

Also $f = \frac{1}{4}\left(\frac{D_e}{L}\right)\left(\frac{P_0 - P_L}{\frac{1}{2}\rho \bar{v}^2}\right)$

Combine $\quad f = \frac{24\eta}{D_e \bar{v} \rho} = \frac{24}{Re}$

3.5 Determine the size of the largest alumina particle in Problem 2.14 that would be expected to obey Stokes' law, remembering that for spheres this law is valid for Re $\leq$ 1.

$$F_S = F_W + F_K$$

$$\frac{4}{3}\pi R^3 \rho_{Fe} g = \frac{4}{3}\pi R^3 \rho_{Al_2O_3} g + 6\pi \eta R v_t$$

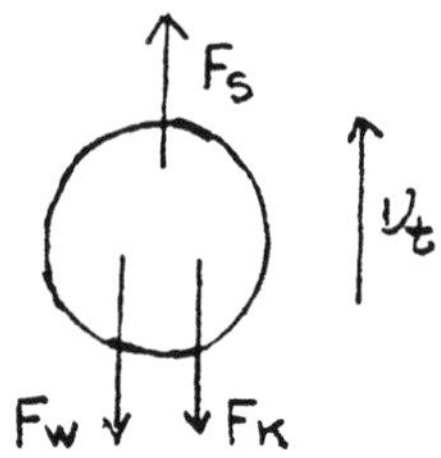

$$v_t = \frac{2}{9}\frac{R^2}{\eta} g \left(\rho_{Fe} - \rho_{Al_2O_3}\right)$$

Laminar to turbulent transition at $Re = 1$

$$Re = \frac{D V_t \rho_{Fe}}{\eta} \quad \therefore \quad \frac{2 R V_t \rho_{Fe}}{\eta} = \frac{4}{9}\frac{R^3}{\eta^2} g\left(\rho_{Fe} - \rho_{Al_2O_3}\right)\rho_{Fe} = 1$$

$$R^3 = \frac{9}{4}\eta^2 \frac{1}{g\left(\rho_{Fe} - \rho_{Al_2O_3}\right)\rho_{Fe}}$$

$\rho_{Fe} = 7600 \text{ kg m}^{-3}$, $\rho_{Al_2O_3} = 3320 \text{ kg m}^{-3}$, $\eta_{Fe}(1600\,°C) \doteq 6.4 \text{ cP} = 6.4 \times 10^{-3} \text{ kg m}^{-1}\text{s}^{-1}$

$$R^3 = \frac{9}{4}\left|\frac{(6.4\times 10^{-3})^2 \text{ kg}^2}{\text{m}^2\,\text{s}^2}\right|\frac{\text{s}^2}{9.8 \text{ m}}\left|\frac{\text{m}^3}{(7600-3320)\text{ kg}}\right|\frac{\text{m}^3}{7600 \text{ kg}} = 2.891 \times 10^{-13} \text{ m}^3$$

$R = 6.61 \times 10^{-5} \text{ m} = 0.066 \text{ mm}$

3.6 A falling-sphere viscometer was used to determine the viscosity of a slag intended for the production of copper. The viscosity of the slag was determined to be 441.2 Poise, using a steel ball as the falling sphere. Is this a valid viscosity? Why or why not? If not, determine the real value of the viscosity and then calculate its kinematic viscosity. The density of the slag may be taken as one-half that of the steel ball.

Data: Radius of steel ball, 88.7 mm; terminal velocity of steel ball, 1.52 m s^{-1}.

$$F_K = F_W - F_S$$

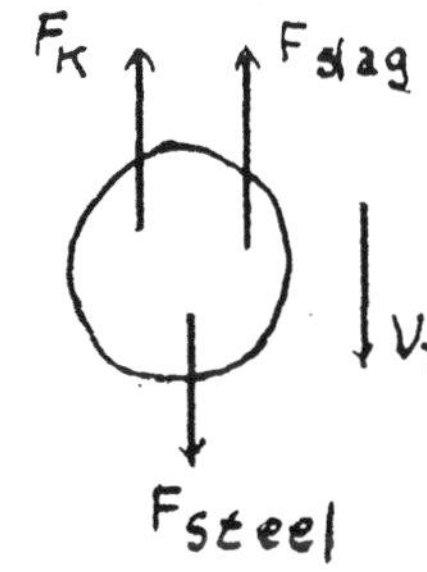

$$F_K = (\pi R^2)\left(\frac{1}{2}\rho_{slag} v_t^2\right) f = \frac{4}{3}\pi R^3 g\left(\rho_{steel} - \rho_{slag}\right)$$

$$\rho_{steel} = 2\rho_{slag}$$

$$f = \left(\frac{8}{3}\right)\frac{Rg}{v_t^2}\,\frac{(2\rho_{slag} - \rho_{slag})}{\rho_{slag}} = \frac{8}{3}\frac{Rg}{V_t^2} = \frac{8}{3}\left|0.0887 \text{ m}\right|\frac{9.8 \text{ m}}{\text{s}^2}\left|\frac{\text{s}^2}{(1.52)^2 \text{ m}^2}\right. = 1.0$$

$f = 1.0$, Fig. 3.8 $Re_D = 100$

$$Re_D = \frac{D V_t \rho_{slag}}{\eta_{slag}} = \frac{D V_t}{\nu}$$

$$\nu_{slag} = \frac{0.1774\ \text{m}}{} \left| \frac{1.52\ \text{m}}{5} \right| \frac{}{100} = 2.70 \times 10^{-3}\ \text{m}^2\ \text{s}^{-1}$$

$$\eta_{slag} = \nu_{slag}\,\rho_{slag} = \frac{\nu_{slag}\,\rho_{st}}{2} = \frac{2.7 \times 10^{-3}\ \text{m}^2}{\text{s}} \left| \frac{7.850 \times 10^{3}\ \text{kg}}{\text{m}^3} \right| \frac{}{2} = 10.6\ \text{kg m}^{-1}\ \text{s}^{-1}$$

$\eta_{slag} = 10.6\ \text{N s m}^{-2} = 106\ \text{P}$. The stated viscosity is incorrect.

3.7 Two spheres of equal density and different diameters fall through a liquid with an unknown density and an unknown viscosity. The diameter of the larger sphere is twice the diameter of the smaller sphere.

a) With an appropriate force balance, derive an equation for the terminal velocity of either sphere. Your equation should be valid for any Reynolds number.

b) Assume that $10^3 < Re_D < 2 \times 10^5$ for both spheres and calculate the ratio of the terminal velocity of the larger to that of the smaller.

c) Assume that $Re_D < 1$ for both spheres, and calculate the ratio of terminal velocities.

a. $F_W = F_s + F_K$

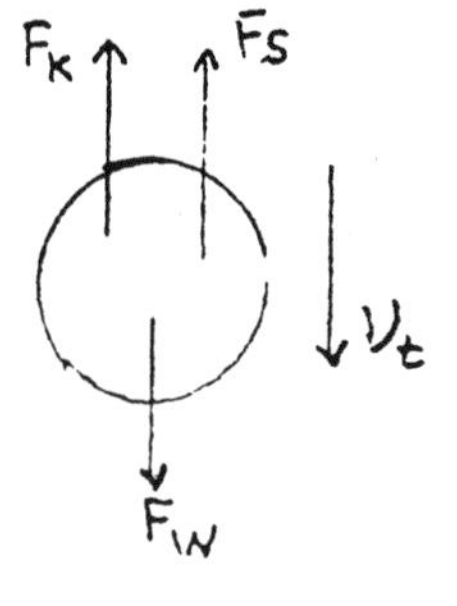

$$F_K = fAK = f(\pi R^2)\left(\tfrac{1}{2}\rho \nu_t^2\right)$$

$$F_s = \tfrac{4}{3}\pi R^3 \rho g$$

$$F_W = \tfrac{4}{3}\pi R^3 \rho_s g$$

$$\tfrac{4}{3}\pi R^3 \rho_s g = \tfrac{4}{3}\pi R^3 \rho g + \tfrac{\pi}{2} f R^2 \rho \nu_t^2$$

$$\nu_t = \left[\frac{8 R (\rho_s - \rho) g}{3 \rho f}\right]^{\frac{1}{2}}$$

b. For these values of Re, $f \simeq 0.42$ (constant)

Let 2 ≑ larger and 1 ≑ smaller

$$\frac{\nu_{t_2}}{\nu_{t_1}} = \left(\frac{R_2}{R_1}\right)^{\frac{1}{2}} = 2^{\frac{1}{2}} = 1.414$$

c. For these values of Re, Stokes' law applies.

$$f = \frac{24}{Re} = \frac{24\eta}{D v_t \rho} = \frac{12\eta}{R v_t \rho}$$

$$\frac{v_{t2}}{v_{t1}} = \left[\frac{\frac{R_2}{f_2}}{\frac{R_1}{f_1}}\right]^{\frac{1}{2}} = \left[\frac{R_2 f_1}{R_1 f_2}\right]^{\frac{1}{2}} = \left[\frac{R_2 R_2 v_{t2}}{R_1 R_1 v_{t1}}\right]^{\frac{1}{2}} \quad \therefore \quad \frac{v_{t2}}{v_{t1}} = \left[\frac{R_2}{R_1}\right]^2 = 4$$

3.8 Bubbles that rise through liquids may be treated as rigid spheres provided they are small enough. Assume that a spherical bubble of air has a diameter of 1 mm at the bottom of a glass melt that is 1 m deep.

a) Calculate the pressure in the bubble as a function of the distance below the top surface of the glass. Neglect added pressure within the bubble because of the surface tension.

b) Calculate the diameter of the bubble as a function of the distance below the top surface of the glass.

c) Neglecting acceleration effects, calculate the velocity of the bubble as a function of distance below the top surface of the glass. *Data for melt*: Temperature is 1700 K; viscosity is 2.0 N s m^{-2}; density is 3000 kg m^{-3}. Assume that air behaves as an ideal gas with a molecular weight of 28.8 kg kmol^{-1}.

a. $P = P_{atm} + \rho g h$ where h is the distance below the melt.

b. $V = \frac{P(\text{bottom})\,V(\text{bottom})}{P} = \frac{4}{3}\pi R^3$

$P(\text{bottom}) = P_{atm} + \rho g h(\text{bottom}) = P_{atm} + \rho g$ since $h = 1\text{m}$

$V(\text{bottom}) = \frac{4}{3}\pi\left[R(\text{bottom})\right]^3$, $R(\text{bottom}) = 0.5 \times 10^{-3}\,\text{m}$

$P = P_{atm} + \rho g h$

$$D = 2\left[\frac{(P_{atm} + \rho g)\left[R(\text{bottom})\right]^3}{P_{atm} + \rho g h}\right]^{\frac{1}{3}}$$

c. Eq. (2.121) $\eta = \frac{2R^2(\rho_L - \rho)g}{9 v_t}$ if $\frac{2R v_t}{\nu} \lesssim 1$; $\quad v_t = \frac{2R^2 g \rho_L}{9\eta}$

The following computer program gives the answers to a, b and c versus the depth of the liquid. The program also calculates the Reynolds numbers to assure the validity of using Eq. (2.121).

```
10 'Problem 3.8
20   P0 = 101330! : G = 9.807
30   T = 1700 : NETA = 2 : RHO = 3000
40   RBOT = .0005 : PI = 3.1416
50   PBOT = P0 + RHO*G : VBOT = (4*PI/3)*RBOT^3
60 '
70   'a) P = P0 + rho*g*h  P0 is 1 stand. atm.  h is distance below top of melt
80   'b) V = P(bottom)*V(bottom)/P
90   'c) f = f(Re)      Vt = (2/9)*R*R*g*rho/neta   valid if Re<1
100         LPRINT
110         LPRINT "  distance     pressure   bubble   velocity  Reynolds"
120        LPRINT "  below, mm      N/m2      dia, mm     mm/s     number"
130     FOR H = 0 TO 1.01 STEP .1
140         P = P0 + RHO*G*H : V = PBOT*VBOT/P
150         R = (3*V/(4*PI))^(1/3) : D = 2*R
160         VT = (2/9)*R*R*G*RHO/NETA   ' Re <= 1
170         RE = D*VT*RHO/NETA
180         LPRINT USING "  ####       ##.###^^^^  ##.###   ##.### ##.###^^^
^ ";H*1000,P,D*1000,VT*1000,RE
190     NEXT H
200 END
```

distance below, mm	pressure N/m2	bubble dia, mm	velocity mm/s	Reynolds number
0	1.013E+05	1.089	0.969	1.582E-03
100	1.043E+05	1.078	0.950	1.537E-03
200	1.072E+05	1.068	0.933	1.495E-03
300	1.102E+05	1.059	0.916	1.455E-03
400	1.131E+05	1.050	0.900	1.417E-03
500	1.160E+05	1.041	0.885	1.381E-03
600	1.190E+05	1.032	0.870	1.347E-03
700	1.219E+05	1.024	0.856	1.315E-03
800	1.249E+05	1.015	0.843	1.284E-03
900	1.278E+05	1.008	0.830	1.254E-03
1000	1.308E+05	1.000	0.817	1.226E-03

3.9 A thermocouple tube lies in a melt that is flowing perpendicular to the axis of the tube. Calculate the force per unit length of tube exerted by the flowing metal. *Data*: Velocity of the melt is 3 m s^{-1}; viscosity is 2 × 10^{-3} N s m^{-2}; density of the melt is 8000 kg m^{-3}; diameter of thermocouple tube is 61 mm.

$F_K = f A K$

$Re_D = \frac{D V_\infty \rho}{\eta} = \frac{(0.061)(3)(8000)}{2\times10^{-3}} = 7.32\times10^5$

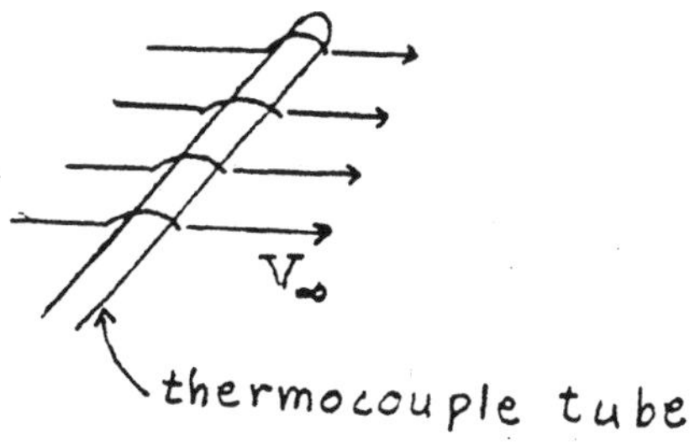

From Fig. 3.9 for infinite circular cylinder $f = 0.30$

$A = DL = 0.061\,L$ with L in m, A is m^2

$$K = \frac{1}{2}\rho V_\infty^2 = \frac{1}{2}\left|\frac{8000\ \text{kg}}{\text{m}^3}\right|\frac{(3)^2\ \text{m}^2}{\text{s}^2} = 3.6\times10^4\ \text{kg m}^{-1}\ \text{s}^{-2}$$

$$F_K = 0.3\left|0.061\,L\ \text{m}^2\right|\frac{3.6\times10^4\ \text{kg}}{\text{m s}^2}\left|\frac{\text{N s}^2}{\text{kg m}}\right| = 659\ N$$

$\frac{F_K}{L} = 659$ N m^{-1} where L is in m.

3.10 A packed bed reduction-reactor, 15.0 m high and 6.0 m in diameter, is packed with spherical metal oxide pellets (D_p = 3 mm). A reducing gas enters the top of the bed at 800 K and at a rate of 95 kg s^{-1} and exits the reactor at the same temperature. What should the pressure at the top of the reactor be if the pressure at the bottom of the reactor is maintained at 1.4 × 10^5 Pa. *Data*: Bed porosity, ω is 0.40; viscosity of the reducing gas at 800 K is 4.13 × 10^{-5} N s m^{-2}; density of the gas at atmospheric pressure and 800 K is 0.5 kg m^{-3}.

Equation (3.49) in differential form:

$$-\frac{dP'}{dx} = K_1 V_0 + K_2 \rho V_0^2$$

where $K_1 = \frac{150\,\eta(1-\omega)^2}{D_p^2\,\omega^3}$

$K_2 = \frac{1.75\,(1-\omega)}{D_p\,\omega^3}$

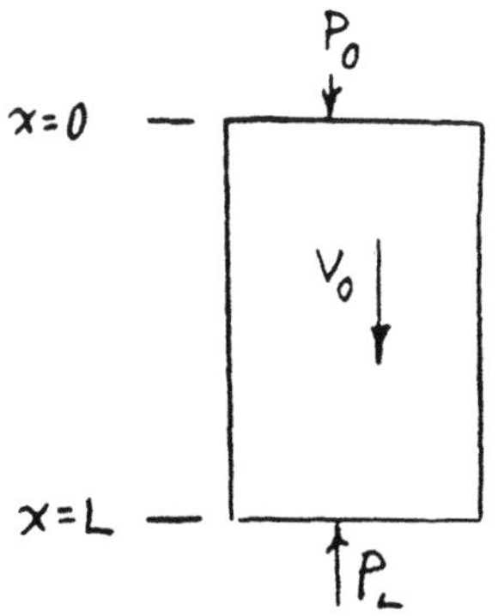

$$P' = P + \rho g(L-x)\,, \quad \frac{dP'}{dx} = \frac{dP}{dx} - \rho g\,.$$

Also substitute $\dot{W} = V_0 A \rho$ (constant mass flow rate)

Then

$$\frac{dP}{dx} = \beta\frac{1}{\rho} + \rho g\,, \quad \text{where } \beta = -K_1\frac{\dot{W}}{A} - K_2\left(\frac{\dot{W}}{A}\right)^2.$$

Density varies with pressure as

$$\rho = \frac{P}{MRT}\,, \text{where } M = \text{molecular mass of the gas.}$$

$$\frac{dP}{dx} = \beta MRT\frac{1}{P} + \frac{g}{MRT}P$$

With $P = P_0$ at $x=0$ and $P = P_L$ at $x = L$, we get

$$\ln\left\{\frac{\beta(MRT)^2 + gP_L^2}{\beta(MRT)^2 + gP_0^2}\right\} = \frac{2gL}{MRT}\,.$$

Evaluate parameters (all SI units):

$$K_1 = \frac{(150)(4.13\times10^{-5})(1-0.40)^2}{(0.003)^2(0.40)^3} = 3.872\times10^3$$

$$K_2 = \frac{(1.75)(1-0.40)}{(0.003)(0.40)^3} = 5.469\times10^3$$

$$\beta = -(3.872\times10^3)\frac{(95)(4)}{(\pi)(6^2)} - 5.469\times10^3\left[\frac{(95)(4)}{(\pi)(6^2)}\right]^2 = -74.57\times10^3$$

$$MRT = \frac{P}{\rho} = \frac{1.0133\times10^5}{0.5} = 2.027\times10^5$$

$g = 9.807$

$P_L = 1.4\times10^5$ Pa

Then

$$\ln\left\{\frac{(-74.57)(2.027\times10^5)^2 + (9.807)(1.4\times10^5)^2}{(-74.57)(2.027\times10^5)^2 + (9.807)P_0^2}\right\} = \frac{(2)(9.807)(15)}{2.027\times10^5}$$

Solve for P_0.

$P_0 = 6.875\times10^5$ Pa

3.11 In a packed bed reactor (diameter = 4.5 m and height = 18 m), metal oxide A forms a central column within the reactor, having a diameter of 3.0 m, while pellets of metal oxide B fill the annulus between metal oxide A and the wall of the reactor. The pressure at the top of the bed is maintained at 6.9×10^4 Pa, while the pressure at the bottom is kept at 1.72×10^5 Pa. Calculate the fraction of reducing gas that passes through metal oxide A. It may be assumed that the temperature and reducing gas density are uniform throughout the reactor. Furthermore, turbulent flow conditions prevail.

Data for A: $\omega = 0.40$, $D_p = 76$ mm.

Data for B: $\omega = 0.25$, $D_p = 19$ mm.

$$A_A = \pi (1.5)^2 = 7.068\ m^2, \quad A_B = \pi (2.25)^2 - \pi (1.5)^2 = 8.836\ m^2$$

$$\frac{\Delta P}{L} = \left.\frac{1.75 \rho V_o^2 (1-\omega)}{D_p \omega^3}\right|_A = \left.\frac{1.75 \rho V_o^2 (1-\omega)}{D_p \omega^3}\right|_B, \quad \rho_A = \rho_B$$

$$V_o = \frac{Q}{A}$$

$$\frac{Q_A^2 (1-\omega_A)}{A_A^2 D_{PA} \omega_A^3} = \frac{Q_B^2 (1-\omega_B)}{A_B^2 D_{PB} \omega_B^3}$$

$$\left(\frac{Q_A^2}{Q_B^2}\right) = \frac{(1-\omega_B)(\omega_A)^3 (D_{PA}) (A_A)^2}{(1-\omega_A)(\omega_B)^3 (D_{PB}) (A_B)^2}$$

$$\frac{Q_A}{Q_B} = \left[\left(\frac{0.75}{0.60}\right)\left(\frac{0.40}{0.25}\right)^3\left(\frac{76}{19}\right)\left(\frac{7.068}{8.836}\right)^2\right]^{\frac{1}{2}} = 3.620$$

$$Q_A = 3.620\ Q_B, \quad Q_A + Q_B = 1 = 3.620\ Q_B + Q_B$$

$$Q_B = 0.216, \quad Q_A = 0.784$$

$$\therefore\ Q_A = 78.4\%$$

$$Q_B = 21.6\%$$

3.12 Preliminary experimental studies have shown that the porosity in a newly developed packed bed reactor is $\omega < 0.6$. The pellets have a diameter of 30.0 mm and the reducing gas flows through the bed at a rate of 0.025 kg s^{-1}. The reactor has 3.0 m $\times$ 3.0 m square cross section and is 15 m in height. A constant pressure difference of 690 Pa is maintained between the inlet and outlet nozzles, and it may be assumed that the temperature is uniform throughout the reactor. You are required to evaluate the bed porosity. The properties of the gas are $\eta = 2.07 \times 10^{-5}$ N s m^{-2} and $\rho = 1.2$ kg m^{-3} (average).

$$Re_E = 6\,Re_C = \frac{D_p \rho V_0}{\eta(1-\omega)} = \frac{D_p \dot{W}}{\eta A(1-\omega)}, \text{ since } \rho V_0 = \frac{\dot{W}}{A}$$

$$Re_E = \frac{0.03\text{ m}}{} \left| \frac{0.025\text{ kg}}{\text{s}} \right| \frac{\text{m}^2}{2.07\times10^{-5}\text{ N s}} \left| \frac{}{3^2\text{ m}^2} \right| \frac{\text{N s}^2}{\text{kg m}} \left| \frac{1}{(1-\omega)} \right. = \frac{4.026}{(1-\omega)}$$

$0 < \omega < 0.6$ - for $\omega = 0$, $Re_E = 4.026$; for $\omega = 0.6$, $Re_E = 10$

$Re_C = \dfrac{Re_E}{6}$; $0.671 \leq Re_C \leq 1.67$ $\therefore$ Blake-Kozeny Eq. is valid.

For spherical particles $\dfrac{\omega^3}{(1-\omega)^2} = \dfrac{150\,\eta\, L\, V_0}{D_p^2\, \Delta P}$; $V_0 = \dfrac{0.025\text{ kg}}{\text{s}} \left| \dfrac{\text{m}^3}{1.2\text{ kg}} \right| \dfrac{}{3^2\text{ m}^2}$

$= 2.315\times10^{-3}$ m s^{-1}

$$\frac{\omega^3}{(1-\omega^3)^2} = \frac{150}{} \left| \frac{2.07\times10^{-5}\text{ N s}}{\text{m}^2} \right| \frac{15\text{ m}}{} \left| \frac{2.315\times10^{-3}\text{ m}}{\text{s}} \right| \frac{}{0.03^2\text{ m}^2} \left| \frac{\text{m}^2}{690\text{ N}} \right. = 1.736\times10^{-4}$$

$\omega = 0.054$

3.13 Molten aluminum is passed through a horizontal filter bed of Al_2O_3 spheres in order to remove drossy oxides from the aluminum. The filter bed comprises two different packings arranged in series.

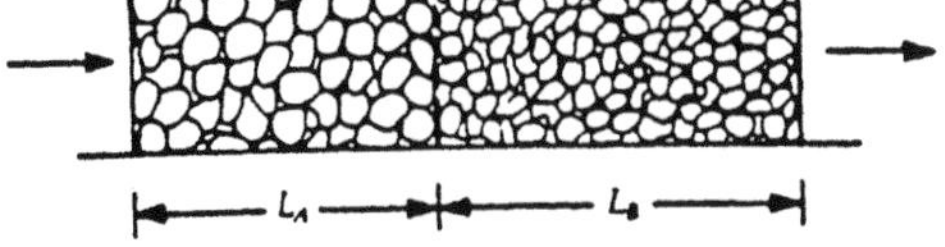

The first packing encountered by the flow captures large drossy particles, and the second packing captures the smaller drossy particles. Given $L_A = 0.7\, L_B$, $\omega_A = \omega_B$, $D_{P,A} = 2D_{P,B}$, compute the ratio of the pressure drop through A to the pressure through B for a) very low Reynolds numbers, and b) very high Reynolds numbers.

a. $\frac{\Delta P}{L} = \frac{150 \eta V_0 (1-\omega)^2}{D_p^2 \omega^3}$, very low Reynolds numbers.

$$\frac{\Delta P_A}{\Delta P_B} = \frac{\frac{L_A}{D_{P,A}^2}}{\frac{L_B}{D_{P,B}^2}} = \left(\frac{D_{P,B}^2}{D_{P,A}^2}\right)\left(\frac{L_A}{L_B}\right) = \frac{0.7}{4} = 0.175 \text{ since } D_{P,A} = 2D_{P,B} \text{ and } L_A = 0.7L_B$$

b. $\frac{\Delta P}{L} = \frac{1.75 \rho V_0^2 (1-\omega)}{D_p \omega^3}$, very high Reynolds numbers.

$$\frac{\Delta P_A}{\Delta P_B} = \frac{\frac{L_A}{D_{P,A}}}{\frac{L_B}{D_{P,B}}} = \left(\frac{D_{P,B}}{D_{P,A}}\right)\left(\frac{L_A}{L_B}\right) = \left(\frac{1}{2}\right)(0.7) = 0.350$$

3.14 Derive an equation for the pressure drop through an isothermal column of a porous medium, that accompanies the flow of a compressible gas. Assume that the gas follows ideal behavior.

Rewrite Eq. (3.46) in differential form:

$$-\frac{dP}{dx} = \frac{4.2 \eta V_0 S_0^2 (1-\omega)^2}{\omega^3} + \frac{0.292 \rho V_0^2 S_0 (1-\omega)}{\omega^3}$$

Let $C_1 \equiv \frac{4.2 \eta S_0^2 (1-\omega)^2}{\omega^3}$ and $C_2 \equiv \frac{0.292 S_0 (1-\omega)}{\omega^3}$

Then $-\frac{dP}{dx} = C_1 V_0 + C_2 \rho V_0^2$; But $V_0 = \frac{Q}{A} = \frac{\dot{W}}{A\rho}$ where $\dot{W}$ = mass flow rate.

At steady state $\dot{W}/A$ is constant but ρ varies with the pressure.

$$-\frac{dP}{dx} = C_1 \frac{\dot{W}}{A}\frac{1}{\rho} + C_2 \rho \left(\frac{\dot{W}}{A}\right)^2 \frac{1}{\rho^2} = \left[C_1 \frac{\dot{W}}{A} + C_2 \left(\frac{\dot{W}}{A}\right)^2\right]\frac{1}{\rho}$$

$$\text{Hence, } -\int_{P_0}^{P_L} \rho\, dP = \left[C_1 \frac{\dot{W}}{A} + C_2 \left(\frac{\dot{W}}{A}\right)^2\right] \int_0^L dx$$

The ideal gas law gives: $\rho = \frac{MP}{RT}$ where M is the molecular weight.

$$\int_{P_0}^{P_L} \rho\, dP = \frac{M}{RT}\int_{P_0}^{P_L} P\, dP = \frac{M}{2RT}\left(P_L^2 - P_0^2\right)$$

$$\text{So } P_0^2 - P_L^2 = \frac{2RT}{M}\left[C_1 \frac{\dot{W}}{A} + C_2 \left(\frac{\dot{W}}{A}\right)^2\right] L$$

We can write: $P_L^2 - P_0^2 = (P_L - P_0)(P_L + P_0) = -2\Delta P \bar{P}$ where $\bar{P} = \frac{1}{2}(P_L + P_0)$, avg. press.

$$\therefore \Delta P = \frac{RT}{M\bar{P}}\left[C_1 \frac{\dot{W}}{A} + C_2\left(\frac{\dot{W}}{A}\right)^2\right] L$$

3.15 For unidirectional flow through a column of a porous medium, show that Eq. (3.61) reduces to Eq. (3.31).

$$\frac{\partial v_x}{\partial t}, \frac{\partial v_x}{\partial x}, \frac{\partial v_x}{\partial y}, \frac{\partial v_x}{\partial z}, \frac{\partial}{\partial x}\left(\frac{1}{\omega}\right); v_y, v_z, \frac{\partial^2 v_x}{\partial x^2}, \frac{\partial^2 v_x}{\partial y^2}, \frac{\partial v_x}{\partial z^2} = 0$$

$$\therefore \omega g_x - \frac{\omega}{\rho}\frac{dP}{dx} - \frac{\nu\omega}{\mathcal{P}} v_x = 0$$

$$\rho g_x - \frac{dP}{dx} = \frac{\eta}{\mathcal{P}} v_x$$

$$v_x = -\frac{\mathcal{P}}{\eta}\left[\frac{dP}{dx} - \rho g_x\right] \text{ which is Eq. (3.31)}$$

3.16 In a *falling head permeameter*, the permeability is determined by measuring the difference in height between two liquid columns. In the apparatus depicted below, H decreases and h increases as liquid flows by gravity through the porous medium of length L. Derive an equation that gives h as a function of time t, assuming that $\mathcal{P}$ is uniform and constant.

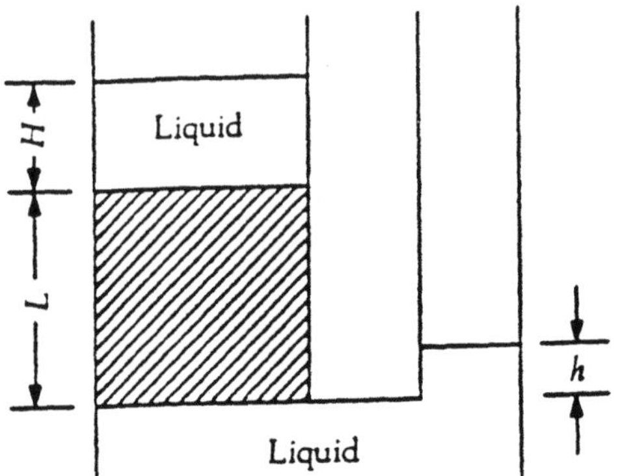

Cross-sectional areas of the larger and smaller columns of liquid are A and a, respectively.

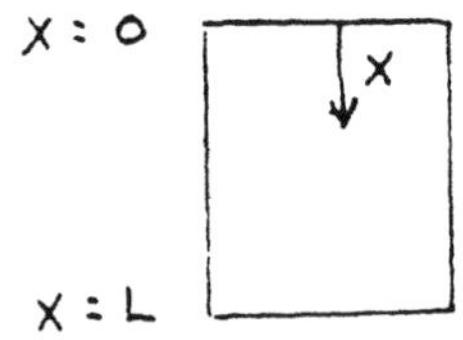

Eq. (3.31) $v_x = -\frac{\mathcal{P}}{\eta}\left[\frac{\partial P}{\partial x} - \rho g_x\right]$, $g_x = g$

$$\frac{dP}{dx} = \frac{P_L - P_0}{L}; \; P_L = \rho g h; \; P_0 = \rho g H$$

$$\frac{dP}{dx} = \frac{\rho g h - \rho g H}{L} = \frac{\rho g}{L}(h - H)$$

$$v_x = -\frac{\mathcal{P}}{\eta}\left[\frac{\rho g}{L}(h-H) - \rho g\right] = -\frac{\mathcal{P}\rho g}{\eta L}\left[h - H - L\right]$$

Also $A v_x = a \frac{dh}{dt}$; $v_x = \frac{a}{A}\frac{dh}{dt}$

$$\frac{a}{A}\frac{dh}{dt} = -\frac{\mathcal{P}\rho g}{\eta L}\left[h - H - L\right]; \; \frac{dh}{dt} = -\frac{\mathcal{P}}{\eta}\frac{A}{a}\frac{\rho g}{L}\left[h - H - L\right]$$

Also $A(H_0 - H) = a(h - h_0)$ where H_0, h_0 = initial values of H, h

$$\frac{dh}{dt} = -\frac{\mathcal{P}A\rho g}{\eta a L}\left[-H_0 + \frac{a}{A}h - \frac{a}{A}h_0 + h - L\right] = -\frac{\mathcal{P}A\rho g}{\eta a L}\left[-H_0 - L - \frac{a}{A}h_0 + \left(1 + \frac{a}{A}\right)h\right]$$

Let $K_1 = \left(-H_0 - L - \frac{a}{A}h_0\right)\frac{\mathcal{P}A\rho g}{\eta a L}$ and $K_2 = -\left(1 + \frac{a}{A}\right)\frac{\mathcal{P}A\rho g}{a L}$

$$\frac{dh}{dt} = -K_1 + K_2 h; \; \int_{h_0}^{h}\frac{dh}{-K_1 + K_2 h} = \int_0^t dt \; \therefore h = \frac{1}{K_2}\left[K_1 + (-K_1 + K_2 h_0)e^{K_2 t}\right]$$

3.17 The tube bundle theory for permeability predicts

$$\mathcal{P} = \frac{\omega^3}{K\, S_0^2\,(1 - \omega)^2}.$$

where K is a constant. Assuming that Eq. (3.55) applies, does the tube bundle theory compare to the empirical result given by Eq. (3.59)?

$\mathcal{P} = \frac{\omega^3}{K\, S_0^2\,(1-\omega)^2}$ under Eq. (3.58) we see that $K = 4.2$ (usually taken to be this value, or could be $K = 5.0$, for grandular materials).

Eq. (3.59): $\phi = \mathcal{P}$ and $\bar{D}_{vs}$ is in μm.

$P = 4.6 \times 10^{-11}\, \bar{D}_{VS}^{\,0.73}\, \omega^{6.8}$, since Eq. (3.55) gives $S_0 = \dfrac{6}{\bar{D}_{VS}}$, so that

$$P = 4.6 \times 10^{-11} \left(\frac{6}{S_0}\right)^{0.73} \omega^{6.8}$$

Obviously the two are not the same, but let's compare in the ranges $59 \le \bar{D}_{VS} \le 715\,\mu m$; $0.25 \le \omega \le 0.56$ (scopes under Eq. 3.60). The following program calculates the dimensionless permeability $-PS_0^2-$ for both equations. The results show that the two equations do not favorably compare. We see that the tube bundle theory predicts much higher permeabilities than the empirical result given by Eq. (3.59).

```
10  Problem 3.17
20   LPRINT "  Dvs                        Dimensionless Permeability"
30   LPRINT " microns   porosity   tube bundle     empirical powders"
40   LPRINT " *******   ********   ***********     *****************"
50   FOR DVS = 100 TO 700 STEP 100  'DVS in microns
60   FOR W = .2 TO .6 STEP .2
70       S00 = 6/DVS          'S00 in 1/microns
80       S0 = 1000000! * S00     'S0 in 1/m
90       Tube bundle theory
100       DENOM = 4.2 * S0^2 * (1-W)^2 : PERM = W^3/DENOM
110       DIMPERM1 = PERM*S0^2
120       Empirical for powders, Eq. (3.59)
130       PERM = 4.6E-11 * (6/S00)^.73 * W^6.8
140       DIMPERM2 = PERM*S0^2
150       LPRINT USING "  ###       #.##       ##.##^^^^          ##.##^^^^";DVS
,W,DIMPERM1,DIMPERM2
160  NEXT W
170  NEXT DVS
180  END
```

Dvs		Dimensionless Permeability	
microns	porosity	tube bundle	empirical powders
*******	********	***********	*****************
100	0.20	2.98E-03	8.43E-05
100	0.40	4.23E-02	9.40E-03
100	0.60	3.21E-01	1.48E-01
200	0.20	2.98E-03	3.50E-05
200	0.40	4.23E-02	3.90E-03
200	0.60	3.21E-01	6.14E-02
300	0.20	2.98E-03	2.09E-05
300	0.40	4.23E-02	2.33E-03
300	0.60	3.21E-01	3.67E-02
400	0.20	2.98E-03	1.45E-05
400	0.40	4.23E-02	1.62E-03
400	0.60	3.21E-01	2.55E-02
500	0.20	2.98E-03	1.09E-05
500	0.40	4.23E-02	1.22E-03
500	0.60	3.21E-01	1.92E-02
600	0.20	2.98E-03	8.67E-06
600	0.40	4.23E-02	9.66E-04
600	0.60	3.21E-01	1.52E-02
700	0.20	2.98E-03	7.13E-06
700	0.40	4.23E-02	7.94E-04
700	0.60	3.21E-01	1.25E-02

3.18 Consider Eqs. (3.62) and (3.63) and assume that the aspect ratio of the fibers is sufficiently large that the fibers can be assumed to have infinite lengths.

a) Derive an equation for the relationship between S_0 and a.

b) Does the equation for permeability given in Problem 3.17 predict the permeabilities for flow through fibrous media?

a. $S_0 = \dfrac{\text{Surface area of fibers}}{\text{Volume of fibers}}$

$$S_0 = \frac{2\pi a L}{\pi a^2 L} = \frac{2}{a}$$

b. Now we compare the tube bundle theory to Eqs. (3.62) and (3.63) for fibers. For convenience, we define a nondimensional permeability as $\frac{P}{a^2}$. The tube bundle theory (with $K=4.2$) gives

$$P S_0^2 = \frac{1}{4.2}\frac{\omega^3}{(1-\omega)^2}; \text{ but } S_0^2 = \frac{4}{a^2}$$

$$\therefore \frac{P}{a^2} = \frac{1}{(16)(4.2)}\frac{\omega^3}{(1-\omega)^2}$$

Eq. (3.62) – Flow parallel to fiber axes.

$$\frac{P}{a^2} = \frac{0.427}{(1-\omega)}\left[1-f(\omega)\right]^4\left[1+0.473\left(f^{-1}(\omega)-1\right)\right], \text{ where } f(\omega) = \left[\frac{2(1-\omega)}{\pi}\right]^{1/2}$$

and $0.21 \leq \omega \leq 0.5$

Eq. (3.63) – Flow perpendicular to fiber axes.

$$\frac{P}{a^2} = \frac{2\sqrt{2}}{9(1-\omega)}\left[1-\sqrt{2}\, f(\omega)\right]^{5/2}, \text{ where } f(\omega) = \left[\frac{2(1-\omega)}{\pi}\right]^{1/2}$$

and $0.21 \leq \omega \leq 0.8$

The following program calculates the nondimensional permeabilities for the three conditions.

```
10 'Problem 3.18, part (b)
20   PI = 3.1416
30   LPRINT "                         Nondimensional Permeabilities  "
40   LPRINT "  Porosity     tube bundle      parallel       perpendicular"
50   LPRINT "  ########     ###########     #########       #############"
60   FOR W = .25 TO .8 STEP .05              'W is porosity
70      'tube bundle theory
80      DIMPERM = W^3 / ( 16*4.2*(1-W)^2 )       'DIMPERM is dimensionless
             permeability, defined as permeability times radius squared.
90      FW = 2*(1-W)/PI : FW = SQR(FW)   'FW is the function under Eq. (3.62).
100     'Eq. (3.62), flow parallel to fibers
110      TERM1 = .427/(1-W) : TERM2 = ( 1 - FW )^4
120      TERM3 = 1 + .473 * ( 1/FW - 1 )
130      DIMPERM1 = TERM1 * TERM2 * TERM3
140     'Eq. (3.63), flow perpendicular to fibers
150      TERM1 = 2 * SQR(2) / ( 9 * (1-W) ) : TERM2 = ( 1 - SQR(2)*FW )^2.5
160      DIMPERM2 = TERM1 * TERM2
170      LPRINT USING "    #.##        ##.##^^^^     ##.##^^^^          ##.##^^^
^ ";W,DIMPERM,DIMPERM1,DIMPERM2
180  NEXT W
185  LPRINT "   ": LPRINT "  Scopes: parallel .21<W<.5, perpendicular .21<W<.8."

190  END
```

Nondimensional Permeabilities

Porosity	tube bundle	parallel	perpendicular
########	###########	#########	#############
0.25	4.13E-04	6.29E-03	3.29E-05
0.30	8.20E-04	9.21E-03	3.32E-04
0.35	1.51E-03	1.34E-02	1.18E-03
0.40	2.65E-03	1.96E-02	2.95E-03
0.45	4.48E-03	2.86E-02	6.15E-03
0.50	7.44E-03	4.21E-02	1.15E-02
0.55	1.22E-02	6.25E-02	2.03E-02
0.60	2.01E-02	9.41E-02	3.45E-02
0.65	3.34E-02	1.45E-01	5.72E-02
0.70	5.67E-02	2.30E-01	9.45E-02
0.75	1.00E-01	3.82E-01	1.58E-01

Scopes: parallel .21<W<.5, perpendicular .21<W<.8.

The tube bundle theory underestimates the permeability for parallel flow by approximately 1/4 to 1/15. It does better for perpendicular flow,

3.19 In the production of titanium (Ti), rutile ore (TiO_2) is fluidized with gaseous chlorine and the following reaction occurs:

$$TiO_2 + 2Cl_2 \rightarrow TiCl_4(g) + O_2$$

The rate of this reaction is controlled by the removal of the oxygen by reaction with coke particles in the reactor, according to

$$C + O_2 \rightarrow CO_2$$

The rutile ore, prior to being placed in the reactor, was analyzed according to size. The following ranges of particle diameters, D_p, were found:

$180 \le D_p < 250\ \mu m$	5%
$150 \le D_p < 180\ \mu m$	6.2%
$106 \le D_p < 150\ \mu m$	77.4%
$75 \le D_p < 106\ \mu m$	11.4%

Using the above data, calculate the possible chlorine mass flow rates (at 1223 K) that are needed to fluidize the ore in a reduction reactor that is 1.2 m in diameter and 10 m in height.

Combining Eqs. (3.53) and (3.54) $\frac{1}{\bar{D}_{vs}} = \sum_{n=1}^{\infty} \frac{\Delta\phi_i}{\bar{D}_{pi}}$

$$\frac{1}{\bar{D}_{vs}} = \frac{0.05}{215} + \frac{0.062}{165} + \frac{0.774}{128} + \frac{0.114}{90.5}; \quad \bar{D}_{vs} = 126\,\mu m = 1.26 \times 10^{-4} m$$

$\eta_{Cl_2}(1223\text{ K}) = 4.49 \times 10^{-5}\ N\,s\,m^{-2}$; $\rho_{Cl_2}(1223\text{ K}) = 0.772\ kg\,m^{-3}$; $\rho_{rutile} = 4250\ kg\,m^{-3}$

Minimum fluidization: $\omega_{mf} = 0.4$, $\bar{D}_{vs} = 1.26 \times 10^{-4} m$

Eq. (3.66) with $\bar{D}_{vs} = D_p$: $(fRe^2)^{1/3} = \dfrac{1.26 \times 10^{-4}}{\left[\dfrac{(3)(4.49 \times 10^{-5})^2}{(4)(9.81)(0.772)(4249)}\right]^{1/3}} = 3.49$

Fig. 3.13 $\left(\frac{Re}{f}\right)^{1/3} \simeq 5 \times 10^{-3}$

Eq. (3.65) with $V_o = V_{mf}$: $V_{mf} = (5 \times 10^{-3})\left[\dfrac{(4)(9.81)(4249)(4.49 \times 10^{-5})}{(3)(0.772)^2}\right]^{1/3} = 8.06 \times 10^{-3}\ m\,s^{-1}$

Mass flow rate $\dot{M} = (8.06 \times 10^{-3}) \frac{\pi(1.2)^2}{4}(0.772) = 7.04 \times 10^{-3}\ kg\ s^{-1}$

Maximum flow rate: Entrainment velocity is based on smallest particles and $\omega = 1$. $(fRe^2)^{1/3} = \frac{75}{126}(3.49) = 2.08$: Fig. 3.13 gives $\left(\frac{Re}{f}\right)^{1/3} \simeq 0.18$

$$V_o = 0.18\left[\frac{(4)(9.81)(4249)(4.49 \times 10^{-5})}{(3)(0.772)^2}\right]^{1/3} = 0.400\ m\,s^{-1}$$

$$\dot{M} = (0.290)\left(\frac{\pi}{4}\right)(1.2)^2(0.772) = 0.253\ kg\ s^{-1}$$

Hence $0.007 < \dot{M} < 0.253\ kg\ s^{-1}$

3.20 A bed of particles of uniform size is fluidized such that the bed voidage is 0.6 when the Reynolds number is 10. A second bed, similar to the first, contains particles with a diameter equal to one-half of the diameter of particles in the first bed. Both beds operate with the same superficial velocity. What is the bed voidage in the bed filled with the smaller particles?

Large particle - A, small particles - B

From Fig. 3.13 with Re = 10 and $\omega = 0.6$

$$\left(\frac{Re}{f_A}\right)^{1/3} = 0.6, \quad \frac{Re}{f} = 0.216, \quad f = 46.30$$

$$\left(f_A Re^2\right)^{1/3} = \left[(46.30)(10)^2\right]^{1/3} = 16.7$$

$$f = \frac{4}{3}\frac{D_p(\rho_p - \rho_f)g}{\rho_f V_o^2}$$

$$D_{pA} = 2D_{pB}$$

$$f_A = K D_{pA} = 2K D_{pB} = 2f_B$$

$$\left(\frac{Re}{2f_B}\right)^{1/3} = 0.6, \quad \left(\frac{Re}{f_B}\right)^{1/3} = \frac{0.6}{(0.5)^{1/3}} = 0.76$$

$$\left(2f_B Re^2\right)^{1/3} = 16.7, \quad \left(f_B Re^2\right)^{1/3} = \frac{16.7}{2^{1/3}} = 13.25$$

From Fig. 3.13, $\omega = 0.54$

3.21 A fan delivers air to two fluidized beds, A and B. Bed A is operating at a minimum volume (i.e., at minimum fluidization) and bed B is fluidized to a volume equal to twice its fixed bed volume.

a) Calculate the superficial velocity through bed A.
b) Calculate the superficial velocity through bed B.
c) Calculate ΔP across bed A.
d) For bed B, prove that $\omega = 0.7$ when it is fluidized to twice the fixed bed volume.

Bed A: $D_p = 91.4\ \mu m$ (uniform); ρ (solid) $= 4808\ kg\ m^{-3}$; $\lambda = 1$.
Bed B: $D_p = 61.0\ \mu m$ (uniform); ρ (solid) $= 4006\ kg\ m^{-3}$; $\lambda = 1$;
ω (fixed bed) $= 0.4$; ω (fluidized) $= 0.7$.
Air: $\rho = 1.28\ kg\ m^{-3}$; $\eta = 2.07 \times 10^{-5}\ N\ s\ m^{-2}$.

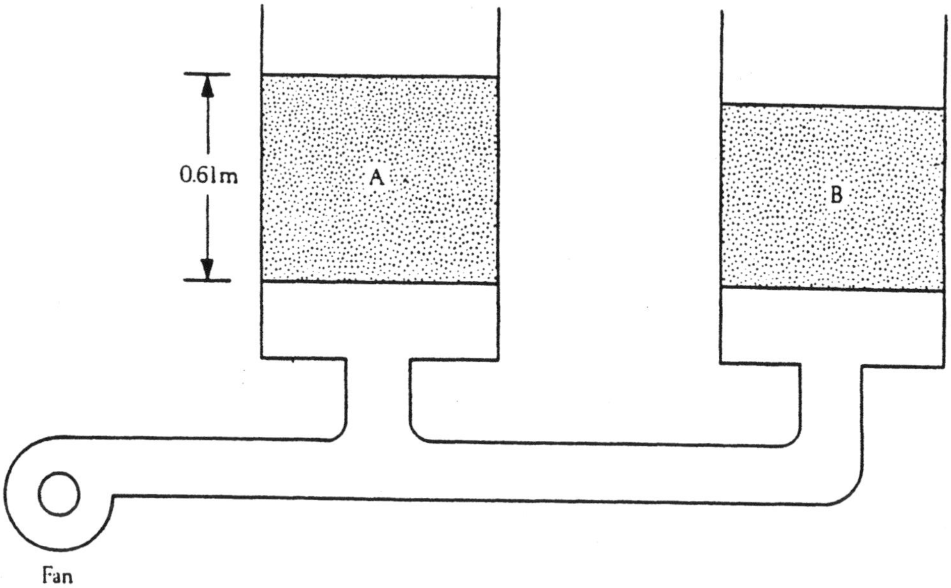

Fan

a. $(f Re^2)^{1/3} = D_p \left[\dfrac{3\eta_f^2}{4g\rho_f(\rho_p-\rho_f)}\right]^{-1/3} = 91.4\times10^{-6}\left[\dfrac{(3)(2.07\times10^{-5})^2}{(4)(9.81)(1.28)(4807)}\right]^{-1/3} = 5.24$

From Fig. 3.13 for $\omega = 0.45$ then $\left(\dfrac{Re}{f}\right)^{1/3} \simeq 0.023$

Eq. (3.65) $V_o = V_{mf} = (0.023)\left[\dfrac{(4)(9.81)(2.07\times10^{-5})(4807)}{(3)(1.28)^2}\right]^{1/3} = 2.1\times10^{-2}\ m\ s^{-1}$

b. $(f Re^2)^{1/3} = 61.0\times10^{-6}\left[\dfrac{(3)(2.07\times10^{-5})^2}{(4)(9.81)(1.28)(4005)}\right]^{-1/3} = 3.29$

From Fig. 3.13 for $\omega = 0.7$ then $\left(\dfrac{Re}{f}\right)^{1/3} \simeq 0.068$

$V_o = 0.068\left[\dfrac{(4)(9.81)(2.07\times10^{-5})(4005)}{(3)(1.28)^2}\right]^{1/3} = 5.9\times10^{-2}\ m\ s^{-1}$

c. By referring to Fig. 3.12, we calculate ΔP as though we have a fixed bed with $\omega = \omega_{mf}$.

Apply Eq. 3.46 $Re_c = \frac{\rho_f V_0 D_P}{6\eta_f(1-\omega)} = \frac{(1.28)(0.021)(91.4\times10^{-6})}{(6)(2.07\times10^{-5})(1-0.45)} = 3.60\times10^{-2}$

Since $Re_c < 2$ we only use the first term of Eq. 3.46

Eq. (3.46) - First term: $\frac{\Delta P}{L} = \frac{(4.2)(2.07\times10^{-5})(0.021, (6)^2(0.55)^2}{(91.4\times10^{-6})^2(0.45)^3} = 2.62\times10^4\,N\,m^{-3}$

$\Delta P = (2.62\times10^4)(0.61) = 1.59\times10^4\,N\,m^{-2}$

d. Let $\mathcal{V}_p$ = Volume of the particles

$\mathcal{V}_f$ = Volume of the fluid

$\mathcal{V}_T$ = Volume total

$\mathcal{V}_T = \mathcal{V}_p + \mathcal{V}_f$

at minimum fluidization: $\omega_{mf} = \frac{\mathcal{V}_{f,mf}}{\mathcal{V}_p + \mathcal{V}_{f,mf}}$: $\mathcal{V}_p = \left[\frac{1-\omega_{mf}}{\omega_{mf}}\right]\mathcal{V}_{f,mf}$

Then $\omega = \frac{\mathcal{V}_f}{\left(\frac{1-\omega_{mf}}{\omega_{mf}}\right)\mathcal{V}_{f,mf} + \mathcal{V}_f} = \frac{1}{\left(\frac{1-\omega_{mf}}{\omega_{mf}}\right)\left(\frac{\mathcal{V}_{f,mf}}{\mathcal{V}_f}\right)+1}$

Also $\omega = \frac{\mathcal{V}_f}{2\mathcal{V}_{T,mf}}$ $\quad \omega_{mf} = \frac{\mathcal{V}_{f,mf}}{\mathcal{V}_{T,mf}}$ So that $\omega = \frac{\mathcal{V}_f\,\omega_{mf}}{2\mathcal{V}_{f,mf}} \Rightarrow \frac{\mathcal{V}_{f,mf}}{\mathcal{V}_f} = \frac{\omega_{mf}}{2\omega}$

Substituting $\omega = \frac{1}{\left(\frac{1-\omega_{mf}}{\omega_{mf}}\right)\frac{\omega_{mf}}{2\omega}+1} = \frac{2\omega}{(1-\omega_{mf})+2\omega}$

$\omega(1-\omega_{mf}) + 2\omega^2 - 2\omega = 0$; $\omega(1-\omega_{mf}-2) + 2\omega^2 = 0$; $-\omega(1+\omega_{mf}) + 2\omega^2 = 0$

$\therefore \omega = 0$ or $2\omega - (1+\omega_{mf}) = 0$

$\omega = \frac{1+\omega_{mf}}{2} = \frac{1+0.4}{2} = 0.7$

3.22 Metal parts are to be annealed at 800 K in a bed of silica sand fluidized by products of combustion of natural gas with 100% excess air. If the sand has a U.S. Standard screen analysis of 20% -30M+50M, 30% -50M+70M, 40%-70M+100M, and 10% -100M+140M, what superficial velocity of gas is required through the bed? (Hint: Think about at what void volume you want to operate.) What is the minimum superficial velocity that you can operate at?

Assume natural gas is CH_4: $CH_4 + 2O_2 \rightarrow CO_2 + 2H_2O$

Basis: 1 mol CH_4 - Products (take air to be $0.79 N_2 - 0.21 O_2$)

$CO_2 = 1$ mol

$H_2O = 2$ mol

$O_2 = 2$ mol (100% excess air)

$N_2 = 7.52$ mol

For mol fractions and properties (Appendix B)

Species	Subscript	X_i	ρ_i, kg m^{-3}	η_i, N s m^{-2}
CO_2	1	0.080	0.661	3.37×10^{-5}
H_2O	2	0.160	0.274	2.79×10^{-5}
O_2	3	0.160	0.481	4.15×10^{-5}
N_2	4	0.600	0.421	3.49×10^{-5}

Viscosity of gases - $\eta_f = \eta_{MIX} \simeq \sum_1^4 X_i \eta_i$ (Law of mixtures is adequate)

$$\eta_f = \left[(0.08)(3.37) + (0.16)(2.79) + (0.16)(4.15) + (0.6)(3.49)\right] \times 10^{-5} = 3.47 \times 10^{-5} \text{ N s m}^{-2}$$

Density of gas - Assume ideal gas - Let $\hat{V}_i$ = molar volume of i.

$$\hat{V}_1 = \frac{\text{m}^3}{0.661 \text{ kg}} \left| \frac{44 \text{ kg}}{1 \text{ kmol}} \right. = 66.56 \text{ m}^3 \text{ kmol}^{-1}; \; \hat{V}_2 = 65.69, \; \hat{V}_3 = 66.53, \; \hat{V}_4 = 66.51$$

$$\overline{MW} = (0.08)(44) + (0.16)(18) + (0.16)(32) + (0.6)(28) = 28.32 \text{ kg kmol}^{-1}$$

$$\rho_f = \frac{\overline{MW}}{\sum_1 X_i \hat{V}_i} = \frac{28.32}{(0.08)(66.56) + (0.16)(65.69 + 66.53) + (0.6)(66.51)} = 0.427 \text{ kg m}^{-3}$$

Density and particle sizes of the silica sand (SiO_2)

$\rho_p = 2840$ kg m^{-3} (Used another source)

For sizes, refer to Appendix D (U.S. Standard)

$\Delta\Phi_i$, %	Minimum D_{pi}	Maximum D_{pi}	$\overline{D}_{pi}$, in.
20	0.0117 in.	0.0232 in	0.01745
30	0.0083	0.0117	0.0100
40	0.0059	0.0083	0.0071
10	0.0041	0.0059	0.0050

Apply Eqs. (3.53) and (3.54)

$$S_w = \frac{6}{\rho_p}\sum_{i=1}^{n}\frac{\Delta\Phi_i}{\overline{D}_{pi}} = \frac{6}{\rho_p}\left[\frac{0.20}{0.01745}+\frac{0.30}{0.0100}+\frac{0.40}{0.0071}+\frac{0.10}{0.0050}\right] = \frac{6}{\rho_p}(117.8)\ \text{in.}^{-1}$$

$$D_{\overline{vs}} = \overline{D}_p = \frac{6}{S_w\rho_p} = \frac{1\ \text{in.}}{117.8}\left|\frac{25.4\ \text{mm}}{\text{in.}}\right|\frac{1\ \text{m}}{1000\ \text{mm}} = 2.16\times10^{-4}\ \text{m}$$

Assume $\omega_{mf} = 0.45$

$$\text{Eq. (3.66)}\quad (fRe^2)^{1/3} = (2.16\times10^{-4})\left[\frac{(3)(3.47\times10^{-5})^2}{(4)(9.81)(0.427)(2840)}\right]^{-1/3} = 5.11$$

$$\left(\frac{Re}{f}\right)^{1/3} = 2.2\times10^{-2}$$

$$V_{mf} = V_o = (2.2\times10^{-2})\left[\frac{(4)(9.81)(3.47\times10^{-5})(2840)}{(3)(0.427)^2}\right]^{1/3} = 4.2\times10^{-2}\ \text{m s}^{-1}$$

3.23 Pellets of polyethylene are to be fluidized in a column 1 meter in diameter and 10 meters high with air at 300 K. A hot steel pipe is lowered into the fluidized bed to bring it into contact with the pellets, which melt onto its surface to form a protective anticorrosion coating. Calculate the total flow of air required, if the pellets are 5 mm in diameter and the desired void fraction of the bed is 0.7. The density of the polyethylene is 920 kg m^{-3}.

At 300 K, $\rho_f = 1.161$ kg m^{-3}; $\eta_f = 1.846\times10^{-5}$ N s m^{-2}

$$\text{Eq. (3.66)}\quad (fRe^2)^{1/3} = D_p\left[\frac{3\eta_f^2}{4g\rho_f(\rho_p-\rho_f)}\right]^{-1/3} = 5\times10^{-3}\left[\frac{(3)(1.846\times10^{-5})^2}{(4)(9.81)(1.161)(919)}\right]^{-1/3} = 172$$

From Fig. 3.13 with $(fRe^2)^{1/3} = 172$ and $\omega = 0.7$, $\left(\frac{Re}{f}\right)^{1/3} = 8$

$$V_o\left[\frac{4g(\rho_p-\rho_f)}{3\rho_f^2}\right]^{-1/3} = 8;\quad V_o = 8\left[\frac{(4)(9.81)(919)}{(3)(1.161)^2}\right]^{1/3} = 166\ \text{m s}^{-1}$$

$$Q = V_oA = \frac{166\ \text{m}}{\text{s}}\left|\frac{\pi(1)^2\ \text{m}^2}{4}\right. = 130.4\ \text{m}^3\ \text{s}^{-1}$$

3.24 During the compaction of metal powders into sheet material (powder rolling), as shown in the figure below, the entrapped air is expelled from the loose powder. This expulsion occurs at the line *AB*. Below *AB*, the powder is coherent, that is, the particles are locked together, but above *AB*, the powder is loose, i.e., a normal packed bed.

If the velocity of expulsion exceeds the minimum fluidization velocity, the powder does not feed properly into the roll gap and the sheet product is not satisfactory. This, in fact, limits the production rate of the process.

The following equation gives the superficial velocity V_0 (cm s^{-1}) of gas expulsion from the coherent zone:

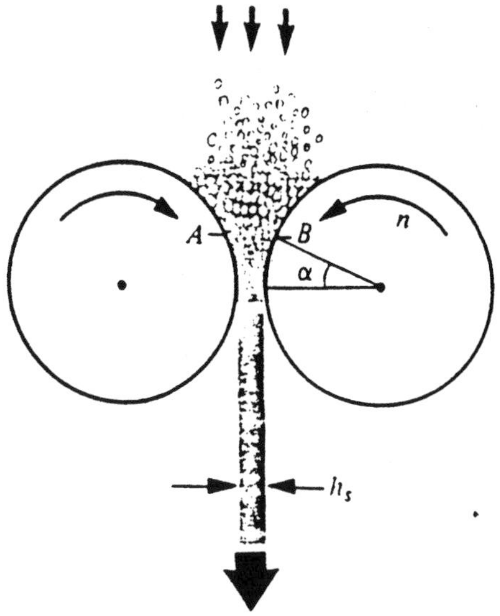

$$V_0 = \frac{2\pi R^2 n(\alpha - \sin\alpha\cos\alpha)}{\alpha\left[2R(1 - \cos\alpha) + h_s\right]},$$

where R = roll radius, cm, n = rolling speed, rev s^{-1}, α = roll-coherent powder contact angle, rad, and h_s = roll gap, cm. For copper powder, α has been found to be 6° (0.1047 radians).

a) Calculate the rolling speed at which copper powder, with properties given below, will just begin to be fluidized at the plane *AB*, for strip thickness 0.5 mm and roll diameter 200 mm.

b) Calculate the corresponding speed of the emerging coherent strip.

c) Discuss the effect that order-of-magnitude changes in particle size and roll radius would have on production rates. Data are given as follows: $\eta = 1.8 \times 10^{-6}$ N s m^{-2}, $\rho_{air} = 1.30$ kg m^{-3}, $D_{P_{Cu}} = 40$ μm, $\rho_{Cu} = 6700$ kg m^{-3}.

a. Assume $\omega_{mf} = 0.45$

$$\text{Eq. (3.66)} - (fRe^2)^{1/3} = \frac{D_p}{\left[\dfrac{3\eta_f^2}{4g\rho_f(\rho_p - \rho_f)}\right]^{1/3}} = \frac{40\times10^{-6}}{\left[\dfrac{3(1.8\times10^{-6})^2}{4(9.8)(1.3)(6700)}\right]^{1/3}} = 13.1$$

$$\text{From Fig. 3.13 @ } \omega = 0.45,\ \left(\frac{Re}{f}\right)^{1/3} = 0.1 = \frac{V_{o,mf}}{\left[\dfrac{4g\eta_f(\rho_p - \rho_f)}{3\rho_f}\right]^{1/3}}$$

$$V_{o,mf} = 0.1\left[\frac{(4)(9.8)(1.8\times10^{-6})(6700)}{3(1.3)^2}\right]^{1/3} = 0.045\ \text{m s}^{-1} = 4.5\ \text{cm s}^{-1}$$

$h = 0.05$ cm, $R = 10.0$ cm – Assume no slip

$$n = \frac{V_2 \alpha \left[2R(1-\cos\alpha) + h_s\right]}{2\pi R^2 (\alpha - \sin\alpha \cos\alpha)} = \frac{(4.5)(0.1047)\left[(2)(10)(1-\cos 0.1047) + 0.05\right]}{2\pi (10)^2 (0.1047 - \sin 0.1047 \cos 0.1047)}$$

$n = 0.157\ s^{-1} \approx 9.4$ rpm

b. Velocity of strip $= \pi D n = (\pi)(20)(9.4) = 590\ cm\ min^{-1} = 9.84\ cm\ s^{-1}$

c. If $D_p = 183\ \mu m$ then $\left(\frac{Re}{f}\right)^{1/3} = 1$ and $V_{o,mf}$ would increase to $0.45\ m\,s^{-1}$

This would increase the rolling speed to 94 rpm.

Vel. of the strip $= \frac{V_2 \alpha \left[2R(1-\cos\alpha) + h_s\right]}{R(\alpha - \sin\alpha \cos\alpha)}$

If roll diameter was increased to 400 mm, $R = 20$ cm, then

vel. of strip $= 8.34\ cm\ s^{-1}$ or a slight decrease in the production rate.

4.1 The reciprocal of β, as defined in Eq. (4.4) is called the kinetic energy correction factor. Derive the kinetic energy correction factors for the following flows:

a) fully developed, laminar flow between infinite parallel plates;

b) Hagen-Poiseuille flow in a circular tube;

c) turbulent flow, in a tube of radius R, described by

$$\frac{\bar{V}}{V_{max}} = \left[\frac{r}{R}\right]^n \quad \text{change to } \frac{V}{V_{max}} = \left(\frac{R-r}{R}\right)^n$$

where $n = 1/7$.

a. $$\frac{1}{\beta_1} = \frac{1}{A_1}\int_0^{A_1}\left[\frac{v_1}{\bar{V}_1}\right]^3 dA_1$$

Fully developed flow between parallel plates.

$$v = \frac{1}{2\eta}(\delta^2 - y^2)\frac{(P_0 - P_L)}{L}$$

$$\bar{V} = \frac{\delta^2}{3\eta}\frac{(P_0 - P_L)}{L}$$

$$\frac{v_1}{\bar{V}_1} = \frac{3}{2\delta^2}(\delta^2 - y^2)$$

$A_1 = 2\delta W$, W = width of parallel plates

$A = 2yW$, $dA = 2W dy$; when $A_1 = 0$, $y = 0$; when $A_1 = A_1$, $y = \delta$

$$\therefore \frac{1}{\beta_1} = \frac{1}{2\delta W}\int_0^{\delta}\left[\frac{3}{2\delta^2}(\delta^2 - y^2)\right]^3 2W dy = \frac{27}{8\delta^7}\int_0^{\delta}(\delta^2 - y^2)^3 dy = \frac{54}{35}$$

$B_1 = 0.648$

b. Hagen-Poiseuille flow in a circular tube.

$$v_z = \left[\frac{P_0 - P_L}{L} + \rho g\right]\left[\frac{R^2}{4\eta}\right]\left[1 - \left(\frac{r}{R}\right)^2\right] \quad : \text{Eq. (2.32)}$$

$$\bar{V} = \left[\frac{P_0 - P_L}{L} + \rho g\right]\frac{R^2}{8\eta} \quad : \text{Eq. (2.33)}$$

$$\frac{v_z}{\bar{V}} = 2\left[1-\left(\frac{r}{R}\right)^2\right]$$

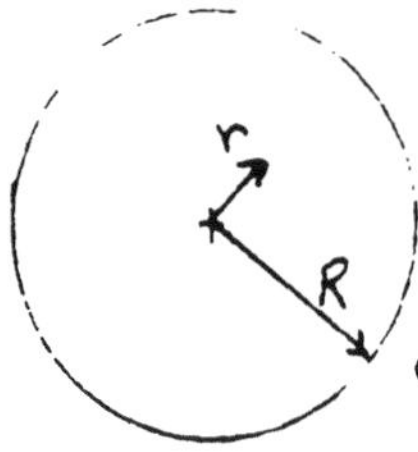

$A = \pi R^2$

$dA' = 2\pi r\,dr$

$$\frac{1}{\beta} = \frac{1}{\pi R^2}\int_0^R 2^3\left[1-\left(\frac{r}{R}\right)^2\right]^3 2\pi r\,dr = \frac{2^4}{R^2}\int_0^R \left[1-\left(\frac{r}{R}\right)^2\right]^3 r\,dr$$

$$= \frac{16}{R^2}\,\frac{R^2}{8} = 2 \quad \therefore \beta = \frac{1}{2}$$

c. At $r=0$, $v = V_{max.}$; $r=R$, $v=0$; $r=\frac{R}{2}$, $v=\left(\frac{1}{2}\right)^n V_{max.}$

$$\frac{1}{\beta} = \frac{1}{A}\int_0^A \left(\frac{v}{\bar{V}}\right)^3 dA'$$

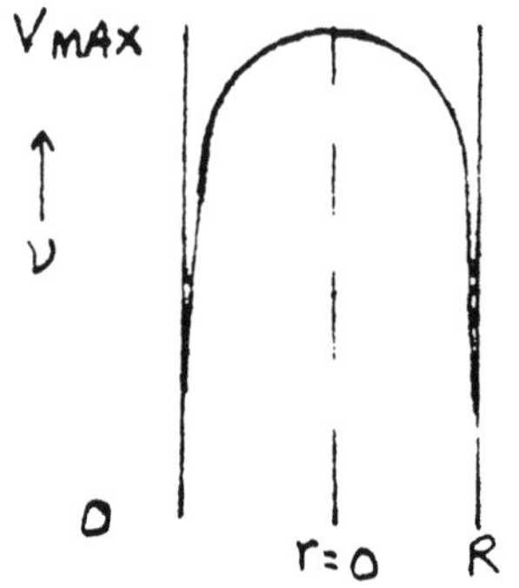

$$\bar{V} = \frac{1}{A}\int_0^A v\,dA' = \frac{2\pi}{\pi R^2}\int_0^R v\,r\,dr$$

$$= \frac{2V_{MAX}}{R^2}\int_0^R \left(\frac{R-r}{R}\right)^n dr = 2V_{MAX}\left(\frac{1}{n+1} - \frac{1}{n+2}\right)$$

with $n = \frac{1}{7}$, $\frac{\bar{V}}{V_{MAX}} = 0.817$

Then $$\frac{v}{\bar{V}} = \frac{V_{max.}\left(\frac{R-r}{R}\right)^n}{2V_{max.}\left(\frac{1}{n+1} - \frac{1}{n+2}\right)} = \left[\frac{1}{2\left(\frac{1}{n+1} - \frac{1}{n+2}\right)}\right]\left(\frac{R-r}{R}\right)^n$$

$$\frac{1}{\beta} = \frac{1}{\pi R^2}\left[\frac{1}{2\left(\frac{1}{n+1} - \frac{1}{n+2}\right)}\right]^3\int_0^R \left(\frac{R-r}{R}\right)^{3n} 2\pi r\,dr$$

With $n = \frac{1}{7}$, $$\frac{1}{\beta} = \frac{2}{R^2}(1.836)\int_0^R \left(\frac{R-r}{R}\right)^{3n} r\,dr = \frac{2}{R^2}(1.836)R^2\left[\frac{1}{3n+1} - \frac{1}{3n+2}\right]$$

$$\frac{1}{\beta} = (2)(1.836)(0.288) = 1.058$$

$$\beta = 0.945$$

4.2 A fan is used to draw exhaust gases from a large hood. For highly turbulent flow, write an equation for the system which gives $-M^*\rho$ as a function of volume flow rate, Q. Neglect potential energy changes because a gas is being exhausted. For friction include only the contraction ($e_f = 0.4$), the expansion ($e_f = 1.0$) and the elbow ($L/D_e = 20, f = 0.001$). Note that $-M^*\rho$ has units of pressure and represents the "total pressure" against which the fan must operate.

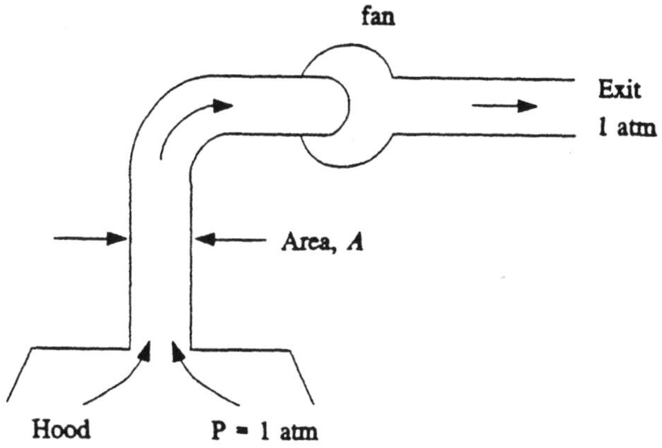

$$\int_{P_1}^{P_2} \frac{dP}{\rho} + \left[\frac{\bar{V}_2^2}{2\beta_2} - \frac{\bar{V}_1^2}{2\beta_1}\right] + g\Delta z + M^* + E_f = 0$$

$\bar{V}_1 = \bar{V}_2$; $P_1 = P_2$, $\Delta z = 0$, $\therefore M^* = -E_f$

E_f = elbow + contraction + expansion

$$E_f(\text{elbow}) = 2f\frac{L_e}{D}\bar{V}^2 = 2(0.001)(20)\bar{V}^2 = 0.04\,\bar{V}^2$$

$$E_f(\text{contraction}) = \frac{1}{2}\bar{V}^2(0.4) = 0.2\,\bar{V}^2$$

$$E_f(\text{expansion}) = \frac{1}{2}\bar{V}^2(1) = 0.5\,\bar{V}^2$$

$$E_f = (0.04 + 0.2 + 0.5)\bar{V}^2 = 0.74\,V^2 = 0.74\frac{Q^2}{A^2}$$

$$-M^*\rho = \frac{0.74\rho}{A^2}Q^2$$

4.3 Cooling water is provided to the mold used in the electroslag remelting process depicted below. For flow through the mold,

$$E_f = K\bar{V}^2$$

where $\bar{V}$ is the average velocity of the water in the lines at the entrance and exit of the mold and K is a turbulent flow constant for the mold. When the pressure gauges read $P_A = 2.76 \times 10^5$ N m^{-2} and $P_B = 2.07 \times 10^5$ N m^{-2}, the volume flow rate is 2.83×10^{-3} m^3 s^{-1}. Calculate $-M^*$ (in N m kg^{-1}) for the pump when the volume flow rate is 5.66×10^{-3} m^3 s^{-1}. Assume that friction losses through the straight lengths of pipes can be ignored. *Data*: D (pipes) = 15.4 mm; $f = 0.001$; $L_e/D = 26$ (elbows); e_f (entrance) = 0.8; e_f (exit) = 1.

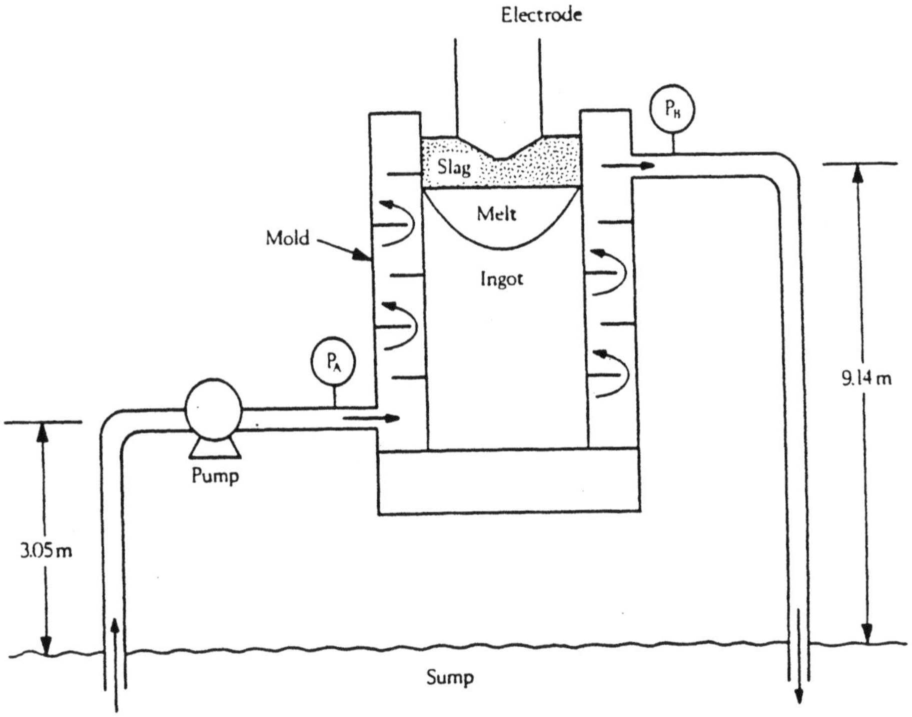

Case 1. Between P_A and P_B is considered to be the mold.

$$\frac{P_B - P_A}{\rho} + g(z_B - z_A) + E_f = 0\ ;\ E_f = \frac{P_A - P_B}{\rho} - g(z_B - z_A)$$

$$E_f = \frac{2.76\times10^5 - 2.07\times10^5}{1\times10^3} - 9.81(9.14 - 3.05);\ E_f = 9.3$$

$$\text{Also } E_f = K\bar{V}^2;\ K = \frac{E_f}{\bar{V}^2} = \frac{E_f(\pi d^2)^2}{4^2 Q^2} = \frac{(9.3)(\pi^2)(0.0154^4)}{(4^2)(2.83\times10^{-3})^2} = 0.0403$$

Case 2. Plane 1 – entrance from sump; Plane 2 – exit to sump.

$$P_2 = P_1,\ \bar{V}_2 = \bar{V}_1,\ z_2 = z_1,\ -M^* = E_f$$

$$-M^* = \left[\frac{1}{2} e_{f,1} + \frac{1}{2} f\left(\frac{L}{D}\right) + \frac{1}{2} e_{f,2} + K\right] \bar{V}^2$$

$$\bar{V} = \frac{(5.66 \times 10^{-3})(4)}{(\pi)(0.0154^2)} = 30.4 \ m \ s^{-1}$$

$$-M^* = \left[\frac{1}{2}(0.8) + \frac{1}{2}(0.001)(26) + \frac{1}{2}(1) + 0.0403\right] 30.4^2 = 881 \ m^2 s^{-2}$$

$$-M^* = 881 \ N m \ kg^{-1}$$

4.4 A blower draws air from a melting area and directs the air to a "bag house" where particulates are filtered before the air is discharged to the environment. The melting area and the environment are at ambient temperature and pressure (289 K and 1.0133×10^5 N m^{-2}, respectively). When the pressure drop across the bag house (ΔP in the diagram) is 5.07×10^3 N m^{-2}, the volume flow rate is 0.944 m^3 s^{-1}. Calculate the work done by the blower in N m per kg of air delivered by the blower. *Conduit Information*: length before elbow, 61 m; length after elbow, 61 m; diameter, 305 mm; L_e/D = 25 (elbows); f = 0.0043; e_f (contraction) = 0.4; e_f (expansion) = 0.8.

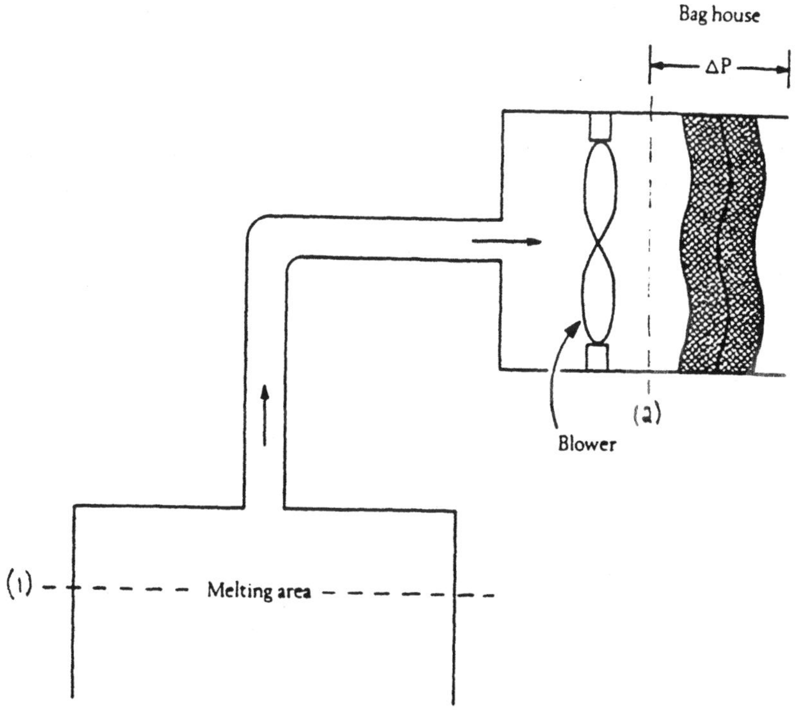

$$\frac{P_2 - P_1}{\rho} + g\Delta z + M^* + E_f = 0; \text{ since } \bar{V}_1^2 \cong \bar{V}_2^2 \cong 0$$

$P_2 - P_1 = 5.07 \times 10^3\ N\ m^{-2}$

$\bar{V}$ (inside 0.305 m dia. conduit) $= \frac{Q}{A} = \frac{4}{\pi} \left| \frac{0.944\ m^3}{s^{-1}} \right| \frac{1}{(0.305)^2\ m^2} = 12.92\ m\ s^{-1}$

$g\Delta z = (9.807\ m\ s^{-2})(61\ m) = 598\ m^2\ s^{-2}$

$\rho_{Air}(1\ atm.,\ 289\ K) = 1.22\ kg\ m^{-3};\ \rho_{1\to 2}(avg.) = \frac{1.03865 \times 10^5}{1.01330 \times 10^5}(1.22) = 1.25\ kg\ m^{-3}$

$E_f = E_f(\text{contraction}) + E_f(\text{elbow}) + E_f(\text{expansion}) + E_f(\text{conduit})$

$$E_f = \frac{1}{2}(0.4)\bar{V}^2 + 2(0.0043)(25)\bar{V}^2 + \frac{1}{2}(0.8)\bar{V}^2 + 2(0.0043)\left(\frac{122}{0.305}\right)\bar{V}^2 = 4.255\bar{V}^2$$

$$E_f = (4.255)(12.92)^2\ m^2\ s^{-2} = 710\ m^2\ s^{-2}$$

$$\frac{P_2 - P_1}{\rho} = \frac{5.07 \times 10^3\ N}{m^2} \left| \frac{m^3}{1.25\ kg} \right| \frac{kg\ m}{N\ s^{-2}} = 4056\ m^2\ s^{-2}$$

$$-M^* = (4056 + 598 + 710)\ m^2\ s^{-1} = 5364\ m^2\ s^{-2} = 5364\ N\ m\ kg^{-1}$$

4.5 Water is pumped from a storage tank to a mold designed to produce nonferrous ingots by the "direct-chill" process. The water supply is at ambient pressure ($1.0133 \times 10^5\ N\ m^{-2}$), and the water leaving the mold impinges upon the surface of the ingot which is also at ambient pressure. A pressure gauge mounted in the manifold portion of the mold (pressure gauge P in the diagram) indicates an absolute pressure of $1.22 \times 10^5\ N\ m^{-2}$, when the volume flow rate is $3.93 \times 10^{-3}\ m^3\ s^{-1}$. The water level in the tank is 3 m, and the vertical length of the pipe is 3 m. Calculate the theoretical power of the pump. Assume that the tank for the water supply has a very large diameter, and that the kinetic energy of the water within the manifold portion of the mold is negligible. *Piping Information*: total length of straight pipe, 9.14 m; diameter, 30.5 mm; $L_e/D = 25$ (elbows); $f = 0.004$; e_f (contraction) $= 0.4$; e_f (expansion) $= 0.8$.

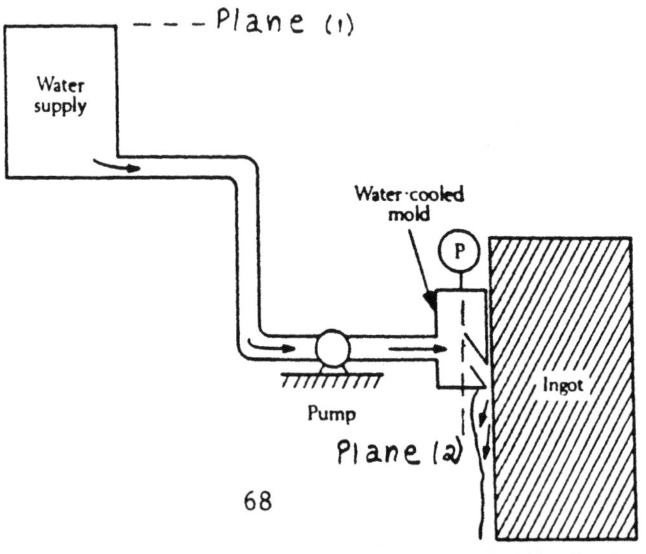

$$\frac{P_2-P_1}{\rho}+\left[\frac{\bar{V}_2^2}{2\beta_2}-\frac{\bar{V}_1^2}{2\beta_1}\right]+g(z_2-z_1)+M^*+E_f=0;\quad \bar{V}_1=\bar{V}_2\approx 0$$

$$\frac{P_2-P_1}{\rho}=\frac{(1.22\times10^5-1.0133\times10^5)\,N}{m^2}\left|\frac{m^3}{1\times10^3\,Kg}\right|\frac{Kg\,m}{N\,s^2}=20.67\,m^2s^{-2}$$

$$g(z_2-z_1)=\frac{9.807\,m}{s^2}\Big|-6\,m=-58.84\,m^2s^{-2}$$

$$\bar{V}=\frac{Q}{A}=\frac{3.93\times10^{-3}\,m^3}{s}\left|\frac{4}{\pi}\right|\frac{}{(0.0305)^2\,m^2}=5.38\,m\,s^{-1}$$

$E_f=E_f\,(Pipe)+2E_f(elbows)+E_f\,(contraction)+E_f\,(expansion)$

$$=\left[(2)(0.004)\left(\frac{9.14}{0.0305}\right)+(4)(0.004)(25)+\left(\frac{1}{2}\right)(0.4)+\left(\frac{1}{2}\right)(0.8)\right]5.38^2=98.34$$

$$-M^*=20.67-58.84+98.34=60.2\,N\,m\,kg^{-1}$$

4.6 Water, maintained at a constant level, is supplied to a long line from a filter tank filled with sand. There are two vertical branches attached to the main line as shown in the accompanying figure; these branches and the tank are open to ambient pressure (i.e., 1.0133×10^5 N m^{-2}). At the end of the line there is a frictionless valve.

a) When the valve is closed, what are the heights h_1 and h_2 in the vertical branches?

b) When the valve is open, what is the mass flow rate?

c) When the valve is open, what is the difference in height between h_1 and h_2?

Data: $\rho = 1{,}000$ kg m^{-3}; $\eta = 1\times10^{-3}$ N s m^{-2}; ε/D (relative roughness) = 0.01; ω (sand) = 0.40; D_p (sand) = 152 μm.

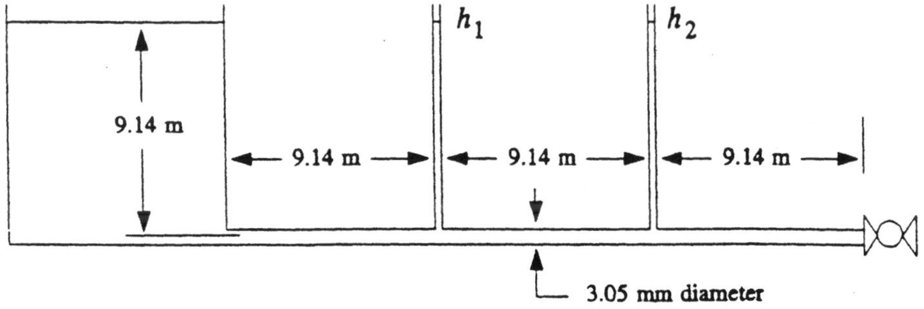

a. $h_1=h_2=9.14\,m$

b. $\frac{P_2-P_1}{\rho}+\frac{\bar{V}_2^2}{2\beta_2}-\frac{\bar{V}_1^2}{2\beta_1}+g(z_2-z_1)+E_f=0;\quad P_2-P_1=0,\ \beta_2=\beta,\ \bar{V}_2=\bar{V},\ \bar{V}_1=0$

$z_2-z_1=-L_b$ (L_b is length of filter bed)

$E_f=E_f(bed)+E_f(entr.)+E_f(Pipe)$

$$E_f(bed) = \frac{\Delta P(bed)}{\rho} = C_1 L_b V_0 + C_2 L_b V_0^2 \quad \text{where } C_1 = \frac{150\,\eta(1-\omega)^2}{\rho D_p^2 \omega^3};\ C_2 = \frac{1.75(1-\omega)}{D_p \omega^3}$$

$$E_f = \left(\frac{C_1 L_b V_0}{\bar{V}^2} + \frac{C_2 L_b V_0^2}{\bar{V}^2} + \frac{1}{2} e_f + 2f\frac{L_b}{D}\right)\bar{V}^2;\ \text{Also } V_0 = \frac{A_p}{A_b}\bar{V} = R\bar{V};\ p \triangleq \text{pipe},\ b \triangleq \text{bed}$$

where R = area ratio $\left(\frac{\text{pipe}}{\text{bed}}\right)$

$$E_f = \left[\frac{C_1 L_b R}{\bar{V}} + C_2 L_b R^2 + \frac{1}{2} e_f + 2f\frac{L_b}{D}\right]\bar{V}^2$$

Substituting into mechanical energy equation.

$$-g L_b + \left[\frac{1}{2\beta} + \frac{C_1 L_b R}{\bar{V}} + C_2 L_b R^2 + \frac{1}{2} e_f + 2f\frac{L_p}{D}\right]\bar{V}^2 = 0$$

Solve for $\bar{V}$: $\bar{V}^2\left[\frac{1}{2\beta} + \frac{C_1 L_b R}{\bar{V}} + C_2 L_b R^2 + \frac{1}{2} e_f + 2f\frac{L_p}{D}\right] = g L_b$

The following program calculates the mass flow rate.

```
10 'Problem 4.6 - solves for velocity in the pipe; SI units throughout
20  VISC = .001 : DENS = 1000: W = .4: DP = .000152: RATIO = .001: G = 9.807
30  LB = 9.140001 : LP = 3*9.140001 : D = .00305  'lengths of bed and pipe, and
                                                  diameter of pipe.
40 'Assume laminar flow because the pipe diameter is so small and the filter bed
    adds a lot of resistance to flow.
50  EF = 1.1 : BETA = .5  'Frict. loss fact. of entrance and beta.laminar flow.
60  C1 = ( 150*VISC*(1-W)^2 )/( DENS*DP^2*W^3 )
70  C2 = ( 1.75*(1-W) )/( DP*W^3 )
80  FOR V = .01 TO 1  STEP .001
90      RE = D*V*DENS/VISC : F = 16/RE      'laminar flow in pipe
100      DENOM = 1/(2*BETA) + C1*LB*RATIO/V + C2*LB*RATIO^2 + EF/2 + 2*F*LP/D
110      NUMER = G*LB
120      SIDE = V*V*DENOM
130     IF SIDE > NUMER THEN 150
140 NEXT V
150 LPRINT "   The velocity in the pipe is";V;"m/s."
160 LPRINT "   The Reynolds no. in the pipe is";RE;"."
170 IF RE > 50000! THEN 210
180 LPRINT "   Reynolds no. is O.K., so assumed friction factor is valid."
190 MASSRATE = (3.1416*D*D/4)*V*DENS
200 LPRINT "   The mass flow rate is";MASSRATE;"kg/s."
210 END
```

```
The velocity in the pipe is .2100003 m/s.
The Reynolds no. in the pipe is 640.5007 .
Reynolds no. is O.K., so assumed friction factor is valid.
The mass flow rate is .0015343 kg/s.
```

:. Mechanical enery equation between the two vertical legs.

$$\frac{P_2 - P_1}{\rho} + E_f = 0;\ E_f = 2f\frac{L}{D}\bar{V}^2 = \frac{(2)(16)}{Re}\left(\frac{L}{D}\right)\bar{V}^2 = \left(\frac{32}{640.5}\right)\left(\frac{9.14}{3.05\times10^{-3}}\right)(0.210^2) = 6.60\ m^2 s^{-2}$$

$$P_2 - P_1 = (1\times10^3)(6.60) = 6.6\times10^3\ N\ m^{-2};\ Also\ P_2 - P_1 = \rho g(h_1 - h_2)$$

$$h_1 - h_2 = \frac{6.6\times10^3}{(1\times10^3)(9.81)} = 0.673\ m$$

4.7 Liquid aluminum contains a small fraction of Al_2O_3 inclusions which are removed by filtering through a bed of ceramic spheres. The refined aluminum (i.e., filtered aluminum) is pumped to a holding vessel from which liquid metal is drawn to cast ingots. For the equipment arrangement shown below, calculate the theoretical power of the pump to process 2.52 kg s^{-1}. The important friction losses are in the filter bed, the transfer line which contains two 90° elbows (medium radius), and the entrance and exit.

Data: Aluminum:
ρ = 2644 kg m^{-3}
η = 1.28 × 10^{-3} N s m^{-2}

Ceramic spheres:
ω = 0.4
D_p = 0.61 mm
Area (bed) = 0.292 m^2

Transfer line:
L (straight portions) = 1.83 m
D = 9.1 mm
ε (roughness) = 9.1 × 10^{-4} mm

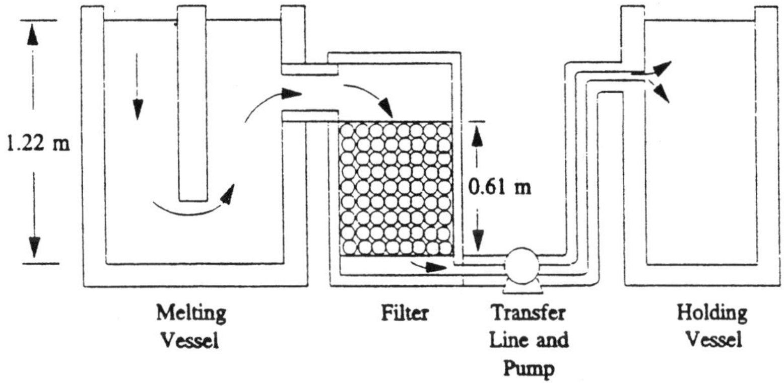

$$\frac{P_2 - P_1}{\rho} + \left[\frac{\bar{V}_2^2}{2B_2} - \frac{\bar{V}_1^2}{2B_1}\right] + g\Delta z + M^* + E_f = 0$$

For the entire system: $\bar{V}_1 = \bar{V}_2 = 0,\ P_1 = P_2,\ z_2 = z_1\ \therefore -M^* = E_f$

$E_f = E_f(\text{filter}) + E_f(\text{transfer line})$

$$E_f(\text{filter}) = \frac{\Delta P}{\rho} = \frac{150\eta V_0 L}{\rho D_p^2}\frac{(1-\omega)^2}{\omega^3} + \frac{1.75\rho V_0^2 L}{\rho D_p}\frac{(1-\omega)}{\omega^3}$$

$$V_0 = \frac{2.52\ kg}{s}\left|\frac{m^3}{2644\ kg}\right|\frac{}{0.292\ m^2} = 3.26\times10^{-3}\ m\ s^{-1}$$

$$E_f(\text{filter}) = \frac{150}{} \left| \frac{1.28\times10^{-3}\ N\ s}{m^2} \right| \frac{3.26\times10^{-3}\ m}{s} \left| \frac{0.61\ m}{} \right| \frac{m^3}{2644\ Kg} \left| \frac{}{(6.1\times10^{-4})^2 m^2} \right| \frac{(1-0.4)^2}{(0.4)^3} \left| \frac{Kg\ m}{N\ s^2} \right.$$

$$+ \frac{1.75}{} \left| \frac{(3.26\times10^{-3})^2 m^2}{s^2} \right| \frac{0.61\ m}{} \left| \frac{}{6.1\times10^{-4} m} \right| \frac{(1-0.4)}{(0.4)^3} = 2.183 + 0.174 = 2.357\ m^2 s^{-2}$$

$$E_f(\text{transfer}) = 2E_f(\text{elbow}) + E_f(\text{contraction}) + E_f(\text{expansion})$$

$$Re = \frac{D\bar{V}\rho}{\eta} = \frac{D\left(\frac{4W}{\pi\rho D^2}\right)\rho}{\eta} = \frac{4W}{\pi\eta D} = \frac{4}{\pi} \left| \frac{2.52\ kg}{s} \right| \frac{m^2}{1.28\times10^{-3}\ N\ s} \left| \frac{}{0.0091\ m} \right. = 2.75\times10^5$$

$$\frac{\varepsilon}{D} = \frac{9.1\times10^{-4}\ mm}{9.1\ mm} = 1\times10^{-4};\ \therefore f = 0.004$$

$$\bar{V} = \frac{4}{\pi} \left| \frac{2.52\ kg}{s} \right| \frac{m^3}{2644\ kg} \left| \frac{}{(0.0091)^2 m^2} \right. = 14.65\ m\ s^{-1}$$

$$2E_f(\text{elbow}) = 2\left[2f\left(\frac{Le}{D}\right)\bar{V}^2\right] = 4(0.004)(26)(14.65)^2 m^2 s^{-2} = 89.3\ m^2 s^{-2}$$

$$E_f(\text{contraction}) = \frac{1}{2}\bar{V}^2 e_f = \frac{1}{2} \left| \frac{(14.65)^2 m^2}{s^2} \right| 0.4 = 42.9\ m^2 s^{-2}$$

where $e_f = 0.4$ is from Fig. 4.4 for $Re = \infty$, $A_2/A_1 = 0$

$$E_f(\text{expansion}) = \frac{1}{2}\bar{V}^2 e_f = \frac{1}{2} \left| \frac{(14.65)^2 m^2}{s^2} \right| 1.0 = 107.3\ m^2 s^{-2}$$

where $e_f = 1.0$ is from Fig. 4.3 for $Re = \infty$, $A_1/A_2 = 0$

$$E_f = (2.4 + 89.3 + 42.9 + 107.3)\ m^2 s^{-2} = 242\ m^2 s^{-2}$$

$$-M^* = 242\ m^2 s^{-2}$$

4.8 Hot-rolled steel sheet is quenched by passing under two water sprays as depicted below. Each spray requires 9.46×10^{-4} m^3 s^{-1} of water at 294 K, and the pressure drop across each nozzle at this flow rate is 1.72×10^5. Calculate the theoretical power required by the pump.

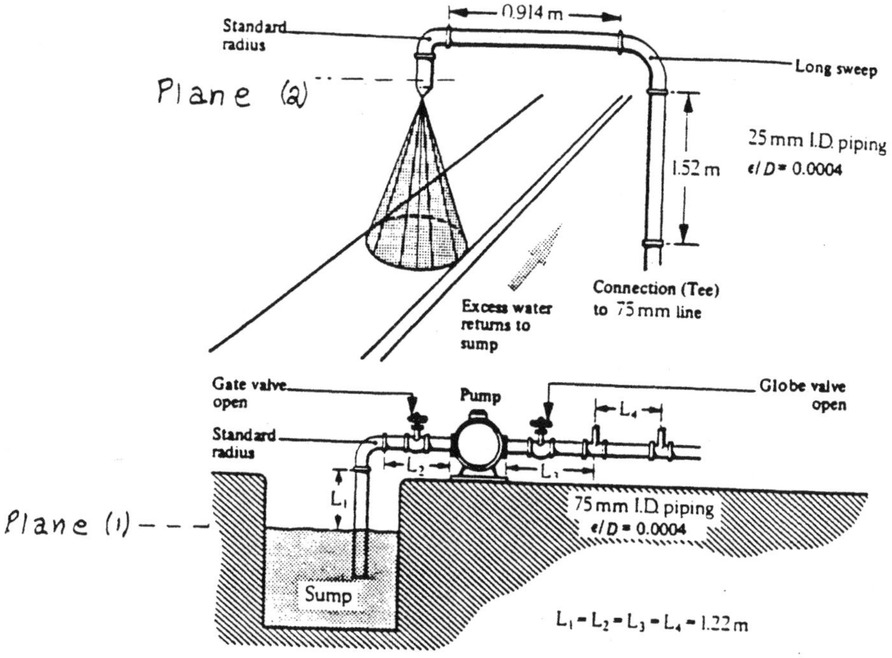

Most of the friction is for flow through the nozzles, so we can assume that the flow rates through the nozzle are equal.

$$\frac{P_2 - P_1}{\rho} + \frac{\bar{V}_2^2}{2\beta_2} - \frac{\bar{V}_1^2}{2\beta_1} + g(z_2 - z_1) + M^* + E_f = 0$$

$P_2 = P_1 + \Delta P \Rightarrow P_2 - P_1 = \Delta P(\text{nozzle})$; $\bar{V}_1 = 0$, $\bar{V}_2 = \bar{V}(25\text{mm pipe})$, $\beta_2 \simeq 1$, $z_2 - z_1 = \Delta z$

$E_f = E_f(75\text{mm}) + 2E_f(25\text{mm})$

$$E_f(75\text{mm}) = \left\{\frac{1}{2}e_{f,1} + 2f\left(\frac{L}{D}\right)_{elbow} + 2\left[2f\left(\frac{L}{D}\right)_{valve}\right] + 2\left[2f\left(\frac{L}{D}\right)_{tees}\right] + 2f\left(\frac{L}{D}\right)_{pipe}\right\}\bar{V}^{*2}$$

where $\bar{V}^*$ = average velocity in 75 mm pipe.

$$\bar{V}^* = \frac{Q}{A} = \frac{(2)(9.46\times10^{-4})}{\left(\frac{\pi}{4}\right)(0.075^2)} = 0.428 \text{ m s}^{-1}; \quad Re^* = \frac{D\bar{V}^*}{\nu} = \frac{(0.075)(0.428)}{9.61\times10^{-7}} = 3.34\times10^4$$

$e_{f,1} = 2(0.4) = 0.8$ (Figs. 4.4 and 4.5); $f = 0.006$ (Fig. 3.2); $\left(\frac{L}{D}\right)_{elbow} = 31$ (Table 4.2)

$$\left(\frac{L}{D}\right)_{valve} = 7 \text{ (Table 4.2)}; \left(\frac{L}{D}\right)_{tee} = 65 \text{ (Table 4.2)}; \left(\frac{L}{D}\right)_{pipe} = \frac{L_1 + L_2 + L_3 + L_4}{D} = \frac{(4)(1.22)}{0.075} = 65$$

$$E_f(75\text{ mm}) = \left[\frac{1}{2}(0.8) + (2)(0.006)(31) + (4)(0.006)(7) + (4)(0.006)(65) + (2)(0.006)(65)\right](0.428^2)$$

$$E_f(75\text{ mm}) = 0.601\ m^2\ s^{-2}$$

$$2E_f(25\text{ mm}) = 2\left\{2f\left(\frac{L}{D}\right)_{tee} + 2f\left[\left(\frac{L}{D}\right)_{elbow\ 1} + \left(\frac{L}{D}\right)_{elbow\ 2}\right] + 2f\left(\frac{L}{D}\right)_{pipe}\right\}$$

$$\bar{V} = \frac{Q}{A} = \frac{9.46 \times 10^{-4}}{\left(\frac{\pi}{4}\right)(0.025^2)} = 1.927\ m\ s^{-1};\ Re = \frac{D\bar{V}}{\nu} = \frac{(0.025)(1.927)}{9.61 \times 10^{-7}} = 5.01 \times 10^4$$

$$f = 0.0055 \text{ (Fig. 3.2)}; \left(\frac{L}{D}\right)_{tee} = 90 \text{ (Table 4.2)}; \left(\frac{L}{D}\right)_{elbow\ 1} = 20 \text{ (Table 4.2)};$$

$$\left(\frac{L}{D}\right)_{elbow\ 2} = 31 \text{ (Table 4.2)}; \left(\frac{L}{D}\right)_{pipe} = \frac{1.52 + 0.914}{0.025} = 97.36$$

$$E_f(25\text{ mm}) = 2\left\{(2)(0.0055)\left[90 + 20 + 31 + 97.4\right]\right\} = 5.24\ m^2 s^{-2}$$

$$E_f = 0.601 + 5.24 = 5.84\ m^2\ s^{-2}$$

$$-M^* = E_f + g\Delta z + \frac{\bar{V}^2}{2} + \frac{\Delta P}{\rho} = 5.84 + 9.81(1.52 + 1.22) + \frac{1.927^2}{2} + \frac{1.72 \times 10^5}{1 \times 10^3} = 207\ m^2 s^{-2}$$

$$-M^* = 207\ N\ m\ kg^{-1}$$

4.9 A fan draws air at rest and sends it through a straight duct 152 m long. The diam. of the duct is 0.61 m and a Pitot-static tube is installed with its impact opening along the center line. The air enters at 300 K and 1 atm and discharges at 1.2 atm. Calculate the theoretical work (in N m kg^{-1}) of the fan if the Pitot-static tube measures a pressure difference of 25.4 mm of water.

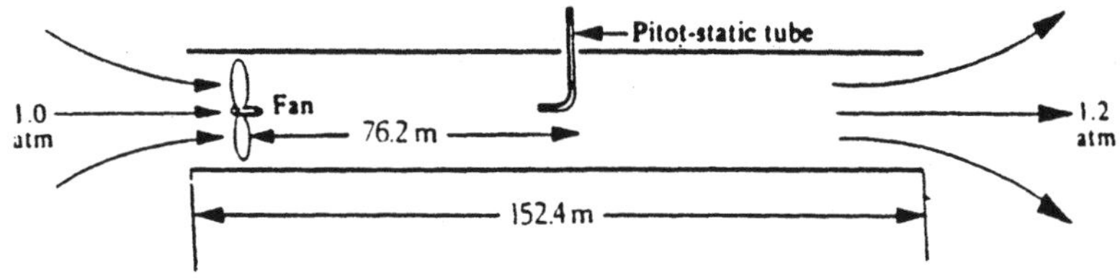

$$V_{max} = C_P\left(\frac{2}{\rho}\Delta P\right)^{\frac{1}{2}} = C_P\left(\frac{2gh\,\rho_{H_2O}}{\rho_{Air}}\right)^{1/2}$$

assume $C_P = 1.0$, $h = 25.4\ mm_{H_2O}$, $1.2\ atm = 1.216\times10^{5}\ N\ m^{-2}$

$$\rho_{Air} = \frac{1\,g\,mol}{0.0224\ m^3}\left|\frac{1.216\times10^{5}\ N\ m^{-2}}{1.0133\times10^{5}\ N\ m^{-2}}\right|\frac{273\ K}{300\ K}\left|\frac{28.8\ g}{1\ g\ mol}\right|\frac{1\ kg}{1000\ g} = 1.404\ kg\ m^{-3}$$

$$V_{max} = 1.0\left[\frac{2}{}\left|\frac{9.807\ m}{s}\right|\frac{0.0254\ m}{}\left|\frac{1\times10^{3}\ kg}{m^3}\right|\frac{m^3}{1.40\ kg}\right]^{\frac{1}{2}} = 18.86\ m\ s^{-1}$$

$$Re_{MAX} = \frac{D\,V_{MAX}\,\rho}{\eta} = \frac{0.61\ m}{}\left|\frac{18.86\ m}{s}\right|\frac{1.404\ kg}{m^3}\left|\frac{m^2}{1.85\times10^{-5}\ N\ s}\right|\frac{N\ m}{kg\ s^2} = 8.73\times10^{5}$$

$$\frac{\bar{V}}{V_{max}} = 0.62 + 0.04\log(Re_{MAX}) = 0.62 + 0.04\log 8.73\times10^{5} = 0.858$$

$$\bar{V} = \frac{0.858}{}\left|\frac{18.86\ m}{s}\right. = 16.2\ m\ s^{-1}$$

$$\int_{P_1}^{P_2}\frac{dP}{\rho} + \left[\frac{\bar{V}_2^{\,2}}{2\beta_2} - \frac{\bar{V}_1^{\,2}}{2\beta_1}\right] + g\Delta z + M^{*} + E_f = 0$$

Plane (1) - just before the inlet; Plane (2) - just after the outlet.

$\therefore \bar{V}_2 = \bar{V}_1 = 0$, $\Delta Z = 0$ since horizontal, $\Delta p = 0.2$ atm $= 2.0266 \times 10^4$ N m^{-2}

$$\frac{\Delta P}{\rho} = \frac{2.0266 \times 10^4 \, N}{m^2} \left| \frac{m^3}{1.404 \, kg} \right| \frac{kg \, m}{N \, s^2} = 1.44 \times 10^4 \, m^2 \, s^{-2}$$

$$E_f = E_1 + E_2 + E_3$$

$E_1 = E_f\,(\text{straight}) = 2f\left(\frac{L}{D}\right)\bar{V}^2$, $Re \simeq Re_{max.} = 8.73 \times 10^5$, $f = 0.003$ From Fig. 3.2

$$E_1 = (2)(0.003)\left(\frac{152.4 \, m}{0.61 \, m}\right)\left(\frac{18.86 \, m}{s}\right)^2 = 533 \quad m^2 s^{-2}$$

$$E_2 = E_f\,(\text{entrance}) = \frac{1}{2}\bar{V}^2 e_f = \left(\frac{1}{2}\right)\left(\frac{18.86 \, m}{s}\right)^2 (2)(0.4) = 142 \, m^2 s^{-2}$$

$$E_3 = E_f\,(\text{exit}) = \frac{1}{2}\bar{V}^2 e_f = \left(\frac{1}{2}\right)\left(\frac{18.86 \, m}{s}\right)^2 (1.0) = 178 \quad m^2 s^{-2}$$

$$E_f = 533 + 142 + 178 = 853 \, m^2 s^{-2}$$

$$-M^* = E_f + \frac{\Delta P}{\rho} = 853 + 14400 = 1.5253 \times 10^4 \, m^2 s^{-2} = 1.5253 \times 10^4 \, N \, m \, kg^{-1}$$

4.10 Compressed air at 6.9×10^5 N m^{-2} and 310 K flows through an orifice plate meter installed in a 75 mm I.D. pipe. The orifice has a 25 mm hole and the downstream pressure tap location is 38 mm from the plate. When the manometer reading is 358 mm Hg,

a) What is the flow rate of air?

b) What is the permanent pressure drop?

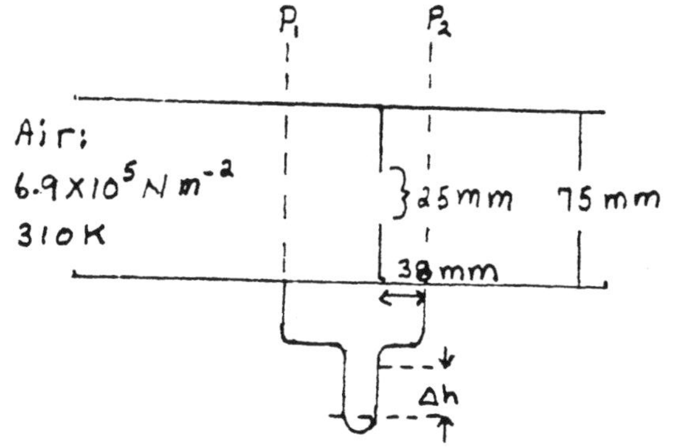

a. Flow rate of air (Kg s^{-1})

$D_o = 25$ mm, $D_1 = 75$ mm, $\gamma = \frac{C_p}{C_v} \simeq \frac{5}{3}$ (ideal gas), $B = \frac{D_o}{D_1} = \frac{1}{3}$, $B^2 = \frac{1}{9}$

$W = KYA_o\,(2\rho\Delta P)^{\frac{1}{2}}$

$\Delta P = (358 \text{ mm Hg})(13.596) = 4.870 \times 10^3 \text{ Kg m}^{-2} \,\Big|\, 9.807 \text{ m s}^{-2} \,\Big|\, \frac{\text{N s}^2}{\text{Kg m}} = 4.776 \times 10^4 \text{ N m}^{-2}$

$P_1 = 6.9 \times 10^5 \text{ N m}^{-2}$, $P_2 = 6.4224 \times 10^5 \text{ N m}^{-2}$

Apply Eq. (4.33)

$$r = \frac{P_2}{P_1} = \frac{6.4224 \times 10^5}{6.90 \times 10^5} = 0.9308$$

$$\frac{1-r}{\gamma} = \frac{(1-0.9308)}{\left(\frac{5}{3}\right)} = 0.042$$

From Fig. 4.11, Y = 0.98

Downstream pressure tap location is 38 mm

Fig. 4.12 gives K = 0.62

$$A_o = \frac{(\pi)(25\text{ mm})^2}{4} = 490.9 \text{ mm}^2$$

ρ_{air} (6.9×10^5 N m^{-2} and 310K)

$$\rho = \frac{1\ \text{gmol}}{0.0224\ \text{m}^3}\left|\frac{6.9\times10^5\ \text{N m}^{-2}}{1.0113\times10^5\ \text{N m}^{-2}}\right|\frac{273\ \text{K}}{310\ \text{K}}\left|\frac{28.8\ \text{g}}{1\ \text{gmol}}\right|\frac{1\ \text{kg}}{1000\ \text{g}} = 7.73\ \text{kg m}^{-3}$$

$$W = (0.62)(0.98)(490.9\ \text{mm}^2)\left[2\left|\frac{7.73\ \text{kg}}{\text{m}^3}\right|\frac{4.776\times10^4\ \text{N}}{\text{m}^2}\left|\frac{\text{kg m}}{\text{N s}^2}\right.\right]^{\frac{1}{2}}\frac{1\ \text{m}^2}{(1\times10^3)^2\ \text{mm}^2} = 0.256\ \text{kg s}^{-1}$$

$W = 0.256$ kg s^{-1} = 923 kg h

b. $\Delta P_{loss} = (1-B^2)(P_1-P_2) = (1-\frac{1}{9})(4.776\times10^4\ \text{N m}^{-2}) = 4.245\times10^4\ \text{N m}^{-2}$

$\Delta P = 4.245\times10^4$ N m^{-2}

4.11 A venturi meter is installed in an air duct of circular cross-section 0.46 m in diameter which carries up to a maximum of 1.18 m^3 s^{-1} of air at 300 K and 1.1×10^{-5} N m^{-2}. The throat diameter is 230 mm.

a) Determine the maximum pressure drop that a manometer must be able to handle, i.e., what range of pressure drops will be encountered? Express the results in mm of water.
b) Instead of the venturi meter, an orifice meter is proposed and the maximum pressure drop to be measured is 50 mm of water. Calculate what diameter of sharp-edged orifice should be installed in order to obtain the full scale reading at maximum flow.
c) Estimate the permanent pressure drop for the devices in parts a) and b).
d) If the air is supplied by a blower operating at 50% efficiency, what is the power consumption associated with each installation?

a. $D_o = 230$ mm

$$\bar{V}_o = \frac{1.18\ \text{m}^3}{\text{s}}\left|\frac{300\ \text{K}}{273\ \text{K}}\right|\frac{4}{\pi(0.23)^2\ \text{m}^2} = 31.2\ \text{m s}^{-1}$$

Assume Y=1; this will be checked later.

$W = KY\sqrt{2\rho(P_1-P_2)}\,A_o$ and $W = \bar{V}_o A_o \rho$

$\therefore \bar{V}_o = KY\sqrt{\frac{2}{\rho}(P_1-P_2)}$, $B = \frac{230\ \text{mm}}{460\ \text{mm}} = 0.5$, $C_o = 0.98$

$$K = \frac{C_D}{\left[1 - B^4\right]^{\frac{1}{2}}} = \frac{0.98}{\left[1 - 0.5^4\right]^{1/2}} = 1.012$$

$$\rho_{Air} = \frac{1 \text{ mol}}{0.0224 \text{ m}^3} \left| \frac{1.1 \times 10^5 \text{ N m}^{-2}}{1.0133 \times 10^5 \text{ N m}^{-2}} \right| \frac{273 \text{ K}}{300 \text{ K}} \left| \frac{28.8}{1 \text{ mol}} \right| \frac{1 \text{ kg}}{1000 \text{ g}} = 1.27 \text{ kg m}^{-3}$$

$$P_1 - P_2 = \frac{1.27 \text{ kg}}{2 \text{ m}^3} \left| \frac{(31.21)^2 \text{ m}^2}{\text{s}^2} \right| \frac{}{1.012^2} \left| \frac{}{1^2} \right| \frac{\text{N s}^2}{\text{kg m}} = 6.04 \times 10^2 \text{ N m}^{-2} = 61.6 \text{ mm } H_2O$$

$\Delta P = 61.6 \text{ mm } H_2O$

Check on value of Y

$$r = \frac{P_2}{P_1} = \frac{1.094 \times 10^5 \text{ N m}^{-2}}{1.1 \times 10^5 \text{ N m}^{-2}} = 0.995$$

$$\frac{1-r}{\gamma} = \frac{0.005}{\left(\frac{5}{3}\right)} = 3 \times 10^{-3}, \quad \therefore Y \simeq 1$$

b. Assume $Y = 1$ and try $D_o = 0.300 \text{ m}$, $P_1 - P_2 = 50 \text{ mm of } H_2O = 4.90 \times 10^2 \text{ N m}^{-2}$

$$A_o = \frac{W}{KY}\left[2\rho(P_1 - P_2)\right]^{-\frac{1}{2}}$$

$$W = Q\rho = \frac{1.18 \text{ m}^3}{\text{s}} \left| \frac{1.27 \text{ kg}}{\text{m}^3} \right. = 1.50 \text{ kg s}^{-1}$$

$$\left[2\rho(P_1 - P_2)\right]^{\frac{1}{2}} = \left[2 \left| \frac{1.27 \text{ kg}}{\text{m}^3} \right| \frac{4.90 \times 10^2 \text{ N}}{\text{m}^2} \left| \frac{\text{kg m}}{\text{N s}^2} \right. \right]^{\frac{1}{2}} = 35.3 \text{ kg m}^{-2} \text{s}^{-1}$$

$B = \frac{0.300}{0.460} = 0.652$

$K = 0.68$ from Fig. 4.12 at Vena contracta

$$A_o = \frac{1.50 \text{ kg}}{\text{s}} \left| \frac{}{0.68} \right| \frac{}{1} \left| \frac{\text{m}^2 \text{ s}}{35.3 \text{ kg}} \right. = 6.25 \times 10^{-2} \text{ m}^2$$

$$D_o = \left(\frac{4A_o}{\pi}\right)^{\frac{1}{2}} = 0.282 \text{ m}, \quad B^2 = \left(\frac{0.282 \text{ m}}{0.46 \text{ m}}\right)^2 = 0.38, \quad r = \frac{\left(1.1 \times 10^5 - 4.9 \times 10^2\right)}{1.1 \times 10^5} = 0.996$$

$$\frac{1-r}{\gamma} = \frac{(1-0.996)}{\left(\frac{2}{3}\right)} = 2.4 \times 10^{-3}, \therefore Y \simeq 1$$

Try $D_o = 0.296\ m$, $B = 0.64$, $K = 0.67$

$$A_o = \frac{1.65\ kg}{s}\bigg|\frac{}{0.67}\bigg|\frac{}{1}\bigg|\frac{m^2\ s}{35.3\ kg} = 6.98 \times 10^{-2}\ m^2;\ \therefore D_o = 0.298\ m - \text{close enough.}$$

c. Venturi $- \Delta P_{loss} \simeq (0.10)(6.11 \times 10^2\ N\ m^{-2}) = 6.11 \times 10^1\ N\ m^{-2}$

Orifice $- \Delta P_{loss} \simeq (1-B^2)(P_1 - P_2) = (1-0.41)(4.90 \times 10^2\ N\ m^{-2}) = 2.89 \times 10^2\ N\ m^{-2}$

d. Power $= \frac{\Delta P}{\rho} w\ eff^{-1}$

$$\text{Power Consumption venturi} = \frac{61.1\ N}{m^2}\bigg|\frac{1\quad m^3}{1.27\ kg}\bigg|\frac{1.65\ kg}{s}\bigg|\frac{}{0.5} = 159\ N\ m\ s^{-1}$$

$$\text{Power Consumption Orifice} = (159)\left(\frac{289}{61.1}\right) = 752\ N\ m\ s^{-1}$$

4.12 Liquid flows through a long straight tube and then through a Venturi meter. Pressure gauges A and B measure the pressure drop for a tube length L, and P_1 and P_2 are the pressures for the manometer used with the Venturi. Derive an equation for the ratio of the two pressure drops, $(P_0 - P_L)/(P_1 - P_2)$ in terms of f, K, D, and D_0 where f is friction factor for the tube, K is the flow coefficient of the Venturi, D is the diameter of the tube, and D_0 is the throat diameter.

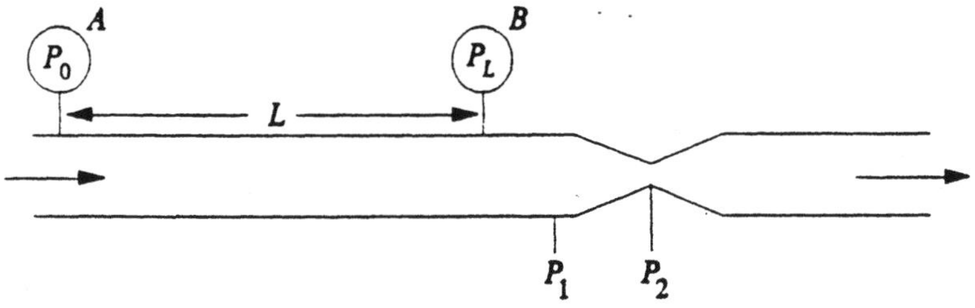

straight tube:

$$\frac{P_2 - P_1}{\rho} + \left[\frac{\bar{V}_2^2}{2\beta_2} - \frac{\bar{V}_1^2}{2\beta_1}\right] + g\Delta z + M^* + E_f = 0$$

$V_1 = V_2$, $\Delta z = 0$, $M^* = 0$, Let $P_2 = P_L$, $P_1 = P_0$

$$E_f = 2f\left(\frac{L}{D}\right)\bar{V}^2$$

$$(P_0 - P_L) = 2\rho f\left(\frac{L}{D}\right)\bar{V}^2$$

Venturi: - Eq. 4.31

$$\bar{V}_0 = K\left[\frac{2}{\rho}(P_1 - P_2)\right]^{1/2}; \qquad \bar{V}_0 = \bar{V}\left(\frac{D}{D_0}\right)^2, \quad (P_1 - P_2) = \frac{\rho}{2}\frac{\bar{V}_0^2}{K^2}$$

$$(P_1 - P_2) = \frac{\rho\bar{V}^2}{2K^2}\left(\frac{D}{D_0}\right)^4$$

$$\frac{(P_0 - P_L)}{(P_1 - P_2)} = \frac{2\rho f\left(\frac{L}{D}\right)\bar{V}^2}{\frac{\rho\bar{V}^2}{2K^2}\left(\frac{D}{D_0}\right)^4} = \frac{4K^2 f L D_0^4}{D^5}$$

4.13 Lacking the funds to purchase a head meter, Mr. Make-do installed a 50 mm dia. tube in a 100 mm dia. line in order to measure flow rate. Pressure taps P_1 and P_2 are connected to a manometer.

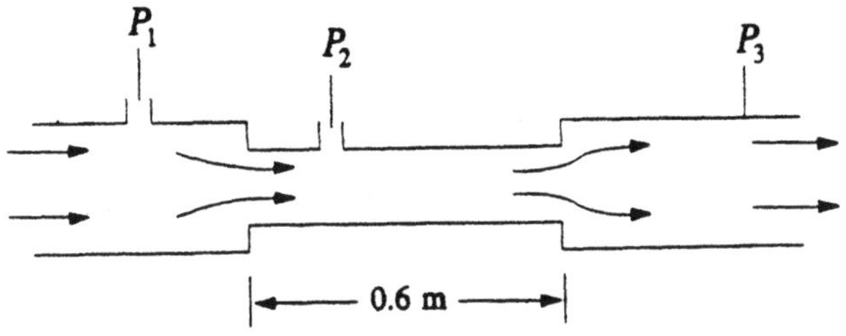

a) For flow which is highly turbulent (Re → ∞), derive an equation which gives mass flow rate through the line in terms of the pressure difference P_1 - P_2.

b) Develop an equation which could be used to calculate the permanent pressure drop (i.e., P_1 - P_3).

a: Plane 1 - at P_1 ; Plane 2 - at P_2

$$\frac{P_2 - P_1}{\rho} + \frac{\bar{V}_2^2}{2} - \frac{\bar{V}_1^2}{2} + E_f = 0 \quad \text{since } \Delta z = 0,\ M^* = 0,\ \beta = 1 \text{ because turbulent}$$

$$\dot{W} = Q\rho = \bar{V}_1 A_1 \rho = \bar{V}_1 \left(\frac{\pi D_1^2}{4}\right)\rho = \bar{V}_1 \left[\frac{\pi}{4}(0.1)^2\right]\rho = 2.5\times10^{-3}\pi \bar{V}_1 \rho$$

$$\therefore \bar{V}_1 = \frac{\dot{W}}{2.5\times10^{-3}\pi\rho}$$

$$\bar{V}_2 = \left(\frac{D_1}{D_2}\right)^2 \bar{V}_1 = \left(\frac{100}{50}\right)^2 \bar{V}_1 = 4\bar{V}_1$$

$$(P_1 - P_2) = \frac{\rho}{2}\left(\bar{V}_2^2 - \bar{V}_1^2\right) + \rho E_f = \frac{\rho}{2}\left(16\bar{V}_1^2 - \bar{V}_1^2\right) + \rho E_f = 7.5\rho\bar{V}_1^2 + \rho E_f$$

$$\frac{A_2}{A_1} = \frac{D_2^2}{D_1^2} = 0.5^2 = 0.25\text{; From Fig. 4.4} - e_f = 0.28$$

$$E_f = \frac{1}{2}\bar{V}_2^2 e_f = \frac{1}{2}\left(16\bar{V}_1^2\right)(0.28)$$

$$(P_1 - P_2) = 7.5\rho\bar{V}_1^2 + 2.24\rho\bar{V}_1^2 = 9.74\rho\bar{V}_1^2$$

$$\bar{V}_1^2 = \frac{\dot{w}^2}{(2.5\times10^{-3})^2\pi^2\rho^2}$$

$$(P_1-P_2) = 9.74\rho\left[\frac{\dot{w}}{(2.5\times10^{-3})\pi\rho}\right]^2$$

$$\dot{w} = \frac{(2.5\times10^{-3})\pi\rho^{\frac{1}{2}}(P_1-P_2)^{\frac{1}{2}}}{(9.74)^{\frac{1}{2}}} = 2.52\times10^{-3}\rho^{1/2}(P_1-P_2)^{\frac{1}{2}}$$

b. $$\frac{P_3-P_1}{\rho} + \left[\frac{\bar{V}_3^2}{2B_3} - \frac{\bar{V}_1^2}{2B_1}\right] + g\Delta z + M^* + E_f = 0$$

$(P_1-P_3) = \rho E_f$ since $V_1 = V_3$, $\Delta z = 0$, $M^* = 0$

$E_f = E_f(\text{contraction}) + E_f(\text{conduit}) + E_f(\text{expansion})$

$E_f(\text{cont.}) = 2.24\,\bar{V}_1^2$ - From part a.

$$E_f(\text{conduit}) = 2f\left(\frac{L}{D}\right)\bar{V}_2^2 = 2f\left(\frac{0.6}{0.05}\right)\bar{V}_2^2 = 2f(12)(16\bar{V}_1^2) = 384f\,\bar{V}_1^2$$

$E_f(\text{expan.}) = \frac{1}{2}\bar{V}_2^2 e_f = \frac{1}{2}(16\bar{V}_1^2)(0.6) = 4.8\,\bar{V}_1^2$; Fig. 4.3 $e_f = 0.6$ since $\frac{A_1}{A_2} = 0.25$

$$P_1 - P_3 = \rho(2.24\bar{V}_1^2 + 384f\bar{V}_1^2 + 4.8\bar{V}_1^2) = \rho\bar{V}_1^2(7.04 + 384f)$$

From part a. - $\bar{V}_1^2 = \dfrac{\dot{w}^2}{(2.5\times10^{-3})^2\pi^2\rho^2}$

$$P_1 - P_3 = (1.62\times10^4)\frac{\dot{w}^2}{\rho}(7.04 + 384f)$$

4.14 In order to *slush* cast seamless stainless-steel pipe, a *pressure pouring* technique is utilized, as depicted below.

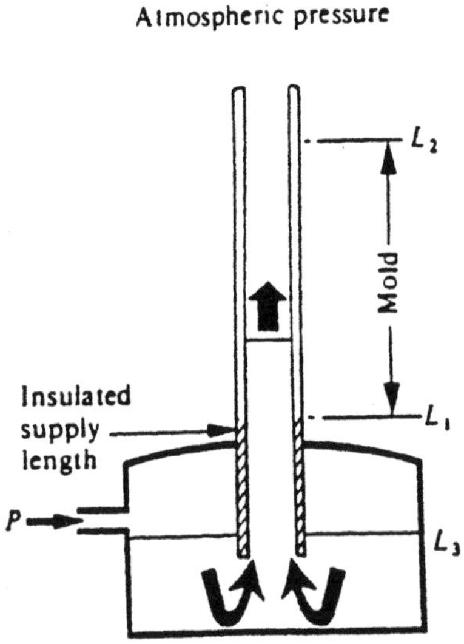

The mold must be filled rapidly up to level L_2 so that no solidification takes place until the entire mold is filled. Determine an expression that we can use to give the time it takes to fill the mold *only*. Consider the mold to be between L_1 and L_2 and open to atmospheric pressure at the top. Neglect the change of metal height L_3 in the ladle; you may also neglect the friction loss associated with the supply tube and mold tube walls, but the entrance loss may *not* be neglected.

Plane A – metal height in the ladle.

Plane B – metal height in the mold.

$$\int_{P_A}^{P_B} \frac{dP}{\rho} + \left[\frac{\bar{V}_B^2}{2\beta_B} - \frac{\bar{V}_A^2}{2\beta_A}\right] + g(z_B - z_A) + M^* + E_f = 0$$

$E_f = \frac{1}{2} e_f V_B^2$ entrance, $P_B - P_A = P_{atm} - P = -(P - P_{atm})$, $M^* = 0$, $\bar{V}_A = 0$

$z_B - z_A = h$

$$\therefore \bar{V}_B = 2^{\frac{1}{2}}\left[\frac{1}{\beta_B} + e_f\right]^{-\frac{1}{2}}\left[\frac{P - P_{atm}}{\rho} - gh\right]^{\frac{1}{2}}, \text{ Let } C_D = \left[\frac{1}{\beta_B} + e_f\right]^{-\frac{1}{2}}$$

$$\bar{V}_B = 2^{\frac{1}{2}} C_D \left[\frac{P - P_{atm}}{\rho} - gh\right]^{\frac{1}{2}}$$

$$\frac{dh}{dt} = \bar{V}_B \quad \text{or} \quad \frac{dh}{dt} = 2^{\frac{1}{2}} C_D \left[\frac{P - P_{atm}}{\rho} - gh\right]^{\frac{1}{2}}$$

$$\int_0^t dt = \frac{1}{2^{\frac{1}{2}} C_D} \int_{L_1 - L_3}^{L_2 - L_3} \frac{dh}{\left[\frac{P - P_{atm}}{\rho} - gh\right]^{\frac{1}{2}}}$$

$$t = \frac{2^{1/2}}{C_D g} \left\{ \left[\frac{P - P_{atm}}{\rho} - g(L_1 - L_3)\right]^{1/2} - \left[\frac{P - P_{atm}}{\rho} - g(L_2 - L_3)\right]^{\frac{1}{2}} \right\}$$

4.15 Calculate the time to fill the mold, as depicted below, with molten metal if the metal level at plane A is maintained constant and the time to fill the runner system (entering *piping*) is ignored. *Data* (all may be taken as constant): $\eta = 1.65 \times 10^{-3}$ N s m^{-2}; $\rho = 6410$ kg m^{-3}; $f = 0.0025$ (runner); e_f(contraction) $= 0.1$; e_f(enlargement for liquid levels below B) $= 0$; e_f(enlargement for liquid levels above B) $= 1.0$; L/D(90° turn) $= 25$; $\beta = 1.0$.

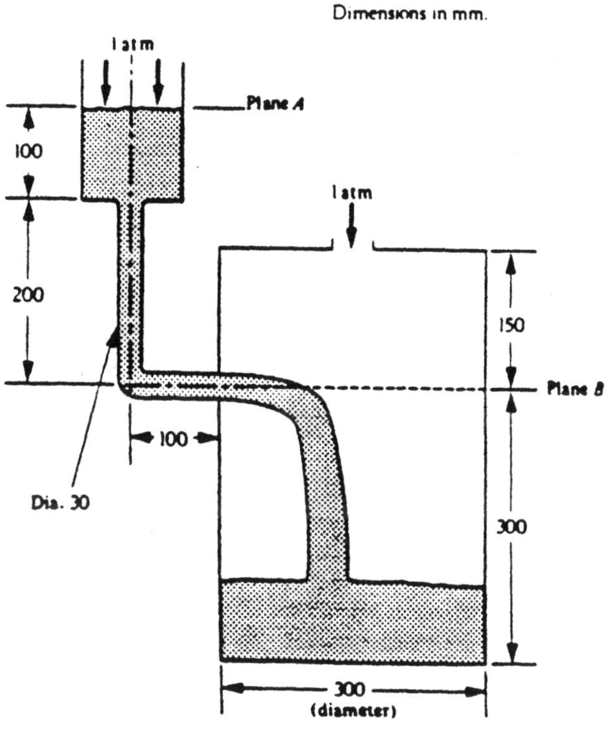

This system is non-interactive until the height of the fluid in the mold is $\geq$ 300 mm.

$$\int_{P_1}^{P_2} \frac{dP}{\rho} + \left[\frac{\bar{V}_2^2}{2\beta_2} - \frac{\bar{V}_1^2}{2\beta_1}\right] + g\Delta z + E_f + W^* = 0$$

$P_1 = P_2$, $\bar{V}_1 = 0$, $\beta = 1.0$, $W^* = 0$: NON-interactive case.

$$\therefore \frac{\bar{V}_B^2}{2} + g\Delta z + E_f = 0 \;,\; \Delta z = -300\,mm$$

$$E_f = E_{f1}\Big|_{contraction} + E_{f2}\Big|_{90°\,elbow} + E_{f3}\Big|_{runner} + E_{f4}\Big|_{enlargement}$$

$$E_{f1} = \frac{1}{2} e_f \bar{V}_B^2 = \frac{1}{2}(0.1)\,V_B^2 = 0.05\,\bar{V}_B^2$$

$$E_{f2} = 2f\left(\frac{Le}{D}\right)\bar{V}_B^2 = 2(0.0025)(25)\,\bar{V}_B^2 = 0.125\,\bar{V}_B^2$$

$$E_{f3} = 2f\left(\frac{L}{D}\right)\bar{V}_B^2 = 2(0.0025)\left(\frac{300}{30}\right)\bar{V}_B^2 = 0.05\,\bar{V}_B^2$$

$E_{f4} = 0$ – Problem Statement.

$$E_f = 0.225\,\bar{V}_B^2$$

$$\frac{\bar{V}_B^2}{2} + 0.225\,\bar{V}_B^2 = -g\,(\Delta z)$$

$$0.725\,\bar{V}_B^2 = -(9.807)(-0.3) = 2.94, \quad \bar{V}_B = 2.01\,m\,s^{-1}$$

$$Q = \bar{V}_B A = \frac{2.01\,m}{s}\left|\frac{\pi}{4}\right|(0.03)^2 m^2 = 1.42\times10^{-3}\,m^3\,s^{-1}, \quad Vol.\ below\ B = \frac{\pi D^2 L}{4} = \frac{\pi(0.3)^2(0.3)}{4}$$

$$Vol. = 2.12\times10^{-2}\,m^3, \quad Time\ to\ fill\ below\ B = \frac{Vol.}{Q} = 15\,s$$

Above Plane B

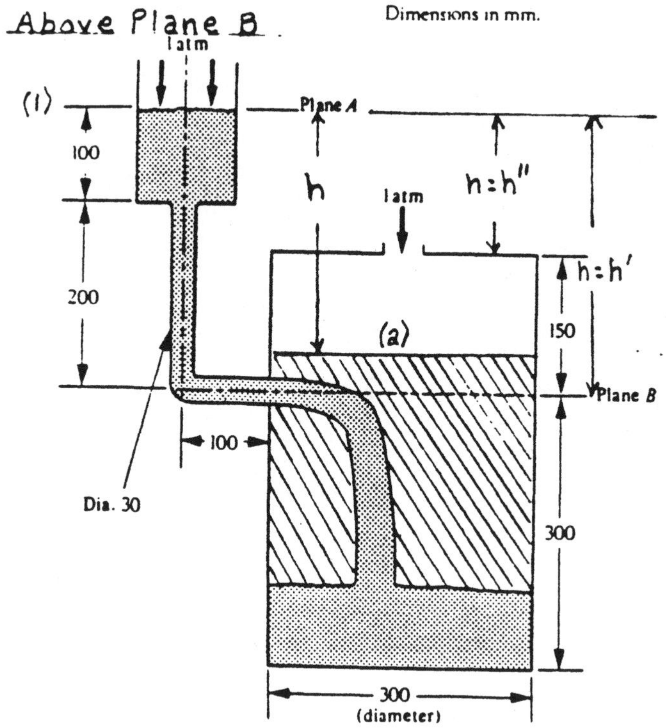

$$\frac{\bar{V}_2^2}{2\beta_2} - \frac{\bar{V}_1^2}{2\beta_1} + g(z_2 - z_1) + E_f = 0 \; ; \; \bar{V}_1^2 \simeq 0 \; ; \; z_2 - z_1 = -h \; ; \; \beta_2 = 1$$

$$E_f = \left[\frac{1}{2}e_{f,c} + \frac{1}{2}e_{f,e} + 2f\left(\frac{L}{D}\right)\right]\bar{V}^2 ; \text{ where } \bar{V} = \text{velocity in gate.}$$

contract. — expansion — elbow + runner

$$\frac{\bar{V}_2^2}{2} - gh + \alpha\bar{V}^2 = 0 \text{ where } \alpha = \frac{1}{2}e_{f,c} + \frac{1}{2}e_{f,e} + 2f\left(\frac{L}{D}\right) : \bar{V} = \left(\frac{A_2}{A_1}\right)\bar{V}_2$$

$$\therefore \frac{\bar{V}_2^2}{2} - gh + \alpha\left(\frac{A_2}{A}\right)^2\bar{V}_2^2 = 0 : \bar{V}_2 = \left[\frac{g}{\frac{1}{2} + \alpha\left(\frac{A_2}{A}\right)^2}\right]^{1/2} h^{1/2} = -\frac{dh}{dt}$$

$$\alpha = \frac{1}{2}(0.1) + \frac{1}{2}(1) + 2(0.0025)\left(\frac{400}{30} + 25\right) = 0.74 \; ; \left(\frac{A_2}{A}\right)^2 = \left(\frac{300}{30}\right)^2 = 100$$

$$\left[\frac{g}{\frac{1}{2} + \alpha\left(\frac{A_2}{A}\right)^2}\right]^{1/2} = \left[\frac{9.81}{0.5 + (0.74)(100)}\right]^{1/2} = 0.36$$

$$dt = -\frac{dh}{0.36\,h^{1/2}} : \int_0^t dt = -\frac{1}{0.36}\int_{h'}^{h''}\frac{dh}{h^{1/2}} \; ; \; h' = 300\,\text{mm}; \; h'' = 150\,\text{mm}$$

$$t = \frac{1}{0.36}\left[2h^{1/2}\right]_{h''}^{h'} = \frac{2}{0.36}\left[300^{1/2} - 150^{1/2}\right] = 28.2 \text{ s}$$

Total time = 15 + 28.2 = 43.2 s

4.16 In planning continuous casting, we use fluid flow analysis. Consider the illustrated configuration of the equipment, which includes in-line vacuum degassing.

a) Determine the tundish and degasser nozzle sizes which are necessary to operate the system at a rate of 6.3 kg s^{-1} per strand. Suppose that for operational reasons it is desirable to maintain tundish and degasser bath depths of 0.76 and 1.83 m, respectively.

b) If only 13 mm diameter degasser nozzles are available, how would their use affect the casting operation?

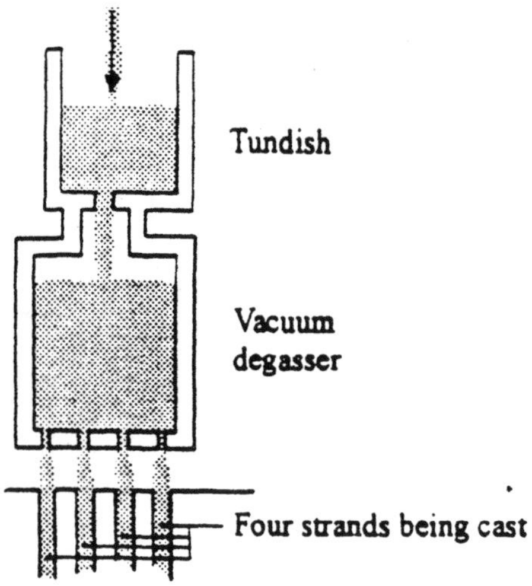

Inside dimensions: tundish, 2.4 m × 2.4 m × 1.2 m; degasser, 1.2 m × 1.2 m × 2.4 m.

Liquid-steel density = 7530 kg m^{-3}.

Discharge coefficients for tundish and vacuum degasser nozzles: $C_D = 0.8$.

Vacuum pressure = 10^{-3} atm (101 N m^{-2}).

a. Tundish calculation

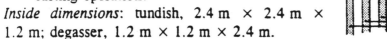

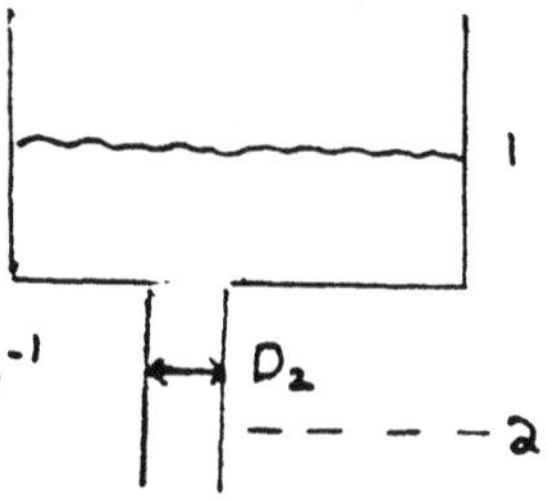

$$A_2 = \pi \frac{D_2^2}{4}, \quad \frac{dw}{dt} = \rho \bar{V}_2 A_2$$

$$\therefore D_2 = \left[\frac{4}{\pi}\left|\frac{1}{\rho}\right|\frac{1}{\bar{V}_2}\left|\frac{dw}{dt}\right.\right]^{1/2}, \text{ For 4 strands } \frac{dw}{dt} = 25.2 \text{ kg s}^{-1}$$

$$\frac{P_2 - P_1}{\rho} + \frac{\bar{V}_2^2}{2\beta_2} + g(z_2 - z_1) + \frac{\bar{V}_2^2}{2} e_f = 0$$

$$\frac{\bar{V}_2^2}{2}\left[\frac{1}{\beta_2} + e_f\right] = \frac{P_1 - P_2}{\rho} + g(z_1 - z_2); \quad \left(\frac{1}{\beta_2} + e_f\right) = C_D$$

$$\bar{V}_2 = \left\{\frac{2}{C_D}\left[\frac{P_1 - P_2}{\rho} + g(z_1 - z_2)\right]\right\}^{1/2}; \quad g(z_1 - z_2) = (9.81 \text{ m s}^{-2})(0.76 \text{ m}) = 7.46 \text{ m}^2 \text{s}^{-2}$$

$$\frac{P_1 - P_2}{\rho} = \frac{(1.0133 \times 10^5 \text{ N m}^{-2} - 101 \text{ N m}^{-2})}{} \left|\frac{m^3}{7530 \text{ kg}}\right|\frac{\text{kg m}}{\text{N s}^2} = 13.44 \text{ m}^2\text{s}^{-2}$$

$$\therefore \bar{V}_2 = \left\{\frac{2}{0.8}\left[13.44 + 7.46\right]\right\}^{1/2} = 7.23 \text{ m s}^{-1}$$

$$D_2 = \left[\frac{4}{\pi}\left|\frac{m^3}{7530 \text{ kg}}\right|\frac{s}{7.23 \text{ m}}\left|\frac{25.2 \text{ kg}}{s}\right.\right]^{1/2} = 0.0243 \text{ m} = 24.3 \text{ mm}$$

Vacuum degasser calculation:

For 1 nozzle

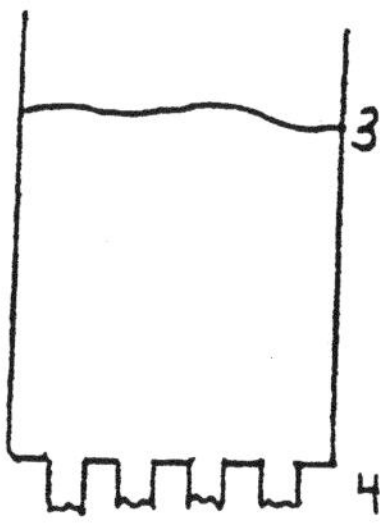

$$D_4 = \left[\frac{4}{\pi}\left|\frac{1}{\rho}\right|\frac{1}{\bar{V}_4}\left|\frac{dW_2}{dt}\right.\right]^{\frac{1}{2}}, \quad \frac{dW_2}{dt} = 6.3\,\text{kg s}^{-1}$$

$$\frac{P_4 - P_3}{\rho} + \frac{V_4^2}{2\beta_4} + g(z_4 - z_3) + \frac{V_4^2}{2} e_f = 0$$

$$\frac{V_4^2}{2}\left[\frac{1}{\beta_4} + e_f\right] = \frac{P_3 - P_4}{\rho} + g(z_3 - z_4); \left(\frac{1}{\beta_i} + e_f\right) = C_D = 0.8$$

$$\bar{V}_4 = \left\{\frac{2}{C_D}\left[\frac{(P_3 - P_4)}{\rho} + g(z_3 - z_4)\right]\right\}^{\frac{1}{2}}$$

$$g(z_3 - z_4) = (9.807\,\text{m s}^{-2})(1.83\,\text{m}) = 17.95\,\text{m}^2\text{s}^{-1}$$

$$\frac{P_3 - P_4}{\rho} = \frac{(101 - 1.0133\times10^5)\,\text{N m}^{-2}}{} \left| \frac{\text{m}^3}{7530\,\text{kg}} \right| \frac{\text{kg m}}{\text{N s}^2} = -13.44\,\text{m}^2\text{s}^{-2}$$

$$\therefore V_4 = \left\{\frac{2}{0.8}\left[-13.44 + 17.95\right]\right\}^{\frac{1}{2}} = 3.36\,\text{m s}^{-1}$$

$$D_4 = \left[\frac{4}{\pi}\left|\frac{\text{m}^3}{7530\,\text{kg}}\right|\frac{\text{s}}{3.36\,\text{m}}\left|\frac{6.3\,\text{kg}}{\text{s}}\right.\right]^{\frac{1}{2}} = 0.018\,\text{m} = 18\,\text{mm}$$

∴ use a 24.3 mm tundish nozzle and four 18 mm vacuum degasser nozzles.

b. If $D_4 = 13\,\text{mm}$, $\bar{V}_4 = \left(\frac{18}{13}\right)^2 3.36 = 6.44\,\text{m s}^{-1}$

$$z_3 - z_4 = \frac{1}{g}\left[0.4\,\bar{V}_4^2 - \frac{P_3 - P_4}{\rho}\right] = \frac{1}{9.807}\left[(0.4)(6.44)^2 - (-13.44)\right] = 3.06\,\text{m}; \quad z_4 = 0$$

∴ $z_3 = 3.06\,\text{m}$, Vacuum degasser would have to be a lot taller.

4.17 Liquid metal flows into a permanent mold through a vertical gating-system. If the mold has a uniform area (A_m) and the metal flows through gates of equal areas (A_g), derive an equation which gives time to fill the mold. Assumptions: (i) no friction, but flow is turbulent; (ii) h_1 is constant; (iii) no flow through the upper gate until h equals h_2.

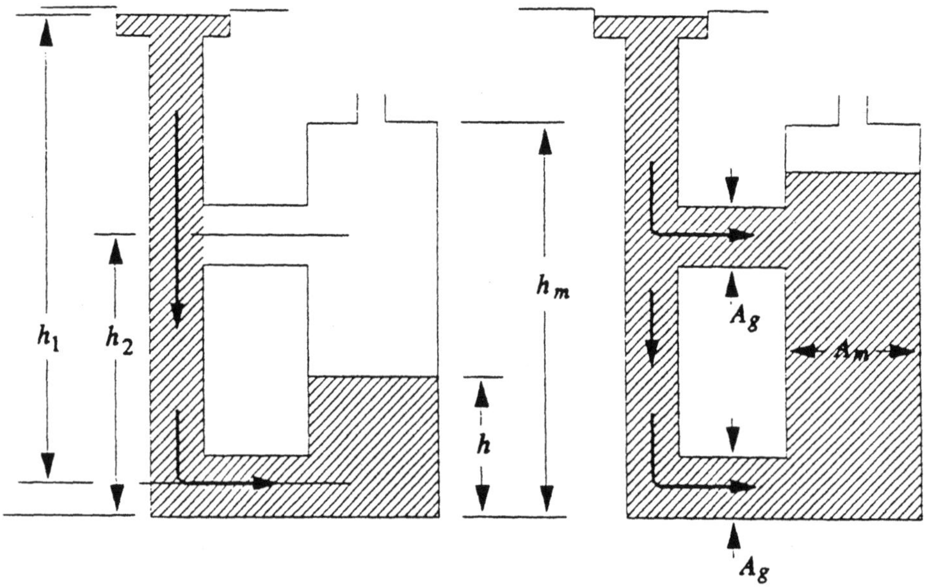

Plane (1) - Top of the vertical gate; Plane (2) - metal height in the mold.

$$\left[\frac{\bar{V}_2^2}{2\beta_2} - \frac{\bar{V}_1^2}{2\beta_1}\right] + g\Delta z = 0; \quad \bar{V}_1 = 0; \quad \beta_2 = 1; \quad z_2 - z_1 = h - h_1 = -(h_1 - h); \quad \bar{V}_2 = \frac{dh}{dt}$$

$$\frac{dh}{dt} = \left[2g(h_1 - h)\right]^{1/2}$$

; when the level rises above the second gate, the mechanical energy is the same. Therefore, integration can be carried out in one step.

$$\int_0^t dt = \frac{1}{(2g)^{1/2}} \int_0^{h_m} \frac{dh}{(h_1 - h)^{1/2}}$$

$$t = \frac{1}{(2g)^{1/2}} \left[-2(h_1 - h)^{1/2}\right]_0^{h_m} = \left(\frac{2}{g}\right)^{1/2} \left[(h_1 - h)^{1/2}\right]_{h_m}^{0} = \left(\frac{2}{g}\right)^{1/2} \left[h_1^{1/2} - (h_1 - h_m)^{1/2}\right]$$

4.18 If the pipes in Fig. 4.15 are of lengths $L_1 = 10$ m, $L_2 = 5$ m, and $L_3 = 1$ m and the nozzles are fan type with a 0.187 inch orifice, and D_1 is 3 cm, determine the diameter of pipes 2 and 3 needed to produce a flow of 3.70 gallons of water per minute through each nozzle. What pressure P_0 needs to be provided? What happens to the flow pattern if a small particle in the water clogs nozzle 1?

$\Delta P_1 = \Delta P_2 = \Delta P_3 = P_o - P_a - \Delta P_N$

For line 1: Assuming $g\,\Delta z \simeq 0$; $\frac{P_1 - P_o}{\rho} + E_f = 0 \Rightarrow E_f = \frac{P_o - P_1}{\rho}$

$E_f = 2f_1 \frac{L_1}{D_1} \bar{V}_1^2$; $2f_1 \frac{L_1}{D_1} \bar{V}_1^2 = \frac{P_o - P_1}{\rho}$; $\bar{V}_1^2 = \left(\frac{P_o - P_1}{\rho}\right) \frac{D_1}{L_1} \frac{1}{2f_1}$

$V_1 = \left(\frac{1}{2f_1}\right)^{1/2} \left(\frac{D_1}{L_1}\right)^{1/2} \left(\frac{P_o - P_1}{\rho}\right)^{1/2}$

$Q_1 = \bar{V}_1 \frac{\pi D_1^2}{4} = \left(\frac{1}{2f_1}\right)^{1/2} \left(\frac{\pi}{4}\right) \frac{D_1^{2.5}}{L_1^{1/2}} \left(\frac{P_o - P_1}{\rho}\right)^{1/2}$

We want $Q_1 = Q_2 = Q_3$: If $f_1 = f_2 = f_3$, then $\frac{D_1^5}{L_1} = \frac{D_2^5}{L_2} = \frac{D_3^5}{L_3}$

$\frac{L_1}{D_1^5} = \frac{L_2}{D_2^5} = \frac{L_3}{D_3^5}$; $L_1 = 10$ m, $L_2 = 5$ m, $L_3 = 1$ m, $D_1 = 3$ cm

$D_2^5 = \frac{L_2}{L_1} D_1^5 = \left(\frac{5}{10}\right)(3^5)\text{ cm}^5 \quad \therefore \; D_2 = 2.61$ cm

$D_3^5 = \frac{L_3}{L_1} D_1^5 = \left(\frac{1}{10}\right)(3^5)\text{ cm}^5 \quad \therefore \; D_3 = 1.89$ cm

Pressure P_o: Use line 1 for calculation.

$Q = 3.70\text{ gal. min}^{-1} = (3.70)(6.309 \times 10^{-5}\text{ m}^3\text{ s}^{-1}) = 2.33 \times 10^{-4}\text{ m}^3\text{ s}^{-1}$

$\bar{V} = \frac{4Q}{\pi D^2} = \frac{(4)(2.33 \times 10^{-4})}{(\pi)(0.03^2)} = 0.330\text{ m s}^{-1}$

$Re = \frac{D\bar{V}\rho}{\eta} = \frac{(0.03)(0.330)(1 \times 10^3)}{1 \times 10^{-3}} = 9.90 \times 10^3$: $f_1 \simeq 0.008$

$E_f = 2f_1 \frac{L}{D} \bar{V}^2 = (2)(0.008)\frac{10}{0.03}(0.330)^2 = 0.581\text{ m}^2\text{s}^{-2}$

Also $\frac{P_1 - P_o}{\rho} + E_f = 0$; $P_o - P_1 = \rho E_f = (1 \times 10^3)(0.581) = 581\text{ N m}^{-2}$

$P_1 = P_{atm} + \Delta P_{nozzle}$; $\Delta P_{NOZ} = 25\text{ psi} = 1.723 \times 10^5\text{ N m}^{-2}$

$P_o = 581 + (1.0133 + 1.723) \times 10^5 = 2.74 \times 10^5\text{ N m}^{-2} = 2.71$ atm

If nozzle 1 becomes clogged, then the flow rate through 2 and 3 will increase.

5.1 Refer to Example 3.5 for the system characteristics and the desired operating point for flow through a sinter bed. The bed has a cross-sectional area of 0.189 m², and V_0 is a superficial velocity. You can select either fan A or fan B to blow the air. Their respective characteristic curves are given to the right.

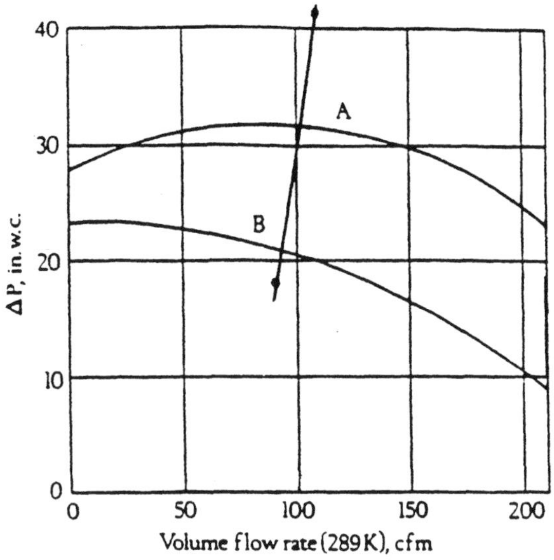

a) If you select fan A, what volume flow rate will be delivered through the sinter bed? Repeat for fan B. You may assume that the pressure drop through the sinter bed is proportional to the square of volume flow rate; i.e., $\Delta P = kQ^2$.

b) What fan, A or B, is better suited for the bed of ore and coal which is discussed? Explain why.

c) What adaptation, if any, must be made to use the fan you have selected?

$$Q = V_oA = \frac{0.25\ m}{s}\left|\frac{0.189\ m^2}{}\right|\frac{60\ s}{min}\left|\frac{3.2808^3\ ft^3}{m^3}\right. = 100\ ft^3\ min^{-1}$$

$Q = 100$ cfm

$$\Delta P = \frac{7.25\times10^3\ N}{m^2}\left|\frac{4.015\times10^{-3}\ in\ H_2O\ m^2}{N}\right. = 29.1\ in\ H_2O$$

To get the system curve, $K = \frac{\Delta P}{Q^2} = \frac{29.1}{(100)^2} = 2.91\times10^{-3}$

To plot the system curve:

ΔP at 80 cfm $= K(80)^2 = 18.62$ in. H_2O

ΔP at 120 cfm $= K(120)^2 = 41.90$ in H_2O

a. Fan A, $Q = 105$ cfm

Fan B, $Q = 85$ cfm

These are the intersections of the system curve with the characteristic curves.

b. Fan A is better suited because it is slightly over-rated. If B is selected, its speed would have to be increased which would require a new motor.

c. The system curve can be shifted by inserting a damper in the system to bring ΔP_{system} = 32 in. H_2O at 100 cfm.

5.2 Refer to Problem 4.4. Suppose the characteristic curve of the fan is as shown to the right.

a) Assume that $\Delta P = kQ^2$ for the entire system and determine the operating point.

b) As particulates are collected, the pressure drop across the bag house increases. When the pressure drop increases by twenty percent, what will be the volume flow rate?

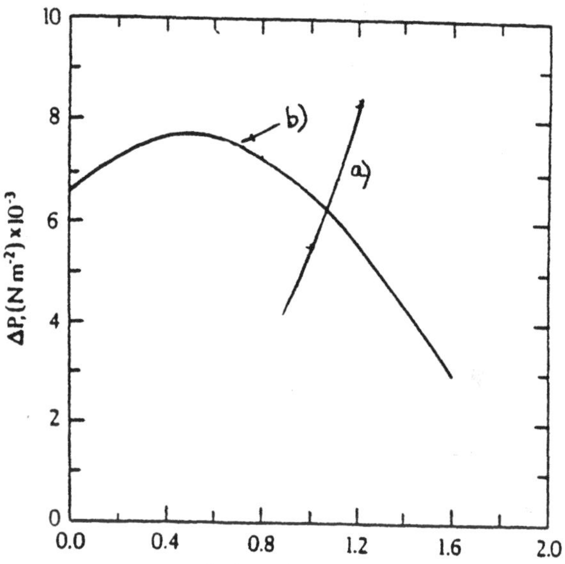

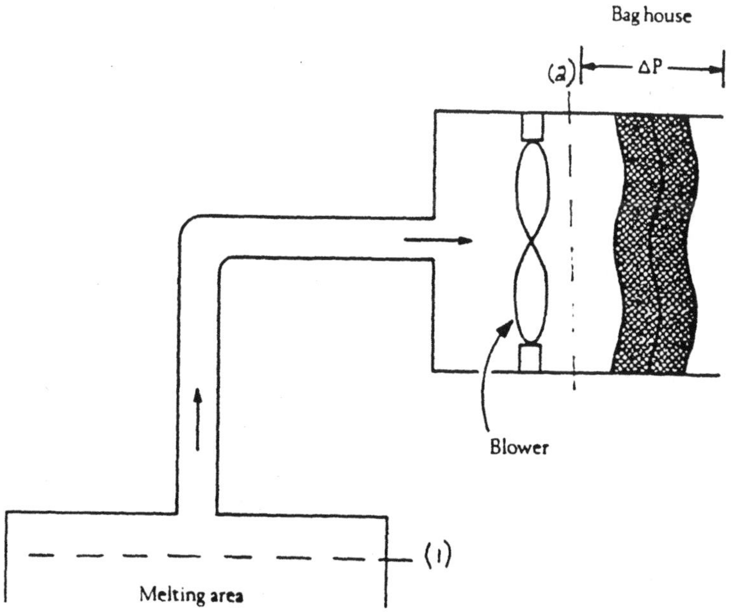

a) From the solution to Problem 4.4, we have

$P_2 - P_1 = 5.07 \times 10^3$ N m^{-2} at $Q = 0.944$ m^3 s^{-1}.

Hence, $k = \frac{\Delta P}{Q^2} = 5.69 \times 10^3$ (SI units).

This intersects the fan curve at $Q = 1.07$ m^3 s^{-1} and $\Delta P =$ 6300 N m^{-2},

b) In effect this is a new system because the characteristics of the bag house have changed. Now $\Delta P = 7.58 \times 10^3$ N m^{-2}. The fan curve gives us $Q = 0.68$ m^3 s^{-1}. The flow has decreased by 36%.

5.3 Two identical pumps are used to pump water from one reservoir to another whose level is 6.1 m higher than the first. When both pumps are operating the flow rate is 0.04 m^3 s^{-1}. What is the flow rate when only one pump operates? Assume highly turbulent flow. The characteristic head curve for the pump is given by the following table.

Flow rate, m^3 s^{-1}	Head, N m kg^{-1}
0	208
0.0057	211
0.0113	212
0.0170	211
0.0227	203
0.0340	178
0.0453	141
0.0566	99
0.0680	39
0.0736	15

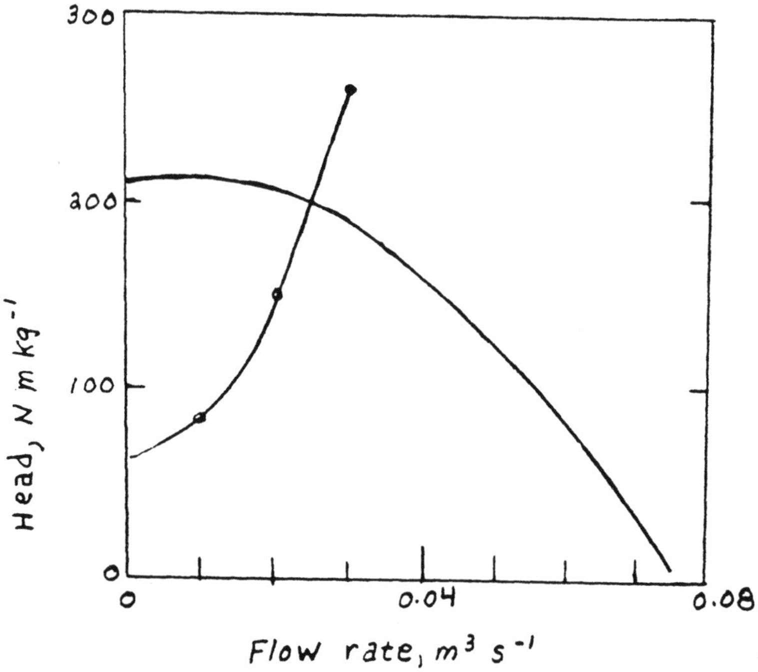

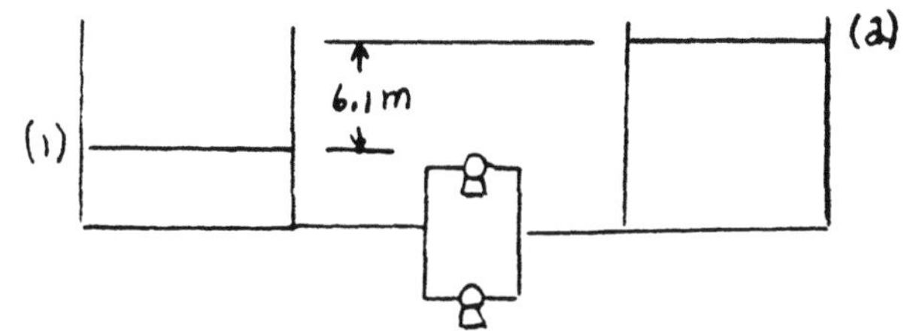

$g(z_2 - z_1) + E_f = -M^*$

Both pumps operating

Q_2 (both pumps) $= 0.04\ m^3 s^{-1}$ so that each pump is operating with a discharge of $0.02\ m^3 s^{-1}$. With Q_1 (each pump) $= 0.02\ m^3 s^{-1}$; $h = 207\ N m\ kg^{-1}$.

$$E_f(\text{one pump}) = -M^* - g(z_2 - z_1) = \Gamma M_p - g(z_2 - z_1) = \left[\frac{207\ N\ m}{kg}\,\frac{kg\ m}{N\ s}\right] - \left[\frac{9.81\ m}{2\,(\text{Pumps})\ s^2}\,\frac{6.1\ m}{}\right]$$

$= 177\ m^2 s^{-2}$

E_f (one pump) $= 177\ N m\ kg^{-1}$; so that E_f (both pumps) $= 354\ N m\ kg^{-1}$

Since $E_f = kQ^2$, then $K = \frac{354}{0.04^2}$

one pump operating

$$h = \Gamma M_p = g(z_2 - z_1) + E_f = g(z^2 - z_1) + KQ^2 = 60\ N m\ kg^{-1} + \frac{354}{0.04^2} Q^2$$

$$h = 60 + 2.2125 \times 10^5 Q^2$$

Q	Q^2	h
0	0	60
0.01	1×10^{-4}	82
0.02	4×10^{-4}	148
0.03	9×10^{-4}	258

A plot of h versus Q gives an intersection with the characteristic head curve for the pump at $Q = 0.025\ m^3 s^{-1}$

5.4 Derive an equation for the conductance of a long straight tube. Assume viscous flow prevails and that the viscosity is given by Eq. (1.13).

a) Write the conductance in terms of viscosity.

b) Show that

$$\eta = \frac{m\bar{V}}{3\pi^{1/6}\sqrt{2}\,d^2},$$

where $\bar{V}$ is the Maxwellian speed of the molecules.

c) Write the conductance in terms of $\bar{V}$.

a. $Q = C(P_1 - P_2)$ Eq. (5.44)

Viscous flow in a long tube Eq. (2.34) $Q = (P_1 - P_2)\left[\frac{\pi R^4}{8L\eta}\right]$

$\therefore\ C = \frac{\pi R^4}{8L\eta}$

b. Eq. (1.13) $\bar{v} = \left[\frac{8K_BT}{\pi m}\right]^{1/2}$; $(K_B T)^{1/2} = \left(\frac{\pi m}{8}\right)^{1/2} \bar{v}$

Eq. (1.13) $\eta = \frac{2}{3\pi^{2/3}} \frac{m^{1/2}}{d^2}(K_BT)^{1/2}$; $(K_BT)^{1/2} = \frac{3\pi^{2/3}}{2} \frac{\eta d^2}{m^{1/2}}$

$$\eta = \frac{2}{3\pi^{2/3}} \frac{m^{1/2}}{d^2}\left(\frac{\pi m}{8}\right)^{1/2} \bar{v} = \frac{m\bar{v}}{3\sqrt{2}\,\pi^{1/6} d^2}$$

c. $C = \frac{3\sqrt{2}}{8} \pi^{7/6} \frac{R^3 d^2}{L m} \bar{v}$

5.5 Compare the conductance for viscous flow in a long straight tube (from Problem 5.4c) to the conductance for molecular flow.

a) How does each vary with $\bar{V}$?
b) How does each vary with temperature?
c) For nitrogen at 300 K, what is the mean free path (see Eq. (1.5)) at normal atmospheric pressure (760 torr).

Viscous flow: $C = \frac{3\sqrt{2}}{8} \pi^{7/6} \frac{R^3 d^2}{L m} \bar{v}$

Molecular flow: Eq. (5.50) $C = \frac{\pi D^3}{12 L} \bar{v}$

a. They each vary linearly with $\bar{v}$

b. $\bar{v} = \left[\frac{8K_BT}{\pi m}\right]^{1/2}$

They each vary with $T^{1/2}$

c. Eq. (1.5) $\lambda = \left(\frac{1}{\sqrt{2}}\right)\left(\frac{1}{\pi d^2 n}\right)$; Assume ideal gas

$PV = NkT$; $n = \frac{N}{V} = \frac{P}{kT} = \frac{1.01\times10^5\ \text{N}}{\text{m}^2} \cdot \frac{\text{K}}{1.38\times10^{-23}\ \text{J}} \cdot \frac{1}{300\ \text{K}} = 2.44\times10^{25}$ molecules m^{-3}

$$\lambda = \frac{1}{\sqrt{2}\,\pi} \cdot \frac{1}{(3.80\times10^{-10})^2\ \text{m}^2} \cdot \frac{\text{m}^3}{2.44\times10^{25}\ \text{mol.}} = 6.39\times10^{-8}\ \text{m}$$

where $d \simeq \sigma = 3.798$ Å (Table 1.1) ; $d = \frac{3.798\ \text{Å}}{1} \cdot \frac{\text{m}}{10^{10}\ \text{Å}} = 3.80\times10^{-10}$ m

5.6 For nitrogen at 300 K, what is the minimum diameter of a long tube for viscous conduction at a) standard atmospheric pressure (760 torr). Repeat for b) 100 torr, c) 10 torr, d) 1 torr and e) 10^{-1} torr. [The criterion is $(\lambda/D) \geq 10$ where λ is the mean free path (see Eq. (1.5)).]

a. From Prob. 5.5c $\lambda = 6.39 \times 10^{-8}$ m

Criterion $(\lambda/D) \leq 10$ ∴ Minimum $D = 6.39 \times 10^{-7}$ m

b. 760 torr. $= 1.0133 \times 10^{5}\ N\,m^{-2}$; 100 torr. $= 1.333 \times 10^{4}\ N\,m^{-2}$

$$n = \frac{1.333 \times 10^{4}}{(1.38 \times 10^{-23})(300)} = 3.22 \times 10^{24} \text{ molecules } m^{-3}$$

$$\lambda = \frac{1}{\sqrt{2}\,\pi (3.8 \times 10^{-10})^{2} (3.22 \times 10^{24})} = 4.84 \times 10^{-7} m$$

minimum $D = 4.84 \times 10^{-6}$ m

c. 10 torr. $= 1.333 \times 10^{-3}\ N\,m^{-2}$

$n = 3.22 \times 10^{23}$ molecules m^{-3}

$\lambda = 4.84 \times 10^{-6}$ m

minimum $D = 4.84 \times 10^{-5}$ m

d. minimum $D = 4.84 \times 10^{-4}$ m

e. minimum $D = 4.84 \times 10^{-3}$ m

5.7 Consult Table 5.2 and obtain the conductance for two chambers connected by a tube with a diameter 250 mm and a length of 750 mm. Compare your result to the approximation given by Eq. (5.52). Assume that the gas is air at 298 K, with a molecular weight of 28.8 kg kmol^{-1}.

Apply Eq. (5.52) $\frac{1}{C} = \frac{1}{C_1} + \frac{1}{C_2} + \frac{1}{C_3}$ Let C_1 and C_2 be the conductances of the openings to the chambers and C_3 the conductance of the tube.

$$C_3 = \frac{\pi D^3}{12L}\bar{V}, \text{ Eq. (5.50)} : C_1 = C_2 = \frac{\bar{V}}{4}A, \text{ Eq. (5.48)}$$

$$\frac{1}{C} = \frac{8}{\bar{V}A} + \frac{12L}{\pi D^3 \bar{V}} = \frac{1}{\pi}\left(\frac{32}{D^2} + \frac{12L}{D^3}\right)\frac{1}{\bar{V}} = \frac{1}{\pi}\left(\frac{32}{D^2} + \frac{12L}{D^3}\right)\left(\frac{\pi m}{8K_BT}\right)^{1/2}$$

$$\frac{1}{C} = \left(\frac{m}{8\pi K_B T}\right)^{1/2}\left(32 + \frac{12L}{D}\right)\frac{1}{D^2} : m = \frac{28.8\text{ kg}}{\text{kmol}}\left|\frac{1\text{ kmol}}{6.022\times10^{26}\text{ molecules}}\right.$$

$$\left(\frac{m}{8\pi K_B T}\right)^{1/2} = \left[\frac{28.8}{(8\pi)(6.022\times10^{26})(1.380\times10^{-23})(298)}\right]^{1/2} = 6.802\times10^{-4}$$

$$C = \left(\frac{1.470\times10^3}{16}\right)\left(\frac{1}{2 + \frac{3L}{4D}}\right)D^2$$

Substituting $L = 0.750$ m; $D = 0.250$ m; $C = 1.35$ m^{3}s^{-1}

Table 5.2 $D = 25$ cm, $L = 75$ cm

$$C = 9.14\frac{D^2}{1 + \frac{3L}{4D}} = 1.76\times10^3 \text{ L s}^{-1}$$

$$C = \frac{1.76\times10^3\text{ L}}{\text{s}}\left|\frac{1000\text{ cm}^3}{1\text{ L}}\right|\frac{1\text{ m}^3}{100^3\text{ cm}^3}$$

$C = 1.76$ m^{3}s^{-1} The results differ by approx. 26 %

5.8 Consider the use of the two-stage pump of Fig. 5.18 that is connected to a chamber of 1 m^3 volume through a duct with an infinitely high conductance. Calculate the time to pumpdown to a) 10^{-2} torr and b) 10^{-4} torr.

The pump speed varies with pressure and the throughput is a volume swept out at a given pressure. A numerical scheme is set up to solve the problem for pressure versus time. The duct has a conductance of ∞. With $C=\infty$, Eq. (5.55) reduces to $S = S_p$. But $S_p = \frac{RT}{P}\frac{dn}{dt}$, which is the volume swept out, based on the ideal gas law, per unit time. This is approximated as $\Delta t = \frac{RT}{\bar{P}}\frac{\Delta n}{\bar{S}_p}$ where $\bar{P}$ is the average pressure and $\bar{S}_p$ is the average speed during one time step. Calculations are done in time steps using the following computer code.

```
10 'Problem 5.8    This does the numerical integration to account for the
12 'variation of pump speed with pressure for the two-stage pump in Fig. 5.18.
20 R = 8.315 : T = 298 : VOL = 1       'R in J/(mol K), T in K, VOL in m^3
30 READ P,SP                    'start at 760 torr (1 atm); P in torrs, SP in L/s
40 POLD = P*101325!/760: SPOLD = SP/1000 :TIME = 0      'P in Pa, SP in m^3/s
50 NO = POLD*VOL/( R*T ) : NOLD = NO       'initial moles of gas in the chamber
60 LPRINT "     Time, s          P, torr
70 LPRINT "     *******       ***************    "
80 FOR I = 1 TO 8
90      READ P,SP                           'P in torrs, SP in L/s
100     PNEW = P*101325!/760   : SPNEW = SP/1000     'P in Pa, SP in m^3/s
110     PAVG = (POLD + PNEW)/2               'average pressure during time step
120     SPAVG = (SPOLD + SPNEW)/2            'average speed during time step
130     NNEW = PNEW*VOL/( R*T )              'moles gas at end of time step
140     DELN = NOLD - NNEW                   'moles gas removed during time step
150     DELTIME = R*T*DELN/(PAVG*SPAVG)      'time step to remove DELN moles gas
160     TIME = TIME + DELTIME
170     POLD = PNEW : SPOLD = SPNEW : PTORR = POLD*760/101325!
180     LPRINT USING "  ##.###^^^^      ##.###^^^^  ";TIME,PTORR
190 NEXT I
200 END
210 'data input as coordinates from Fig. 5.18
220 DATA 760, 6.5
230 DATA 100, 6.2
240 DATA 10, 5.8
250 DATA 1, 5.5
260 DATA .1, 4.5
270 DATA .01, 4
280 DATA .001, 3.2
290 DATA .0005, 3
300 DATA .0001, 1.8
```

Time, s	P, torr
*******	**************
2.417E+02	1.000E+02
2.514E+03	1.000E+01
2.694E+04	1.000E+00
3.033E+05	1.000E-01
3.555E+06	1.000E-02
4.194E+07	1.000E-03
3.688E+08	5.000E-04
1.424E+09	1.000E-04

The results indicate that to achieve 10^{-2} torr, almost 1000 h would be required. Obviously the throughput of the pump is significantly undersized. It would be practically unrealistic to expect the pump to get the chamber to 10^{-4} torr. The pump has too small of a throughput.

5.9 A heat of steel (5×10^4 kg) is to be vacuum degassed from 5 ppm H_2 to 1 ppm H_2 and from 100 ppm N_2 to 75 ppm N_2 in 15 min. The steel is at 1873 K, and the chamber has 9 m^3 of space occupied by air after the top is closed with the ladle inside. At what pressure would you recommend operating the system? Calculate the throughputs of air, hydrogen and nitrogen that must be removed from the chamber. Consult Fig. 5.22 and specify a steam ejector to do the job.

H_2 removal — $(5\times10^4 \text{ kg})(5 \text{ ppm} - 1 \text{ ppm}) = 0.2 \text{ kg } (H_2) = 100 \text{ mol } H_2$

N_2 removal - $(5\times10^4 \text{ kg})(100 \text{ ppm} - 75 \text{ ppm}) = 1.25 \text{ kg } (N_2) = 44.6 \text{ mol } N_2$

Air (initial): $n = \dfrac{PV}{RT} = \dfrac{(1.01325\times10^5)(9)}{(8.315)(300)} = 365.6 \text{ mol air}$

Total amount of gas to be evacuated: $n = 365.6 + 44.6 + 100 = 510.2 \text{ mol}$

We must select a pressure that is low enough to achieve the stated goals for dissolved H_2 and N_2, according to thermodynamics of these gasses dissolving in steel. Let's use configuration AB–X–B–X–C in Fig. 5.22, which has the greatest throughput up to 5 torrs. Let's operate at 0.4 torr.

Then $Q = \dfrac{10 \text{ lbm}}{\text{h}} \cdot \dfrac{1 \text{ kg}}{2.205 \text{ lbm}} \cdot \dfrac{1 \text{ h}}{3600 \text{ s}} \cdot \dfrac{1 \text{ kmol}}{28.8 \text{ kg}} \cdot \dfrac{1000 \text{ mol}}{1 \text{ kmol}} = 0.0437 \text{ mol s}^{-1}$

We can calculate an approximate time as

$t = \dfrac{510.2 \text{ mol s}}{0.0437 \text{ mol}} = 1.166\times10^4 \text{ s} = 3.24 \text{ h}$ This is too long for maintaining molten steel.

let's operate at 4 torr. Then $Q = 40\ lb_m\ h^{-1} = 0.175\ mol\ s^{-1}$ and

$$t = \frac{510.2}{0.175} = 2.919 \times 10^3\ s = 0.81\ h$$ This is still rather a long time, so we would probably go to 20 torr, and switch to configuration A-X-B-X-C.

Then $Q = 75\ lb_m\ h^{-1} = 0.328\ mol\ s^{-1}$ and $t = \frac{510.2}{0.328} = 1.555 \times 10^3\ s = 0.432\ h$

This is about the best we can do, otherwise if the operating pressure gets too high we will not effectively refine the steel.

5.10 An ultrahigh vacuum chamber (300 liters) is equipped with two pumping modes, one to achieve 10^{-4} torr and a titanium sublimation pump (Fig. 5.23) to achieve pressures below 10^{-4} torr. Assuming that nitrogen must be removed from the chamber, which pump(s) of Fig. 5.23 can be used if 10 minutes is an acceptable pumpdown time to go from 10^{-4} torr to 10^{-8} torr?

Assume that the chamber is a short cylinder with height (H) equal to diameter (D). Then the volume (V) is $V = \frac{\pi D^2}{4} H = \frac{\pi D^3}{4}$ and

$$D = \sqrt[3]{\frac{4V}{\pi}} = \sqrt[3]{\frac{(4)(300 \times 1000)cm^3}{\pi}} = 72.56\ cm = 28.6\ in.$$

Based on this diameter, let's restrict ourselves to the maximum pump diameter of 16 inches. This problem is similar to 5.8, so we use the same program and take data as a set of coordinates on the curve for the 16 inch diameter pump (0.10 g h^{-1} sublim. rate) in Fig. 5.23, starting at 10^{-4} torr.

```
10 'Problem 5.10  This does the numerical integration to account for the
   variation of pump speed with pressure for the two-stage pump in Fig. 5.23.
20 R = 8.315 : T = 298 : VOL = .3     'R in J/(mol K), T in K, VOL in m^3
30 READ P,SP                    'start at .0001 torr ; P in torrs, SP in L/
40 POLD = P*101325!/760: SPOLD = SP/1000 :TIME = 0      'P in Pa, SP in m^3/s
50 NO = POLD*VOL/( R*T ) : NOLD = NO      'initial moles of gas in the chamber
60 LPRINT "   Time, s        P, torr
70 LPRINT "   *******     **************  "
80 FOR I = 1 TO 8
90      READ P,SP                       'P in torrs, SP in L/s
100     PNEW = P*101325!/760  : SPNEW = SP/1000    'P in Pa, SP in m^3/s
110     PAVG = (POLD + PNEW)/2            'average pressure during time step
120     SPAVG = (SPOLD + SPNEW)/2         'average speed during time step
130     NNEW = PNEW*VOL/( R*T )           'moles gas at end of time step
140     DELN = NOLD - NNEW                'moles gas removed during time step
150     DELTIME = R*T*DELN/(PAVG*SPAVG)   'time step to remove DELN moles gas
160     TIME = TIME + DELTIME
170     POLD = PNEW : SPOLD = SPNEW : PTORR = POLD*760/101325!
180     LPRINT USING "  ##.###^^^^      ##.###^^^^ ";TIME,PTORR
190 NEXT I
200 END
210 'data input as coordinates from Fig. 5.23
220 DATA 1e-04,60
230 DATA 5e-05,100
240 DATA 1e-05,530
250 DATA 5e-06,1000
260 DATA 1e-06,3200
270 DATA 5e-07,4000
280 DATA 1e-07,4900
290 DATA 5e-08,4900
300 DATA 1e-08,4900
```

Time, s	P, torr
*******	**************
2.500E+00	5.000E-05
5.357E+00	1.000E-05
1.032E+01	5.000E-06
1.504E+01	1.000E-06
2.609E+01	5.000E-07
4.354E+01	1.000E-07
1.301E+02	5.000E-08
3.342E+02	1.000E-08

This pump requires 334 s, assuming that conductance in the system is infinite. It should be suitable for a pumpdown of 10 minutes.

5.11 Supersonic nozzles are arranged circumferentially around a central orifice through which liquid metal is fed. The argon gas jets are focused on a point below the exit of the orifice, where they impinge on the metal stream to break it into fine droplets that solidify to microstructures of particular interest. It has been found that nozzle exit velocities on the order of Mach 3 are desirable. For a Mach 3 nozzle calculate the reservoir pressure P_0 needed, if the desired exit pressure is 1.0 atm and the flow rate of argon is 0.1 kg s^{-1}. What should the throat diameter, exit diameter and length of diverging section be? Assume $\gamma_{Ar} = 1.67$.

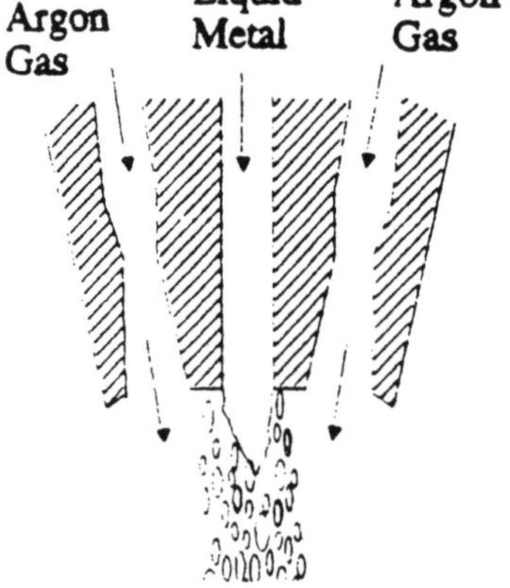

Eq. (5.33): $M_e^2 = \frac{2}{\gamma-1}\left\{\left[\frac{P_0}{P_e}\right]^{(\gamma-1)/\gamma} - 1\right\}$: $\frac{P_0}{P_e} = \left[M_e^2\left(\frac{\gamma-1}{2}\right)+1\right]^{\gamma/\gamma-1}$

$$\therefore P_0 = 1\left[3^2\left(\frac{1.67-1}{2}\right)+1\right]^{1.67/1.67-1} = 31.97\,atm = 3.24\times10^6\,N\,m^{-2}$$

$$\rho_{0Ar}(294\,K) = \frac{P_0}{RT} = \frac{32.4\times10^5 N}{m^2}\Big|\frac{K\,mol}{8.31441\,J}\Big|\frac{39.948\,g}{mol}\Big|\frac{}{294\,K}\Big|\frac{J}{N\,m}\Big|\frac{kg}{1000\,g} = 52.9\ kg\,m^{-3}$$

Eq. (5.31): $W_t^* = A_t\left[\rho_0 P_0 \gamma\left(\frac{2}{\gamma+1}\right)^{(\gamma+1)/(\gamma-1)}\right]^{1/2}$: $0.1 = A_t\left\{(52.9)(3.27\times10^6)(1.67)\left(\frac{2}{2.67}\right)^{3.985}\right\}^{1/2}$

$A_t = 10.46\times10^{-6}\ m^2 \quad (d_t = 3.65\ mm)$

Eq. (5.35): $\left[\frac{A_t}{A_e}\right]^2 = \left(\frac{2}{\gamma-1}\right)\left(\frac{\gamma+1}{2}\right)^{(\gamma+1)/(\gamma-1)}\left(\frac{P_e}{P_0}\right)^{2/\gamma}\left[1-\left(\frac{P_e}{P_0}\right)^{(\gamma-1)/\gamma}\right]$

$\left(\frac{A_t}{A_e}\right)^2 = \left(\frac{2}{0.67}\right)\left(\frac{2.67}{2}\right)^{1.67/0.67}\left(\frac{1}{32.3}\right)^{2/1.67}\left[1-\left(\frac{1}{32.3}\right)^{0.67/1.67}\right]$; $\left(\frac{A_t}{A_e}\right)^2 = 7.19\times10^{-2}$

$A_e = 39.02\times10^{-6}\ m^2 \quad (d_e = 7.05\ mm)$

Length from throat to exit (based on 7°) = 13.8 mm.

5.12 Derive Eq. (5.19).

Eq. 4.11, with $\beta_t = 1$ and $\bar{V}_0^2 = 1$, gives

$$\bar{V}_t^2 = -2\int_{P_t}^{P_0}\frac{dP}{\rho}.$$

Adiabatic, reversible compression of ideal gas gives

$$P\left(\frac{1}{\rho}\right)^\gamma = A^\gamma = \text{constant}$$

so that

$$\frac{1}{\rho} = \frac{A}{P^{1/\gamma}}.$$

Then

$$\int_{P_t}^{P_0}\frac{dP}{\rho} = A\left(\frac{\gamma}{\gamma-1}\right)\left[P_t^{(\gamma-1)/\gamma}\ P_0^{(\gamma-1)/\gamma}\right]$$

$$= P_0^{1/\gamma}\left(\frac{1}{\rho_0}\right)\frac{\gamma}{\gamma-1}\left[P_t^{(\gamma-1)/\gamma} - P_0^{(\gamma-1)/\gamma}\right].$$

Finally

$$\bar{V}_t = \left\{\frac{2P_0}{\rho_0}\left(\frac{\gamma}{\gamma-1}\right)\left[1-\left(\frac{P_t}{P_0}\right)^{\frac{\gamma-1}{\gamma}}\right]\right\}^{1/2}$$

6.1 In the same system described in Problem 1.2, the temperature profile at $x = x_1$ is given by

$$T - T_0 = 6 \sin\left[\frac{\pi y}{2}\right], \qquad 0 \le y \le 0.1 \text{ m},$$

where T is temperature (K) in the water, T_0 is temperature at $y = 0$, and y is distance from the flat plate in m. Find the heat flux to the wall at $x = x_1$. (The thermal conductivity of water is 0.62 W m^{-1} K^{-1}, and the heat capacity is 4.19×10^3 J kg^{-1} K^{-1}.)

$$T = T_o + 6 \sin\left(\frac{\pi Y}{2}\right)$$

$$\frac{dT}{dY} = 6\left(\frac{\pi}{2}\right)\cos\left(\frac{\pi Y}{2}\right)$$

at $y = 0$: $\frac{dT}{dY} = \frac{6\pi}{2} = 3\pi \text{ K m}^{-1}$

$$q\Big|_{\substack{Y=0\\X=X_1}} = -\frac{0.62 \text{ W}}{\text{m K}}\left|\frac{3\pi \text{ K}}{\text{m}}\right. = -5.84 \text{ W m}^{-2}$$

6.2 Determine the thermal conductivity of a test panel 150 mm × 150 mm and 12 mm thick, if during a two-hour period 8.4×10^4 J are conducted through the panel when the two faces are at 290 K and 300 K.

$$q = -\frac{Q}{A} = -k\frac{\Delta T}{\Delta X}$$

$$k = \frac{Q}{A\left|\frac{\Delta T}{\Delta X}\right|} = \frac{8.4 \times 10^4 \text{ J}}{(0.150 \text{ m})^2}\left|\frac{0.012 \text{ m}}{(300-290)\text{K}}\right|\frac{\text{h}}{2 \text{ h}}\left|\frac{\text{h}}{3600 \text{ s}}\right. = 0.622 \text{ J s}^{-1}\text{ m}^{-1}\text{ K}^{-1}$$

$$k = 0.622 \text{ W m}^{-1}\text{ K}^{-1}$$

6.3 At steady state, the temperature profile in a laminated system appears thus:

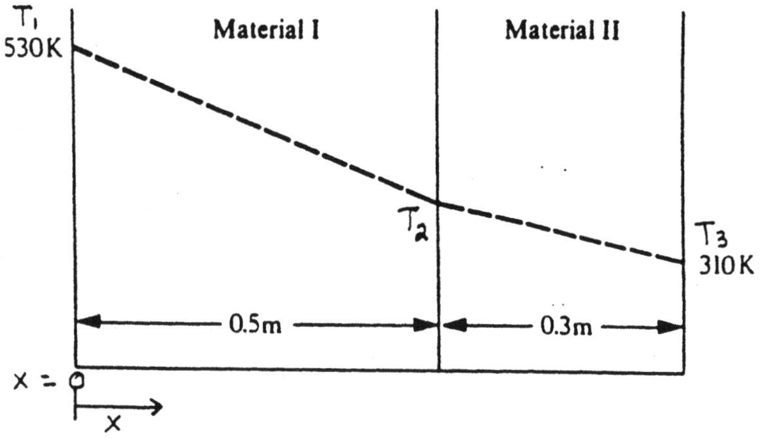

Determine the thermal conductivity of II if the steady-state heat flux is 12.6×10^3 W m^{-2} and the conductivity of I is 52 W m^{-1} K^{-1}.

Material I: $q_x = -k_I \frac{dT}{dx} = -k_I \frac{T_2 - T_1}{L_1}$

$$T_2 - T_1 = -q_x \frac{L_1}{k_1} = -\frac{12.6\times10^3\ \text{W}}{\text{m}^2}\left|\frac{0.5\ \text{m}}{}\right|\frac{\text{m K}}{52\ \text{W}} = -121.2\ \text{K}$$

$$T_2 = -121.2 + 530 = 408.8\ \text{K}$$

Material II: $-k_I \frac{T_2 - T_1}{L_1} = -k_{II} \frac{T_3 - T_2}{L_2}$

$$k_{II} = k_I \left[\frac{T_2 - T_1}{T_3 - T_2}\right]\frac{L_2}{L_1} = \frac{52\ \text{W}}{\text{m K}}\left|\frac{(408.8 - 530)\text{K}}{(310.0 - 408.8)\text{K}}\right|\frac{0.3\ \text{m}}{0.5\ \text{m}} = 38.3\ \text{W m}^{-1}\text{K}^{-1}$$

6.4 Show that Fourier's law can be written (for constant ρC_p) as

$$q_y = -\alpha \frac{d}{dy} (\rho C_p T)$$

for one-dimensional heat flow. In addition, show that Newton's law, for constant ρ, is

$$\tau_{yx} = -\nu \frac{d}{dy} (\rho v_x).$$

Discuss the analogies between the fluxes, constants, and gradients as they appear in these equations.

Fourier's Law: $q_Y = -k\frac{dT}{dY}$; $\alpha = \frac{k}{\rho C_p}$ = thermal diffusivity; $\therefore k = \alpha \rho C_p$;

$q_Y = -\alpha \rho C_p \frac{dT}{dY}$: For constant ρC_p ; $q_Y = -\alpha \frac{d(\rho C_p T)}{dY}$

Newton's Law of Viscosity: $\tau_{YX} = -\eta \frac{dv_x}{dY}$; $\nu = \frac{\eta}{\rho}$ = momentum diffusivity

$\therefore \eta = \nu \rho$; $\tau_{YX} = -\nu \rho \frac{dv_x}{dY}$: For constant ρ ; $\tau_{YX} = -\nu \frac{d(\rho v_x)}{dY}$

Both fluxes represent the flow of a quanity/(area x time).

In both cases the fluxes are proportional to the gradients.

For energy, the gradient is $\frac{d(\rho C_p T)}{dY}$.

For momentum, the gradient is $\frac{d(\rho v_x)}{dY}$.

Finally, the proportionality constant in each is a diffusivity.

α = energy diffusivity, $m^2 s^{-1}$

ν = momentum diffusivity, $m^2 s^{-1}$

Of course, the more common names are:

α = thermal diffusivity

ν = kinematic vicosity

6.5 The thermal conductivity of helium at 400 K is 0.176 W m^{-1} K^{-1}. Knowing only this datum, estimate the thermal conductivity of helium at 800 K. Compare your estimate to the value obtained from Fig. 6.2. What do you conclude about the equation that you used for your estimate?

Eq. (6.10) $k = \frac{1}{d^2}\left[\frac{K_B^3 T}{\pi^3 m}\right]^{\frac{1}{2}}$

$\therefore \frac{k_2}{k_1} = \left(\frac{T_2}{T_1}\right)^{\frac{1}{2}}$; $k_2 = \frac{0.176\,W}{m\,K}\left|\frac{(800K)^{\frac{1}{2}}}{(400K)^{\frac{1}{2}}}\right. = 0.249\ W m^{-1} K^{-1}$ (estimated)

From Fig. 6.2 $k(800K) = 0.303\ W\,m^{-1}K^{-1}$

Conclusion: The thermal conductivity of He increases more strongly with temperature than indicated by Eq. (6.10).

6.6 Repeat Problem 6.5 but use the following equation to estimate the thermal conductivity of helium at 800 K:

$$k = \frac{15R}{4M}\eta$$

where R is the gas constant, M is the molecular weight, and η is the viscosity.

Using viscosities directly from Fig. 1.7

$\eta_2 = \eta(800K) = 0.0385$ cP

$\eta_1 = \eta(400K) = 0.0250$ cP

$k_2 = (0.176)\left(\frac{0.0385}{0.0250}\right) = 0.271\ W\,m^{-1}K^{-1}$

Conclusion: Thermal conductivity of helium increases more strongly with temperature than indicated by the given equation.

6.7 Calculate the thermal conductivity of carbon dioxide at 800 K and compare your result to that given in Fig. 6.2. The heat capacity of CO_2 at 800 K is 1.17 kJ kg^{-1} K^{-1}.

Eq. (6.11) $k = \eta \left[c_p + \frac{1.25 R}{M} \right]$

$M = 44.01 \frac{g}{mol}$, $R = 8.31467 \frac{J}{mol\ K}$, From Fig. 1.7 $\eta_{CO_2}(800K) = 3.5 \times 10^{-5}\ N\ s^{-1} m^{-2}$

$$k = \frac{3.5\times10^{-5}\ N}{s\ m^2} \frac{kg\ m}{N\ s^2} \left[\frac{1.17\ kJ}{kg\ K} + \frac{1.25}{} \frac{8.315\ J}{mol\ K} \frac{mol}{44.01\ g} \frac{1\times10^3\ g}{kg} \frac{kJ}{1\times10^3\ J} \right] = 4.92\times10^{-5}\ kJ\ s^{-1} m^{-1} K^{-1}$$

$k = 4.92\times10^{-2}\ J\ s^{-1} m^{-1} K^{-1} = 4.92\times10^{-2}\ W\ m^{-1} K$

From Fig. 6.2 $k_{CO_2}(800K) = 5.0\times10^{-2}\ W\ m^{-1} K$

The results from Eq. 6.11 compare very well with the actual value from Fig. 6.2 at 800K.

6.8 a) Calculate the thermal conductivity of a gas containing 40 mol% He, 40 mol% H_2, and 20 mol% N_2 at 1400 K.
b) Assume that the concentration of He is constant but that the concentrations of H_2 and N_2 vary as much as ±5 mol% in a process. What is the variation in the thermal conductivity of the gas?

Eq. (6.12) $k_{mix} = \frac{\sum_i x_i k_i M_i^{1/3}}{\sum_i x_i M_i^{1/3}}$

Fig. 6.2 at 1400K; $k_{He} = 0.515$, $k_{H_2} = 0.565$, $k_{N_2} = 0.0753\ W\ m^{-1} K^{-1}$

```
10 ' problem 6.8
20   KHE = .515 : KH2 = .565 : KN2 = .0753          ' thermal conductivities
30   MHE = 4.003 : MH2 = 2.016 : MN2 = 28.014       ' molecular weights
40   ' Eq. (6.12) is applied
50   CHE = KHE*MHE^(1/3) : CH2 = KH2*MH2^(1/3) : CN2 = KN2*MN2^(1/3)
60   AHE = MHE^(1/3) : AH2 = MH2^(1/3) : AN2 = MN2^(1/3)
70   LPRINT "       Select thermal conductivities for X(He) = 0.4 " : LPRINT
80   LPRINT "                                    Therm. Cond."
90   LPRINT "       X(He)    X(H2)    X(N2)      W/(m K)"
100   FOR XH2 = .35 TO .451 STEP .05
110       FOR XN2 = .15 TO .25 STEP .05
120           XHE = 1 - XH2 - XN2
130           NUMER = XHE*CHE + XN2*CN2 + XH2*CH2
140           DENOM = XHE*AHE + XN2*AN2 + XH2*AH2
150           KMIX = NUMER/DENOM
160           LPRINT USING "     #.##     #.##     #.##        #.### ";XHE,XH2,
XN2,KMIX
170       NEXT  XN2
180  LPRINT " "
190  NEXT XH2
200  END
```

Select thermal conductivities for X(He) = 0.4

X(He)	X(H2)	X(N2)	Therm. Cond. W/(m K)
0.50	0.35	0.15	0.410
0.45	0.35	0.20	0.376
0.40	0.35	0.25	0.345
0.45	0.40	0.15	0.411
0.40	0.40	0.20	0.377
0.35	0.40	0.25	0.345
0.40	0.45	0.15	0.411
0.35	0.45	0.20	0.377
0.30	0.45	0.25	0.346

$k_{Mix} = 0.377^{+0.034}_{-0.032}$ W m^{-1} K^{-1}

6.9 Refer to Fig. 6.3 and explain the variation of the thermal conductivity of MgO with temperature.

Conductivity decreases with increasing temperature for T<1366 K because the meanfree path for phonons decreases with increasing temperature. At the higher temperatures, T>1366 K, the conductivity increases because both electronic and photonic contributions become important. At lower temperatures, these contributions are not important.

6.10 Refer to Fig. 6.5 and comment on the effect of impurity scattering of phonons in dielectric solid solutions. Assume that the inverse mean free path for different scattering processes are additive, so that

$$\frac{1}{\lambda_{ph}} = \frac{1}{\lambda_t} + \frac{1}{\lambda_i},$$

where λ_t is the thermal mean free path and λ_i is the impurity mean free path.

Fig. 6.5 shows that the thermal conductivity of the solid-solution is less than that of the pure dielectric compounds And as shown by Eq. 6.14 the thermal conductivity decreases with temperature.

If $\frac{1}{\lambda_{ph}} = \frac{1}{\lambda_t} + \frac{1}{\lambda_i}$

λ_t decreases as T increases.

λ_i decrease as the impurities increase.

$\therefore \lambda_{ph}$ will decrease as the temperature and the impurities increase.

6.11 Electrical resistivities of Ti-Al alloys at 800 K are given in the table below. Aluminum is an "α-stabilizer."

At. pct. Al	Resistivity μohm cm
0	112
3	140
6	165
11	190
33	210

Estimate the thermal conductivity for each alloy at 800 K.

Apply the Smith-Palmer Eq. (6.19) $K = A L \sigma_e T + B$

where $A = 0.997$; $B = 2.7 \text{ W m}^{-1} \text{K}^{-1}$; $L = 2.45 \times 10^{-8} \text{ W ohm K}^{-2}$

For the first case, $\rho = 112 \times 10^{-8}$ ohm m

$$K = \frac{0.997 \mid 2.45 \times 10^{-8} \text{ W ohm} \mid \text{m} \mid 800\text{K}}{\mid \text{K}^2 \mid 112 \times 10^{-8} \text{ ohm} \mid} + 2.7 \text{ W m}^{-1}\text{K}^{-1} = 20.1 \text{ W m}^{-1}\text{K}^{-1}$$

Similar calculations for the other cases are given in the following table.

At. pct. Al	k, W m^{-1}K^{-1}
0	20.1
3	16.7
6	14.5
11	13.0
33	12.0

6.12 The electrical conductivity of molten Pb-Sn alloys at 673 K is

$$\sigma = (100 - 48X)^{-1}$$

where σ is in μohm^{-1} cm^{-1} and X is the atom fraction of Sn. Estimate the thermal conductivities of 90Pb-10Sn, 50Pb-50Sn and 10Pb-90Sn alloys (compositions in mol pct).

use Eq. (6.18) $L = \frac{k_{el}}{\sigma_e T} = 2.45 \times 10^{-8}$ W ohm K^{-2}; assume $k_{el} = k$

$k = \sigma_e (673)(2.45 \times 10^{-8}) = 1.649 \times 10^{-5} \sigma$ where $\sigma_e = \sigma$.

$M_{Pb} = 186.47$; $M_{Sn} = 118.69$

90 Pb - 10 Sn, $X = 0.1$

$\sigma = [100 - (48)(0.10)]^{-1} = 1.050 \times 10^{-2}$ μohm^{-1}cm^{-1} $= 1.050 \times 10^{6}$ ohm^{-1}m^{-1}

$k = (1.649 \times 10^{-5})(1.050 \times 10^{6}) = 17.3$ W m^{-1}K^{-1}

50 Pb - 50 Sn, $X = 0.5$

$\sigma = 1.32 \times 10^{-2}$ μohm^{-1}cm^{-1} $= 1.32 \times 10^{6}$ ohm^{-1}m^{-1}

$k = 21.8$ W m^{-1} K^{-1}

10 Pb - 90 Sn, $X = 0.9$

$\sigma = 1.76 \times 10^{-2}$ μohm^{-1}cm^{-1} $= 1.76 \times 10^{-6}$ ohm^{-1}m^{-1}

$k = 29.0$ W m^{-1} K^{-1}

6.13 Use the Maxwell-Eucken equation and predict the thermal conductivity of a two-phase solid (*A* plus *B*) as a function of composition (wt.pct.). *A* and *B* are insoluble in each other, and the following data apply: $k_A = 13\ \mathrm{W\ m^{-1}\ K^{-1}}$; $k_B = 7\ \mathrm{W\ m^{-1}\ K^{-1}}$, $\rho_A = 4 \times 10^3\ \mathrm{kg\ m^{-3}}$; $\rho_B = 3 \times 10^3\ \mathrm{kg\ m^{-3}}$.

Eq. (6.26) $$\frac{k_{mix.}}{k_A} = \frac{k_B/k_A + 2 - 2V_B(1 - k_B/k_A)}{k_B/k_A + 2 + V_B(1 - k_B/k_A)}$$

$$= \frac{\left(\frac{13}{7}\right) + 2 - 2V_B\left(1 - \left(\frac{13}{7}\right)\right)}{\left(\frac{13}{7}\right) + 2 + V_B\left(1 - \left(\frac{13}{7}\right)\right)}$$

$$K_{MIX} = 13\left[\frac{2.538 - 0.924V_B}{2.538 + 0.462V_B}\right]$$

$$V_B = \frac{\frac{\text{Wt.\%B}}{\rho_B}}{\frac{\text{wt.\%B}}{\rho_B} + \frac{(100 - \text{wt.\%B})}{\rho_A}} = \frac{3\ \text{Wt.\%B}}{3\ \text{Wt.\%B} + 4(100 - \text{Wt.\%B})}$$

```
10 'Problem 6.13  Calculates the thermal conductivity of two phase mixture.
20 'Assume that the continuous phase is A to 50% B and B for greater than 50%B.
30 LPRINT "  wt pct B   vol frac B  k(mix), W/(m K)
40 LPRINT "  ********   **********  ***************"
50 KA = 13 : KB = 7                 'thermal conductivities in W/( m K )
60 DENSA = 4000 : DENSB = 3000      'densities in kg/m^3
70 '
80 KCONT = KA : KDIS = KB           'A is set as the continous phase
90 DCONT = DENSA : DDIS = DENSB     'A is set as the continuos phase
100 FOR CB = 0 TO 50 STEP 5          'CB is wt. pct. B
110      VOLA = (100 - CB)/DENSA     'volume of A
120    VOLB = CB/DENSB : VOLTOT = VOLA + VOLB   'volume of B and total volume
130    FRACB = VOLB/VOLTOT          'volume fraction of B
140    FRACD = FRACB                'keeping A the continuous phase
150    RATIO = KDIS/KCONT : NUMER = RATIO + 2 -2*FRACD*( 1-RATIO )
160    DENOM = RATIO + 2 + FRACD*( 1-RATIO )
170    KMIX = KCONT*( NUMER/DENOM )
180    LPRINT USING "    ###.       #.####       ###.## "; CB,FRACB,KMIX
190 NEXT CB
200 '
210 LPRINT "                 switch continuous phase from A to B"
220 KCONT = KB : KDIS = KA           'B is set as the continous phase
230 DCONT = DENSB : DDIS = DENSA     'B is set as the continuos phase
240 FOR CB = 50 TO 100 STEP 5        'CB is wt. pct. B
250    VOLA = (100 - CB)/DENSA      'volume of A
260    VOLB = CB/DENSB : VOLTOT = VOLA + VOLB   'volume of B and total volume
270    FRACB = VOLB/VOLTOT          'volume fraction of B
280    FRACD = 1 - FRACB            'keeping B the continuous phase
290    RATIO = KDIS/KCONT : NUMER = RATIO + 2 -2*FRACD*( 1-RATIO )
300    DENOM = RATIO + 2 + FRACD*( 1-RATIO )
310    KMIX = KCONT*( NUMER/DENOM )
320    LPRINT USING "    ###.       #.####       ###.## "; CB,FRACB,KMIX
330 NEXT CB
340 END
```

wt pct B	vol frac B	k(mix), W/(m K)
********	**********	***************
0	0.0000	13.00
5	0.0656	12.54
10	0.1290	12.11
15	0.1905	11.69
20	0.2500	11.30
25	0.3077	10.93
30	0.3636	10.58
35	0.4179	10.25
40	0.4706	9.93
45	0.5217	9.62
50	0.5714	9.33
		switch continuous pha
50	0.5714	9.21
55	0.6197	8.94
60	0.6667	8.68
65	0.7123	8.43
70	0.7568	8.20
75	0.8000	7.98
80	0.8421	7.76
85	0.8831	7.56
90	0.9231	7.37
95	0.9620	7.18
100	1.0000	7.00

6.14 A flat heater is sandwiched between two solids of equal areas (0.1 m^2) with different thermal conductivities and thicknesses. The heater operates at a uniform temperature and provides a constant power of 290 W. The external surface temperature of each solid is 300 K, and there is perfect thermal contact at each internal interface.

a) Calculate the heat flux through each solid.
b) What is the operating temperature of the heater?

Solid	Thermal Conductivity, W m^{-1} K^{-1}	Thickness, mm
A	35	60
B	9	30

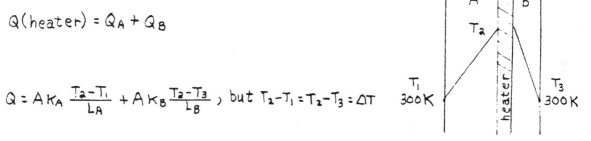

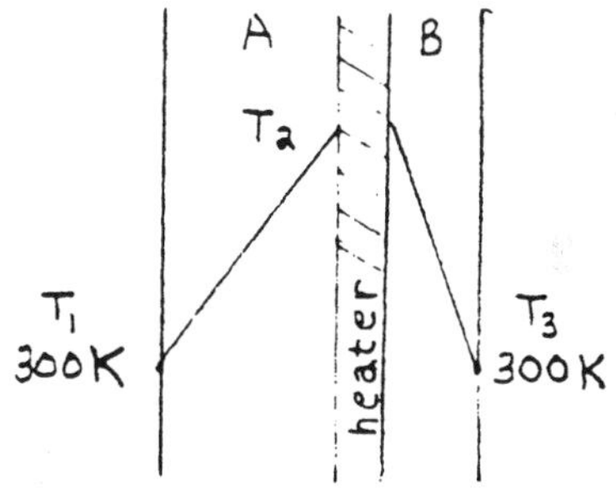

a. $Q(\text{heater}) = Q_A + Q_B$

$$Q = A K_A \frac{T_2 - T_1}{L_A} + A K_B \frac{T_2 - T_3}{L_B}, \text{ but } T_2 - T_1 = T_2 - T_3 = \Delta T$$

$$\therefore Q = \left(\frac{K_A}{L_A} + \frac{K_B}{L_B}\right) A \Delta T$$

$$\frac{Q_A}{Q} = \frac{\frac{K_A}{L_A}}{\frac{K_A}{L_A} + \frac{K_B}{L_B}} = \frac{\frac{35}{60}}{\frac{35}{60} + \frac{9}{30}} = 0.660$$

$$\therefore Q_A = (0.660)(290) = 191.5 \text{ W}; \quad Q_B = (0.340)(290) = 98.5 \text{ W}$$

$$q_A = \frac{Q_A}{A} = \frac{191.5 \text{ W}}{0.1 \text{ m}^2} = 1915 \text{ W m}^{-2}; \quad q_B = \frac{Q_B}{A} = \frac{98.5 \text{ W}}{0.1 \text{ m}^2} = 985 \text{ W m}^{-2}$$

b. $q_A = K_A \frac{T_2 - T_1}{L_A}$

$$T_2 - T_1 = \frac{q_A L_A}{K_A} = \frac{1915 \text{ W}}{\text{m}^2}\left|\frac{\text{m K}}{35 \text{ W}}\right|0.060 \text{ m} = 3.3 \text{ K} \quad \therefore T_2 = 303.3 \text{ K}$$

Obviously, we could run this heater with more power.

7.1 For laminar flow, calculate the results given in Table 7.1 for Nu_∞ for slug flow (v_x = uniform) and uniform heat flux in a circular tube.

$Nu_\infty \equiv \frac{hD}{k}$, Prove $Nu_\infty = 8.00$; $h \equiv \frac{q_o}{T_R - T_m}$

Slug flow and uniform q_o

$$v_z \frac{\partial T}{\partial z} = \frac{k}{\rho C_p} \frac{1}{r} \frac{\partial}{\partial r}\left(r \frac{dT}{dr}\right); \quad h = -\frac{k}{R} \frac{\partial}{\partial\left(\frac{r}{R}\right)} \left(\frac{T_R - T}{T_R - T_m}\right)_{r=R}$$

$$v_z \left(\frac{dT}{dz}\right) \int_0^r r\,dr = \frac{k}{\rho C_p} \int_0^{r\,\partial T/\partial r} d\left(\frac{r\,\partial T}{\partial r}\right); \quad \text{Let } \alpha = \frac{k}{\rho C_p}$$

$$v_z \left(\frac{\partial T}{\partial z}\right) \frac{r^2}{2} = \alpha r \frac{\partial T}{\partial r}$$

Also at $r = R$, $T = T_R$

$$\frac{v_z}{2}\left(\frac{\partial T}{\partial z}\right) \int_R^r r\,dr = \alpha \int_{T_R}^T \partial T; \quad \frac{v_z}{2}\left(\frac{\partial T}{\partial z}\right)\left(\frac{R^2 - r^2}{2}\right) = \alpha (T_R - T)$$

$\therefore$ Temperature distribution: $T_R - T = \frac{v_z}{4}\left(\frac{\partial T}{\partial z}\right)\left(\frac{1}{\alpha}\right)(R^2 - r^2)$

$$T_R - T_m = \frac{2\pi \int_0^R v_z (T_R - T)\, r\,dr}{2\pi \int_0^R v_z\, r\,dr}$$

$$T_R - T_m = \left(\frac{2}{R^2}\right)\left(\frac{v_z}{4}\right)\left(\frac{\partial T}{\partial z}\right)\left(\frac{1}{\alpha}\right) \int_0^R (R^2 r - r^3)\,dr = \frac{2}{R^2} c^* \left[\frac{R^2 r^2}{2} - \frac{r^4}{4}\right]_0^R = \frac{2}{R^2} c^* \left(\frac{R^4}{4}\right) = \frac{c^*}{2} R^2$$

where $c^* = \left(\frac{v_z}{4}\right)\left(\frac{dT}{dz}\right)\left(\frac{1}{\alpha}\right)$

$$\frac{T_R - T}{T_R - T_m} = 2\left(\frac{R^2 - r^2}{R^2}\right) = 2\left(1 - \left(\frac{r}{R}\right)^2\right)$$

$$\left.\frac{d\left(\frac{T_R - T}{T_R - T_m}\right)}{d\left(\frac{r}{R}\right)}\right|_{r=R} = \left. -4 \frac{r}{R}\right|_{r=R} = -4$$

$$h = \frac{k}{R} 4 = \frac{8k}{D} \quad \therefore Nu_\infty = \frac{hD}{k} = 8$$

7.2 A liquid film at T_0 flows down a vertical wall at a higher temperature T_s. Consider heat transfer from the wall to the liquid for such contact times that the liquid temperature changes appreciably only in the immediate vicinity of the wall.

a) Show that the energy equation can be written (state assumptions):

$$\rho C_p v_z \frac{\partial T}{\partial z} = k \frac{\partial^2 T}{\partial y^2}.$$

b) The energy equation contains v_z. What would you use for v_z?

c) Write appropriate boundary conditions.

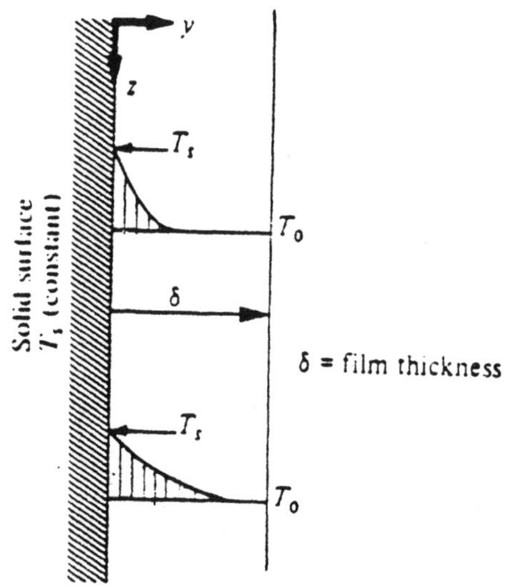

a. Use Eq. (A) in Table 7.5. Assume steady state and fully developed flow:

$$\frac{\partial T}{\partial t} = 0, \frac{\partial v_z}{\partial z} = 0, \frac{\partial T}{\partial x} = v_x = v_y = \frac{\partial^2 T}{\partial x^2} = \frac{\partial^2 T}{\partial z^2} = 0;$$ neglect viscous dissipation.

$$\therefore \rho C_p v_z \frac{\partial T}{\partial z} = k \frac{\partial^2 T}{\partial y^2}$$

b. Eq. (2.12) for fully developed flow applies with $\cos\beta = 1$.

Using appropriate change of notation to correspond to the coordinate system, Eq. (2.12) gives $v_z = \frac{\rho g \delta^2}{2\eta}\left[1 - \left(\frac{y}{\delta}\right)^2\right]$

c. Boundary Conditions – For temperature: $T(y,0) = T_0$; $T(0,z) = T_s$; $T(\delta, z) = T_0$

7.3 A gap of thickness L exists between two parallel plates of porous solids. Fluid is forced to flow through the bottom plate, across the gap, and then through the upper plate. Assume that the fluid flows with a constant velocity V in laminar flow with straight streamlines across the gap. The system is at steady state with the upper and lower plates at T_L and T_0, respectively. a) Write an appropriate energy equation and boundary conditions for the fluid in the gap. b) Solve for the temperature in the gap. c) Derive an equation for the heat flux across the gap.

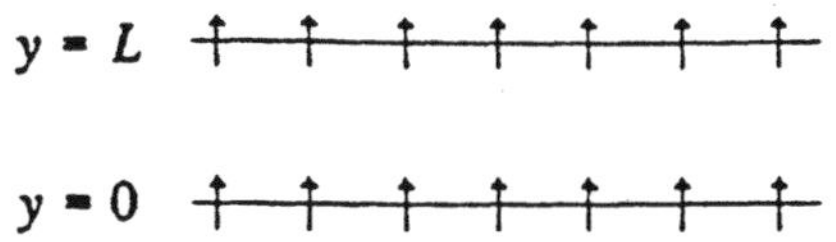

a. $$\frac{\partial^2 T}{\partial Y^2} - \frac{V}{\alpha}\frac{\partial T}{\partial Y} = 0$$

at $Y=0,\ T=T_0;\ Y=L,\ T=T_L$

b. $$\frac{\partial\left(\partial T/\partial Y\right)}{\partial T/\partial Y} = \frac{V}{\alpha}\, dY$$

$$\frac{\partial T}{\partial Y} = c \exp\left(\frac{V}{\alpha} Y\right)$$

$$T = C_1 \exp\left(\frac{V}{\alpha} Y\right) + C_2$$

APPLY B.C. $T_0 = C_1 + C_2 \Rightarrow C_2 = T_0 - C_1$

$$T_L = C_1 \exp\left(\frac{V}{\alpha} L\right) + C_2 = C_1 \exp\left(\frac{V}{\alpha} L\right) - C_1 + T_0$$

$$\left[\frac{T_L - T_0}{\exp\left(\frac{V}{\alpha} L\right) - 1}\right] = C_1 \qquad C_2 = T_0 - \left[\frac{T_L - T_0}{\exp\left(\frac{V}{\alpha} L\right) - 1}\right]$$

$$\therefore\ T = C_1 \exp\left(\frac{V}{\alpha} Y\right) - C_1 + T_0$$

$$T - T_0 = C_1\left[\exp\left(\frac{V}{\alpha} Y\right) - 1\right] \quad \therefore\ \frac{T - T_0}{T_L - T_0} = \frac{\exp\left(\frac{V}{\alpha} Y\right) - 1}{\exp\left(\frac{V}{\alpha} L\right) - 1}$$

c. $$q_Y = -C_1 k \frac{V}{\alpha} \exp\left(\frac{V}{\alpha} Y\right);\ q_Y\big|_{Y=0} = -C_1 k \frac{V}{\alpha} = -\frac{k(T_L - T_0)}{\exp\left(\frac{V}{\alpha} L\right) - 1}\frac{V}{\alpha}$$

7.4 A liquid of constant density and viscosity flows upward in the annulus ($R_1 \le r \le R_2$) between two very long and concentric cylinders. Assume that both the flow and the temperature are fully developed. The inner cylinder is electrically heated and supplies a constant and uniform flux, q_I, to the liquid. The outer cylinder is maintained at a constant temperature, T_0.

a) Solve for v_z.
b) Write the energy equation and state your assumptions.
c) Write appropriate boundary conditions.

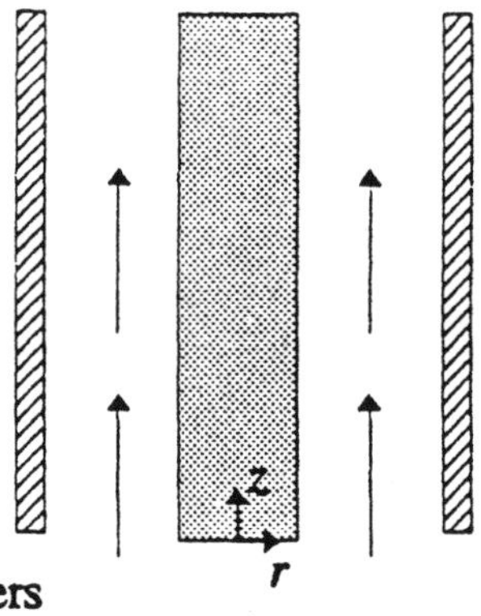

a. From Problem 2.13

$$v_z = \frac{1}{4\eta}\left(\frac{dP}{dz} - \rho g\right) r^2 + C_1 \ln r + C_2$$

B.C. at $r = R_1$, $v_z = 0$; $r = R_2$, $v_z = 0$

$$C_1 = \frac{1}{4\eta}\left(\frac{dP}{dz} - \rho g\right)\frac{\left[R_2^2 - R_1^2\right]}{\ln\left(\frac{R_1}{R_2}\right)}$$

$$C_2 = -\frac{1}{4\eta}\left(\frac{dP}{dz} - \rho g\right)\left[R_1^2 + \frac{(R_2^2 - R_1^2)}{\ln\left(\frac{R_1}{R_2}\right)}\ln R_1\right]$$

$$v_z = \frac{1}{4\eta}\left(\frac{dP}{dz} - \rho g\right)\left[r^2 - R_1^2 + \frac{(R_1^2 - R_2^2)}{\ln\left(\frac{R_1}{R_2}\right)}\ln\left(\frac{R_1}{r}\right)\right]$$

b. Assumptions: $c_v = c_p$; $\frac{\partial T}{\partial t} = 0$; $v_r = v_\theta = 0$, $\frac{\partial^2 T}{\partial \theta^2} = 0$, $\frac{\partial^2 T}{\partial z^2} = 0$, neglect viscous dissipation.

$\therefore$ Eq. B - Table 7.5 $\quad v_z \frac{\partial T}{\partial z} = \alpha \frac{1}{r}\frac{\partial}{\partial r}\left(r \frac{\partial T}{\partial r}\right)$

c. Boundary Conditions for v_z are in part a.

For temperature:

at $r = R_1$, $q = q_I$ $\qquad$ at $z = 0$, $T = T_0$

$r = R_2$, $T = T_0$

7.5 Air at 0.3 m s^{-1} and 365 K flows parallel to a flat plate at 310 K. a) Calculate the distance from the leading edge to where the momentum boundary layer thickness is 6 mm. b) At the same distance from the leading edge, what is the thermal boundary layer thickness? c) Up to the same distance from the leading edge, how much heat is transferred to the plate (one side) in 600 s, if the plate is 100 mm wide?

a. $V_\infty = 0.3\,m\,s^{-1}$; $T_\infty = 365\,K$; $T_o = 310\,K$

Evaluate the properties at the average temp. (called the film temp.)

$$T_f = \frac{1}{2}(T_\infty + T_o) = 338\,K$$

Appendix B gives the properties; $\nu(338\,K) = 19.89 \times 10^{-6}\,m^2\,s^{-1}$

Eq. (2.101) $\dfrac{\delta}{x} = \dfrac{5.0}{\sqrt{V_\infty x/\nu}}$; $x = \dfrac{\delta^2 V_\infty}{5^2 \nu} = \dfrac{0.006^2\,m^2 \;|\; 0.3\,m \;|\; 1 \;|\; s}{s \;|\; 25 \;|\; 19.89 \times 10^{-6}} = 0.022\,m$

$x = 0.022\,m = 22\,mm$

b. Eq. (7.29) $\dfrac{\delta_T}{\delta} = 0.975\,Pr^{-1/3}$; $Pr(338\,K) = 0.701$; $\delta_T = (0.006)(0.975)(0.701)^{-1/3}$

$\delta_T = 0.00658\,m = 6.58\,mm$

c. $Nu_L = 0.664\,Pr^{0.343} Re_L^{0.5}$ $(L = 22 \times 10^{-3}\,m)$; $Re_L = Re_x = \dfrac{x V_\infty}{\nu} = \dfrac{(0.022)(0.3)}{19.89 \times 10^{-6}}$

$= 331.8$

$Nu_L = (0.664)(0.701)^{0.343}(331.8)^{0.5} = 10.71$

$Nu_L = \dfrac{hL}{k}$; $k(338\,K) = 29.2 \times 10^{-3}\,W\,m^{-1}\,K^{-1}$

$$h = \frac{Nu_L k}{L} = \frac{(10.71)(29.2 \times 10^{-3})}{(0.022)} = 14.22\,W\,m^{-2}\,K^{-1}$$

$Q = hA(T_\infty - T_o) = (14.22)(0.1)(0.022)(365 - 310) = 1.72\,W$

In 600 s, $Q\Delta t = (1.72)(600) = 1032\,J$

7.6 Consider natural convection between parallel vertical plates maintained at T_1 and T_0, respectively. Assume that the plates are very long and the convection is fully developed. For constant properties: a) Write the energy equation and boundary conditions for temperature. b) Write the momentum equation with the Boussinesq approximation and boundary conditions for velocity.

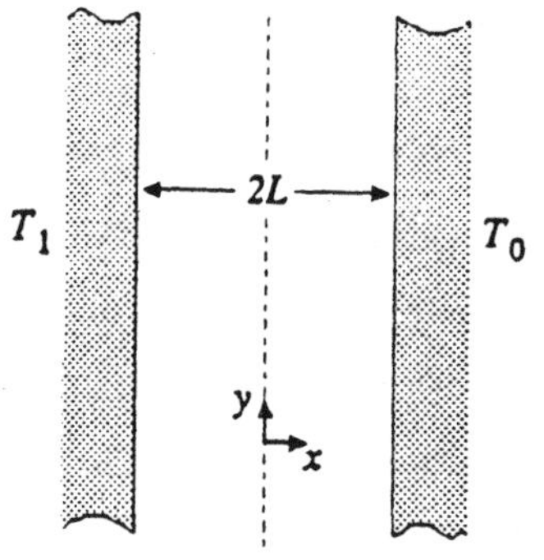

a. $T = T(x)$ because the plates are very long.

Energy equation $\frac{d^2T}{dx^2} = 0$

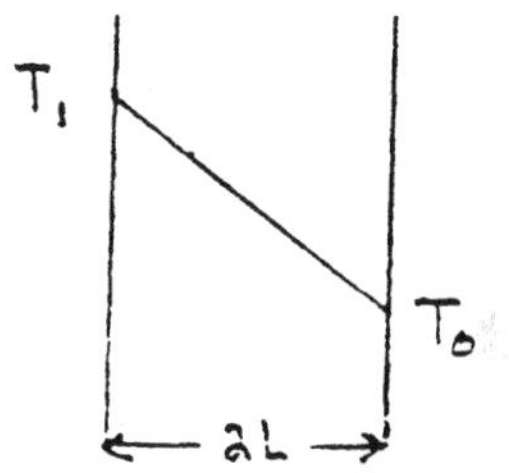

B.C. $T = T_1$ at $x = -L$; $T = T_0$ at $x = L$

b. $v_y = v_y(x)$, $v_x = 0$ because flow is fully developed.

$$0 = -\frac{dP}{dy} + \eta\frac{d^2v_y}{dx^2} + \rho g_y \quad \text{Momentum Eq.}$$

But $\frac{dP}{dy} \simeq -\rho_m g$ and $g_y = -g$ where ρ_m is the density at the mean temperature T_m.

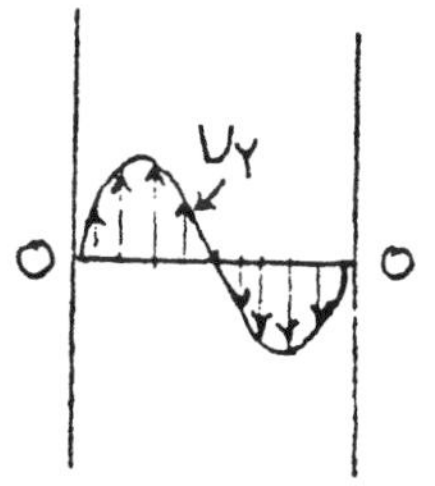

$$\therefore -\frac{dP}{dy} + \rho g_y = \rho_m g - \rho g = g(\rho_m - \rho)$$

$$\beta = -\frac{1}{\rho}\left(\frac{\rho_m - \rho}{T_m - T}\right) \quad \text{so that} \quad -\frac{dP}{dy} + \rho g_y = g\rho\beta(T - T_m)$$

$$0 = \eta\frac{d^2v_y}{dx_2} + g\rho\beta(T - T_m) \quad \text{or} \quad 0 = \nu\frac{d^2v_y}{dx^2} + g\beta(T - T_m) \quad \text{Momentum Eq.}$$

B.C. $v_y = 0$ at $x = 0$; $v_y = 0$ at $x = L$

7.7 The surface temperature of a vertical plate is maintained at 390 K. At 0.24 m from the bottom of the plate, calculate the heat transfer coefficient to: a) air at 290 K; b) helium at 290 K.

a. Evaluate the properties at $T_f = \frac{1}{2}(390+290) = 340$ K.

$Pr(340K) = 0.700$; $k(340K) = 29.3 \times 10^{-3}$ W m^{-1}K^{-1}

$$Gr_x = \frac{g\beta(T_0 - T_\infty)x^3}{\nu^2}; \quad \nu(340K) = 20.10 \times 10^{-6} \text{ m}^2 \text{s}^{-1}$$

$$\beta = -\frac{1}{\rho}\frac{d\rho}{dT}\bigg|_p = \frac{1}{V}\frac{\partial V}{\partial T}\bigg|_p = \frac{1}{T} \text{ (for ideal gas)}$$

$\beta(340K) = 2.941 \times 10^{-3}$ K^{-1}

$$Gr_x = \frac{9.807 \text{ m}}{\text{s}^2}\bigg|\frac{2.941\times10^{-3}}{\text{K}}\bigg|(390-290)\text{K}\bigg|(0.24)^3\text{m}^3\bigg|\frac{\text{s}^2}{(20.10\times10^{-6})^2\text{m}^4} = 9.869\times10^7$$

$Gr_x Pr = 6.91 \times 10^7$

$$\text{Eq. (7.44) } Nu_x = \left(\frac{Gr_x}{4}\right)^{\frac{1}{4}}\frac{0.676\,Pr^{\frac{1}{2}}}{(0.861+Pr)^{\frac{1}{4}}} = \left(\frac{9.869\times10^7}{4}\right)^{\frac{1}{4}}\frac{(0.676)(0.7)^{\frac{1}{2}}}{(0.861+0.700)^{\frac{1}{4}}} = 35.7$$

$$h_x = 35.7\frac{k}{x} = 35.7\frac{(29.3\times10^{-3})}{(0.24)} = 4.35 \text{ W m}^{-2}\text{K}^{-1} \text{ (Air)}$$

b. $Pr(340K) = 0.678$, $\nu(340K) = 153\times10^{-6}$ m^{2}s^{-1}, $k(340K) = 166\times10^{-3}$ W m^{-1}K^{-1}

$$Gr_x = \frac{(9.807)(2.941\times10^{-3})(390-290)(0.24)^3}{(153\times10^{-6})^2} = 1.703\times10^6$$

$$Nu_x = \left(\frac{1.703\times10^6}{4}\right)^{\frac{1}{4}}\frac{(0.676)(0.678)^{\frac{1}{2}}}{(0.861+0.678)^{\frac{1}{4}}} = 12.72$$

$$h_x = (12.72)\frac{(166\times10^{-3})}{(0.24)} = 8.80 \text{ W m}^{-2}\text{K}^{-1} \quad (h(He) > h(air))$$

7.8 Liquid metal flows through a channel with a rectangular cross-section. Two walls are perfectly insulated and two are at a constant temperature of T_w. The metal has temperature T_0 as it enters the channel, and $T_w > T_0$. Assume steady state, fully developed flow and no solidification.

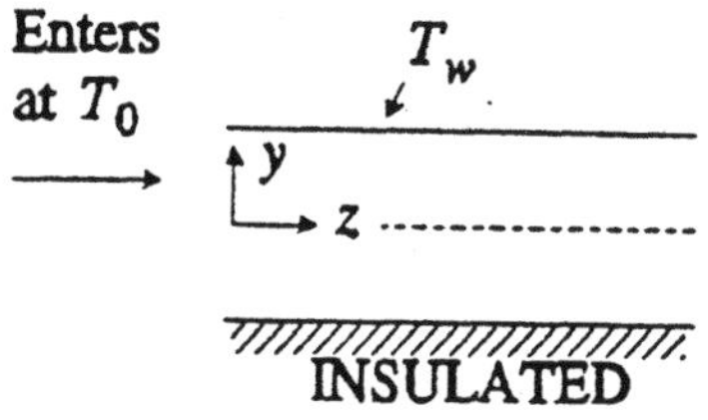

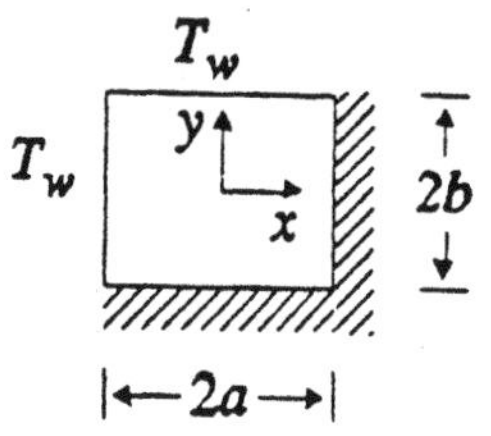

a) Write the energy equation in terms of temperature for constant thermal properties.
b) Write the boundary conditions.

a. $V_Y = V_X = 0$, $V_z = f(x,Y)$, $T = f(x,Y,z)$

Eq. (A) - p. 242 - neglect vicous dissipation; $C_p = C_v$.

$$\rho c_p\left(V_z \frac{\partial T}{\partial z}\right) = k\left[\frac{\partial^2 T}{\partial x^2} + \frac{\partial^2 T}{\partial Y^2} + \frac{\partial^2 T}{\partial z^2}\right] \quad \text{or} \quad V_z \frac{\partial T}{\partial z} = \alpha\left[\frac{\partial^2 T}{\partial x^2} + \frac{\partial^2 T}{\partial Y^2} + \frac{\partial^2 T}{\partial z^2}\right]$$

b. Boundary Conditions

$T(x, Y, 0) = T_0$

$T(x, Y, \infty) = T_w$

$\frac{\partial T}{\partial x}(a, Y, z) = 0$ ⇐ Insulated surface

$T(-a, Y, z) = T_w$

$\frac{\partial T}{\partial Y}(x, -b, z) = 0$ ⇐ Insulated surface

$T(x, +b, z) = T_w$

$\frac{\partial T}{\partial z}(x, Y, \infty) = 0$

7.9 Consider the creeping flow of a fluid about a rigid sphere as illustrated by Fig. 2.9. The sphere is maintained at T_0 and the fluid approaches from below with a temperature T_∞ and velocity V_∞. a) Write the energy equation which applies to the fluid in the vicinity of the sphere. Assume steady-state conditions. b) Write appropriate boundary conditions for part a). c) What other equations or results would you use in order to solve the system described by parts a) and b)?

a. $T = f(r, \theta)$

start with Eq. (c) in Table 7.5

Ignore viscous dissipation, $C_p = C_v$, $v_\phi = 0$

$$\rho C_p\left(v_r \frac{\partial T}{\partial r} + \frac{v_\theta}{r}\frac{\partial T}{\partial \theta}\right) = k\left[\frac{1}{r^2}\frac{\partial}{\partial r}\left(r^2 \frac{\partial T}{\partial r}\right) + \frac{1}{r^2 \sin\theta}\frac{\partial}{\partial \theta}\left(\sin\theta \frac{\partial T}{\partial \theta}\right)\right]$$

b. For the r direction:

At $r = R$, $T = T_0$; $r = \infty$, $T = T_\infty$

For the θ direction:

At $\theta = 0$, $\frac{\partial T}{\partial \theta} = 0$ (symmetry) ; $\theta = \pi$, $\frac{\partial T}{\partial \theta} = 0$ (symmetry)

c. The energy equation contains v_r and v_θ, so these would have to be known or the momentum equation would have to be solved.

7.10 A very long fiber of glass (radius $= R$) is extracted from a hole in the bottom of a crucible. It is extracted with a constant velocity V into a gas at T_∞; assume slug flow.

a) For uniform properties write the energy equation for temperature in the fiber. Do not ignore conduction in the direction of flow.

b) Write boundary conditions. [Hint: At $r = R$, the flux to the surface must equal the flux to the surrounding gas "via h."]

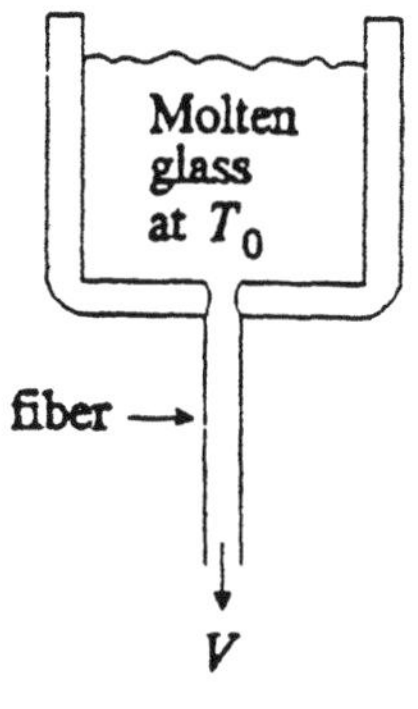

a. $v_z = V$, $C_v = C_p$, $v_r = v_\theta = 0$, $\frac{\partial^2 T}{\partial \theta^2} = 0$

Eq. (8) Table 7.5

$$\rho C_p \left(V \frac{\partial T}{\partial z} \right) = k \left[\frac{1}{r} \frac{\partial}{\partial r} \left(r \frac{\partial T}{\partial r} \right) + \frac{\partial^2 T}{\partial z^2} \right]$$

b. For the r direction:

At $r = 0$, $\frac{\partial T}{\partial r} = 0$; $r = R$, $\frac{\partial T}{\partial r} = -\frac{h}{k}(T - T_\infty)$

At $z = 0$, $T = T_0$; $z = \infty$, $T = T_\infty$ or $\frac{\partial T}{\partial z} = 0$

7.11 Starting with Eq. (7.44), derive Eq. (7.45) and define the dimensionless numbers in Eq. (7.45).

Let $C = \frac{0.676\, Pr^{\frac{1}{2}}}{(0.861 + Pr)^{\frac{1}{4}}}$; then Eq. (7.44) is $Nu_x = \left(\frac{Gr_x}{4} \right)^{1/4} C$

$$\therefore h_x = \frac{k}{x} \left[\frac{g \beta (T_0 - T_\infty) x^3}{4 \nu^2} \right]^{\frac{1}{4}} C = k \left[\frac{g \beta (T_0 - T_\infty)}{4 \nu^2} \right]^{\frac{1}{4}} x^{-\frac{1}{4}} C$$

For the average h, we need $h = \frac{1}{L} \int_{x=0}^{L} h_x \, dx$

$$h = \frac{k}{L} \left[\frac{g \beta (T_0 - T_\infty)}{4 \nu^2} \right]^{\frac{1}{4}} C \int_{x=0}^{L} x^{-\frac{1}{4}} dx = \frac{k}{L} \left[\frac{g \beta (T_0 - T_\infty)}{4 \nu^2} \right]^{\frac{1}{4}} L^{\frac{3}{4}} \left(\frac{4}{3} C \right)$$

$$\frac{hL}{k} = \frac{4}{3} C \left[\frac{g \beta (T_0 - T_\infty) L^3}{4 \nu^2} \right]^{\frac{1}{4}}$$

$$\frac{hL}{k} \equiv Nu_L, \qquad \frac{g \beta (T_0 - T_\infty) L^3}{4 \nu^2} \equiv \frac{Gr_L}{4}$$

$$\therefore \text{Eq. (7.45)} \quad \frac{Nu_L}{4\sqrt{Gr_L/4}} = \frac{0.901\, Pr^{\frac{1}{2}}}{(0.861 + Pr)^{\frac{1}{4}}}$$

7.12

a) Determine an expression that gives the heat flow Q (W) through a solid spherical shell with inside and outside radii of r_1 and r_2, respectively.

b) Examine the results regarding what happens as the shell thickness becomes larger compared with the inside radius.

a. $T = f(r) \quad v_r = v_\theta = v_\phi = 0$

Eq. (C) – p. 242

$$\frac{d}{dr}\left(r^2 \frac{dT}{dr}\right) = 0$$

at $r = R_1,\ T = T_1$

$r = R_2,\ T = T_2$

$$r^2 \frac{\partial T}{\partial r} = C_1$$

$$T = -\frac{C_1}{r} + C_2$$

$$T_1 = -\frac{C_1}{r_1} + C_2 \ ; \ T_2 = -\frac{C_1}{r_2} + C_2$$

$$\therefore C_1 = \frac{T_1 - T_2}{\frac{1}{R_2} - \frac{1}{R_1}} \ ; \ C_2 = T_1 + \frac{C_1}{R_1} \ ; \ T = -\frac{C_1}{r} + T_1 + \frac{C_1}{R_1}$$

$$q = -k\frac{\partial T}{\partial r} \ ; \ Q = 4\pi r^2 q = 4\pi r^2\left(-k\frac{\partial T}{\partial r}\right)$$

$$\frac{\partial T}{\partial r} = \frac{C_1}{r^2} = \frac{1}{r^2}\,\frac{(T_1 - T_2)}{\frac{1}{R_1} - \frac{1}{R_2}}$$

$$Q = \frac{4\pi k (T_2 - T_1)}{\frac{1}{R_2} - \frac{1}{R_1}}$$

b. $$\lim_{R_2 \to \infty} Q = \frac{4\pi k (T_2 - T_1)}{-\frac{1}{R_1}} = -4\pi k R_1 (T_2 - T_1)$$

7.13 A sphere of radius R is in a motionless fluid (no forced or natural convection). The surface temperature of the sphere is maintained at T_R and the bulk fluid temperature is T_∞. a) Develop an expression for the temperature in the fluid surrounding the sphere. b) Determine the Nusselt number for this situation. Such a value would be the limiting value for the actual system with convection as the forces causing convection become very small.

a. $T = f(r) \qquad v_r = v_\theta = v_\phi = 0$

Eq. (c) - p. 242

$$\frac{d}{dr}\left(r^2 \frac{dT}{dr}\right) = 0$$

$$\frac{dT}{dr} = \frac{C_1}{r^2} \; ; \; T = -\frac{C_1}{r} + C_2$$

at $r = R$, $T = T_R$

$r = \infty$, $T = T_\infty$

$\therefore C_2 = T_\infty$ and $C_1 = -R(T_R - T_\infty)$

$$\frac{T - T_\infty}{T_R - T_\infty} = \frac{R}{r}$$

b. $$h = \frac{-k \left.\frac{dT}{dr}\right|_R}{(T_R - T_\infty)} = \frac{-k C_1}{R^2 (T_R - T_\infty)} = \frac{k R (T_R - T_\infty)}{R^2 (T_R - T_\infty)}$$

$$h = \frac{k}{R} = \frac{2k}{D} \text{ or } Nu_D = \frac{hD}{k} = 2$$

7.14 For the system in Fig. 2.1 develop an expression for the temperature distribution in the falling film. Assume fully developed flow, constant properties, and fully developed temperature profile. The free liquid surface is maintained at $T = T_0$ and the solid surface at $T = T_s$, where T_0 and T_s are constants. a) Ignore viscous heating effects. b) Include viscous heating effects.

Answer b)

$$\frac{T - T_0}{T_s - T_0} = \frac{x}{\delta}\left\{1 + \tfrac{3}{4}\,\mathrm{Br}\left[1 - \left(\frac{x}{\delta}\right)^3\right]\right\},$$

where $\mathrm{Br} = \dfrac{\eta \bar{V}^2}{k(T_s - T_0)}$, Brinkman number.

a. $\frac{\partial T}{\partial t} = 0$, $v_x = 0$, $v_y = 0$, $\frac{\partial T}{\partial z} = 0$, $\frac{\partial T}{\partial y} = 0$, $\Phi = 0$

$\therefore$ Eq. A, Table 7.5 reduces to; $k\frac{d^2T}{dx^2} = 0$

so that a linear temperature profile exists: $\frac{T - T_0}{T_\delta - T_0} = \frac{x}{\delta}$

b. Include viscous effects, $\Phi \neq 0$

$$k\frac{d^2T}{dx^2} + \eta\left(\frac{dv_z}{dx}\right)^2 = 0$$

From Chap. 2; Eqs. (2.13) and (2.14)

$$v_z = \frac{3\bar{v}_z}{2}\left[1 - \left(\frac{x}{\delta}\right)^2\right] \quad \text{where } \bar{v}_z = \frac{\rho g \delta^2 \cos\beta}{3\eta}$$

$$\frac{dv_z}{dx} = \frac{-3\bar{v}_z x}{\delta^2}\,; \quad \left(\frac{dv_z}{dx}\right)^2 = \frac{9\bar{v}_z^2 x^2}{\delta^4}$$

$$k\frac{d^2T}{dx^2} + \eta\frac{9\bar{v}_z^2 x^2}{\delta^4} = 0\,; \quad \frac{d^2T}{dx^2} + Ax^2 = 0 \text{ where } A = \frac{9\bar{v}^2\eta}{k\,\delta^4}$$

$$T = -\frac{A}{12}x^4 + C_1x + C_2$$

at $x = 0$, $T = T_0$

$C_2 = T_0$

at $x = \delta$, $T = T_\delta$

$$C_1 = \frac{1}{\delta}(T_\delta - T_0) + \frac{A}{12}\delta^3$$

$$T - T_0 = -\frac{A}{12}x^4 + (T_\delta - T_0)\frac{x}{\delta} + \frac{A}{12}\delta^3 x = \frac{A}{12}\delta^4\left[\frac{x}{\delta} - \left(\frac{x}{\delta}\right)^4\right] + (T_\delta - T_0)\frac{x}{\delta}$$

$$\frac{T - T_0}{T_\delta - T_0} = \frac{9\bar{v}_z^2 \eta \delta^4}{12k\,\delta^4(T_\delta - T_0)}\left[1 - \left(\frac{x}{\delta}\right)^3\right]\left(\frac{x}{\delta}\right) + \left(\frac{x}{\delta}\right)$$

$$\frac{T - T_0}{T_\delta - T_0} = \left(\frac{x}{\delta}\right)\left\{1 + \frac{3}{4}Br\left[1 - \left(\frac{x}{\delta}\right)^3\right]\right\} \quad \text{where } Br = \frac{\eta \bar{v}_z^2}{k(T_\delta - T_0)}$$

7.15 Consider heat conduction through a plane wall of thickness Δx, and T_1 and T_2 are the surface temperatures. Derive the steady-state heat flux in terms of T_1, T_2 and Δx if the thermal conductivity varies according to

$$k = k_0(1 + aT)$$

where k_0 and a are constants.

Steady state $\frac{d}{dx}\left(k\frac{dT}{dx}\right) = 0$

$-k\frac{dT}{dx} = q$, constant

$$-\int_{T_1}^{T_2} k\,dT = \int_{x_1}^{x_2} q\,dx = q\int_{x_1}^{x_2} dx$$

$$-\int_{T_1}^{T_2}(k_0 + k_0 aT)\,dT = q(x_2 - x_1) = q\Delta x$$

$$-\int_{T_1}^{T_2} k_0\,dT - \int_{T_1}^{T_2} k_0 aT\,dT = q\Delta x \qquad q = -\frac{k_0}{\Delta x}\left[(T_2 - T_1) + \frac{a}{2}(T_2^2 - T_1^2)\right]$$

7.16 A liquid at a temperature T_0 continuously enters the bottom of a small tank, overflows into a tube, and then flows downward as a film on the inside. At some position down the tube ($z = 0$) when the flow is fully developed, the pipe heats the fluid with a uniform flux q_R. The heat loss from the liquid's surface is sufficiently small so that it may be neglected.

a) For steady-state laminar flow with constant properties, develop by shell balance or show by reducing an equation in Table 7.5 the pertinent differential energy equation that applies to the falling film.

b) Write the boundary conditions for the heat flow.

c) What other information must complement parts a) and b) in order to solve the energy equations?

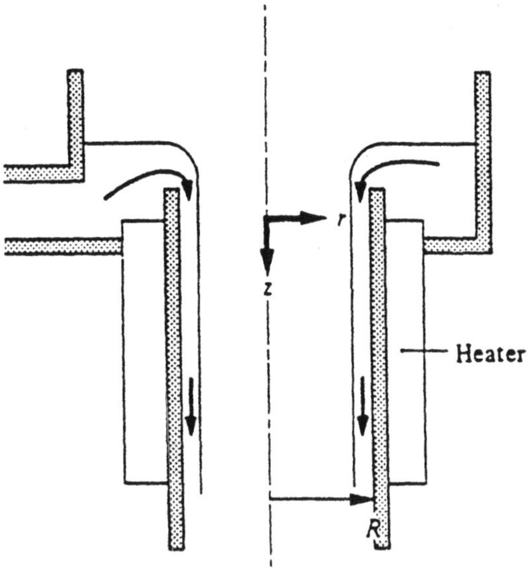

a. $v_r = v_\theta = 0$; $\frac{\partial T}{\partial t} = 0$ steady state; $\frac{\partial T}{\partial \theta} = 0$

Eq.(B) - p.242 - neglect vicous dissipation; $C_p = C_v$

$$\rho C_p v_z \frac{\partial T}{\partial z} = k\left[\frac{1}{r}\frac{\partial}{\partial r}\left(r\frac{\partial T}{\partial r}\right) + \frac{\partial^2 T}{\partial z^2}\right]; \quad \frac{\partial T}{\partial r} >> \frac{\partial T}{\partial z} \text{ so we can ignore } \frac{\partial^2 T}{\partial z^2}$$

$$\therefore \; v_z \frac{\partial T}{\partial z} = \alpha\left[\frac{1}{r}\frac{\partial}{\partial r}\left(r\frac{\partial T}{\partial r}\right)\right]$$

b. Boundary Conditions:

at $z = 0$, $T = T_0$, $aR \le r \le R$

at $r = R$, $\frac{\partial T}{\partial r} = \frac{+q_R}{k}$, $z > 0$

at $r = aR$, $\frac{\partial T}{\partial r} = 0$, $z > 0$

c. To solve this problem an expression for v_z must be obtained to substitute into the energy equation. Thus we would have to solve an appropriate "conservation of momentum" equation. We could use the result for a falling film down the inside of a cylindrical wall.

8.1 Hot gases flow inside an insulated horizontal tube with dimensions shown to the right. Determine the heat transfer coefficients for both the inside and outside surfaces. The gas at 1370 K flows with an average velocity of 4 m s^{-1}. The environment surrounding the tube is air at 300 K, and the outside surface temperature is 330 K. The gas is ideal so that $\beta = 1/T$, and at these high temperatures η for the gas does not change appreciably with temperature. *Data for gas inside tube*:

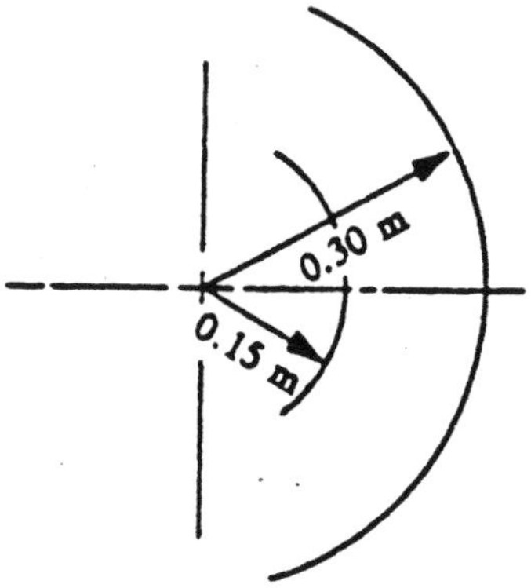

T, K	ρ, kg m^{-3}	η, kg s^{-1} m^{-1}	k, W m^{-1} K^{-1}	C_p, J kg^{-1} K^{-1}
1370	0.30	4.1×10^{-5}	0.086	1.0×10^3

Inside heat transfer coefficient $- h_i$

Use Fig. 8.2 $- \quad Re_m = \dfrac{D\bar{V}\rho}{\eta_m} = \dfrac{0.30\ \text{m}}{} \Big| \dfrac{4.0\ \text{m}}{\text{s}} \Big| \dfrac{0.30\ \text{kg}}{\text{m}^3} \Big| \dfrac{\text{m s}}{4.1 \times 10^{-5}\ \text{kg}} = 8.78 \times 10^3$

$$Nu_m Re_m^{-1} Pr_m^{-1/3} \left(\frac{\eta_m}{\eta_o}\right)^{-0.14} = 0.004 ; \quad \left(\frac{\eta_m}{\eta_o}\right)^{-0.14} \simeq 1$$

$$Pr_m = \left(\frac{\eta C_p}{k}\right) = \frac{4.1 \times 10^{-5}\ \text{kg}}{\text{m s}} \Bigg| \frac{1.0 \times 10^3\ \text{J}}{\text{kg K}} \Bigg| \frac{1\ \text{m K}}{0.086\ \text{W}} \Bigg| \frac{\text{W s}}{\text{J}} = 0.477$$

$$Nu_m = 0.004\, Re_m Pr^{1/3} \left(\frac{\eta_m}{\eta_o}\right)^{0.14} = (0.004)(8.78 \times 10^3)(0.477^{1/3})(1) = 27.4$$

$$h_i = \frac{Nu_m k_m}{D} = \frac{(27.4)(0.086)}{(0.30)} = 7.85\ \text{W m}^{-2}\ \text{K}^{-1}$$

Outside heat transfer coefficient $- h_o$

Use Fig. 8.8 $- \quad Gr_L Pr = \dfrac{g\beta}{\nu^2}(T_o - T_\infty)\left(\dfrac{\pi D}{2}\right)^3 \left(\dfrac{\eta C_p}{k}\right); \quad \beta = \dfrac{1}{T_f} = \dfrac{1\ (2)}{(330+300)} = 3.17 \times 10^{-3}\ \text{K}^{-1}$

$C_p(315) = 1047$ J kg^{-1} K^{-1}, $k_{Air}(315\ \text{K}) = 0.0294$ W m^{-1} K^{-1}

$\eta_{Air}(315\ \text{K}) = 2.1 \times 10^{-5}$ N s m^{-2}, $\rho_{Air}(315\ \text{K}) = 1.041$ kg m^{-3}, $\nu = 2.02 \times 10^{-5}$ N s m kg^{-1}

$$Gr_L Pr = \frac{9.81\ \text{m}}{\text{s}^2} \Bigg| \frac{3.17 \times 10^{-3}}{\text{K}} \Bigg| \frac{\text{kg}^2}{(2.02 \times 10^{-5})^2\ \text{N}^2 \text{s}^2 \text{m}^2} \Bigg| (330-300)\ \text{K} \Bigg| \left(\frac{\pi\, 0.60}{2}\right)^3 \text{m}^3 \Bigg| \frac{2.1 \times 10^{-5}\ \text{N s}}{\text{m}^2}$$

$$\Bigg| \frac{1047\ \text{J}}{\text{kg K}} \Bigg| \frac{\text{m K}}{0.0294\ \text{W}} \Bigg| \frac{\text{W s}}{\text{J}} \Bigg| \frac{\text{N s}^2}{\text{kg m}} = 1.43 \times 10^9$$

$\log_{10}(Gr_L Pr) = 9.16$: From Fig. 8.8, $\log_{10} Nu_L = 1.87$; $Nu_L = 74$

$$h_o = \frac{Nu_L k 2}{\pi D} = \frac{74}{\pi} \Bigg| 2 \Bigg| \frac{0.0294\ \text{W}}{\text{m K}} \Bigg| \frac{}{0.60\ \text{m}} = 2.31\ \text{W m}^{-2}\text{K}^{-1}$$

8.2 Molten aluminum is preheated while being transferred from a melting furnace to a casting tundish by pumping through a heated tube, 50 mm in diameter, at a flow rate of 1.3 kg s^{-1}. The tube wall is kept at a constant temperature at 1030 K. a) Calculate the heat transfer coefficient between the tube wall and the aluminum. b) Using this value of the heat transfer coefficient, how long would the tube have to be to heat the aluminum from 950 K to 1025 K? *Data for aluminum*: $k = 86$ W m^{-1} K^{-1}; $\rho = 2560$ kg m^{-3}; $C_p = 1050$ J kg^{-1} K^{-1}; $\eta = 1.2 \times 10^{-3}$ kg s^{-1} m^{-1}.

a. From Exam. 8.1 we can apply: $Re = \frac{D\bar{v}\rho}{\eta} = \frac{D\dot{W}}{\eta\left(\frac{\pi D^2}{4}\right)} = \frac{4\dot{W}}{\pi\eta D}$

$$Re = \frac{4}{\pi}\,\frac{1.3\ \text{kg}}{\text{s}}\,\frac{\text{s m}}{1.2\times10^{-3}\ \text{kg}}\,\frac{1}{0.05\ \text{m}} = 2.76\times10^{4}$$

$$Pr = \frac{\nu}{\alpha} = \frac{\eta C_p}{k} = \frac{1.2\times10^{-3}\ \text{kg}}{\text{s m}}\,\frac{1050\ \text{J}}{\text{kg K}}\,\frac{\text{m K}}{86\ \text{W}}\,\frac{\text{W s}}{\text{J}} = 0.0147$$

$$\text{Eq.}(8.7) - Nu_g = 6.7 + 0.0041\left[(2.76\times10^{4})(0.0147)\right]^{0.793}\exp\left[(41.8)(0.0147)\right] = 7.53$$

Correct for Nu_T with Fig. 8.4: $\frac{Nu_T}{Nu_g} \simeq 0.78$ ∴ $Nu_T = 5.92$

$$h = 5.92\,\frac{k}{D} = \frac{5.92 \mid 86\ \text{W}}{\text{m K} \mid 0.05\ \text{m}} = 1.02\times10^{4}\ \text{W m}^{-2}\ \text{K}^{-1}$$

b. $\ln\frac{\Delta T_L}{\Delta T_o} = -\frac{hL\pi D}{\dot{W} C_p}$; where ΔT_L = Temp. difference at exit; ΔT_o = Temp. difference at entrance; L, D = Length and diameter of tube; $\dot{W}$ = Mass flow rate

$$L = \frac{\dot{W} C_p}{\pi h D}\ln\frac{\Delta T_o}{\Delta T_L} = \frac{1.3\ \text{kg}}{\pi\ \text{s}}\,\frac{1050\ \text{J}}{\text{kg K}}\,\frac{\text{s m}^2\text{K}}{1.02\times10^{4}\ \text{J}}\,\frac{1}{0.05\ \text{m}}\ln\frac{(1030-950)\ \text{K}}{(1030-1025)\ \text{K}} = 2.36\ \text{m}$$

<u>Added note</u> - This solution ignores conduction in the direction of flow. At sufficiently high flow rates (high Reynolds number), this conduction may be neglected. The student is invited to ascertain whether conduction is important in this case.

8.3 A sheet of glass (1 m length) is cooled from an initial temperature of 1250 K by blowing air at 300 K parallel to the surface of the glass. The free stream velocity of the air is 30 m s^{-1}. Calculate the initial heat transfer coefficient and when the glass has cooled to 400 K.

Glass at 1250 K

$T_f = \frac{1}{2}(1250 + 300) = 775\ K$

Thermal properties of air at 775 K: $\nu = 80.28 \times 10^{-6}\ m^2\ s^{-1}$; $Pr = 0.706$;

$k = 56.08 \times 10^{-3}\ W\ m^{-1}\ K^{-1}$

$$Re_L = \frac{L V_\infty}{\nu} = \frac{(1)(30)}{80.28 \times 10^{-6}} = 3.74 \times 10^5$$

According to Fig. 8.7, this is laminar flow so we resort to Eq. (7.27).

$$Nu_L = 0.664\ Pr^{0.343} Re_L^{0.5} = 0.664(0.706)^{0.343}(3.74 \times 10^5)^{0.5} = 360$$

$$h = \frac{k}{L} Nu_L = \left(\frac{56.08 \times 10^{-3}}{1}\right)(360) = 20.2\ W\ m^{-2}\ K^{-1}$$

Glass at 400 K

$T_f = \frac{1}{2}(400 + 300) = 350\ K$

Thermal properties at 350 K: $\nu = 21.15 \times 10^{-6}\ m^2\ s^{-1}$; $Pr = 0.698$;

$k = 30.0 \times 10^{-3}\ W\ m^{-1}\ K^{-1}$

$$Re_L = \frac{(1)(30)}{21.15 \times 10^{-6}} = 1.42 \times 10^6$$

According to Fig. 8.7, this is turbulent flow so we use Eq. 8.17.

$Nu_x = 0.026\ Re^{0.8} Pr^{1/3}$. After we put on a Nu_L basis,

$$Nu_L = 0.0325\ Re_L^{0.8} Pr^{1/3} = 0.0325(1.42 \times 10^6)^{0.8}(0.698)^{1/3} = 2.40 \times 10^3$$

$$h = \left(\frac{30 \times 10^{-3}}{1}\right)(2.40 \times 10^3) = 72.2\ W\ m^{-2}\ K^{-1}$$

(This calculation neglects short distances from the leading edge where there could be laminar flow).

8.4 A long cylindrical bar of steel (30 mm diameter) is heated in a tempering furnace to 810 K. It is then cooled in a cross-stream of moving air at 300 K with a free stream velocity of 30 m s^{-1}. Calculate the heat transfer coefficient that applies when the bar begins to cool.

$$T_f = \frac{(810+300)}{2} = 555\,K$$

Properties of air at 555 K: $\nu_f = 46.44 \times 10^{-6}\ m^2\,s^{-1}$; $Pr = 0.685$;

$$k = 44.11 \times 10^{-3}\ W\,m^{-1}\,K^{-1}$$

$$Re_f = \frac{D\,V_\infty}{\nu_f} = \frac{(0.03)(30)}{46.44 \times 10^{-6}} = 1.94 \times 10^4$$

use Eq. (8.9). $Nu_f\,Pr^{-0.3} = 0.26\,Re_f^{0.60}$; $Nu_f = 0.26\,Re_f^{0.60}\,Pr^{0.3}$

$$Nu_f = (0.26)(1.94\times10^4)^{0.60}(0.685)^{0.3} = 86.7$$

$$h = \frac{Nu_f\,k}{D} = \frac{86.7 \left| \frac{0.0441\ W}{m\,K} \right|}{0.03\,m} = 127\ W\,m^{-2}\,K$$

8.5 Refer to Example 8.1, in which a tube with a uniform surface temperature was considered. Now consider flow through a tube which is electrically heated so that the heat flux along the length of the tube is uniform. The fluid has a mixed mean temperature of T_0 as it enters the heated length of the tube. Assume constant and uniform thermal properties and steady state. a) For a small length of tube, Δx, write an energy balance on the fluid (see Example 8.1). b) Take the limit as $\Delta x \to 0$. c) Integrate part b) and obtain an equation which gives the mixed mean temperature of the fluid as it leaves the heated section of the tube.

a.

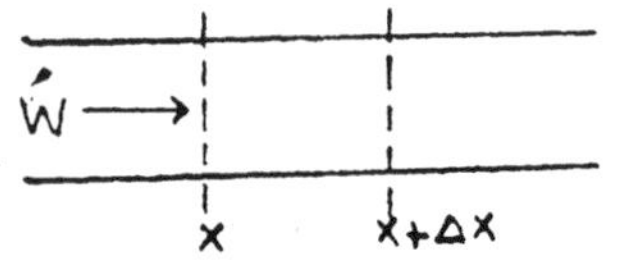

$\dot{W}$ = kg s^{-1}

$$\dot{W} C_p T_m \Big|_x + q_w \pi D \Delta x = \dot{W} C_p T_m \Big|_{x+\Delta x}$$ where q_w = uniform flux at wall.

b. $$\lim_{\Delta x \to 0} \frac{T_m|_{x+\Delta x} - T_m|_x}{\Delta x} = \frac{\pi D q_w}{\dot{W} C_p}$$

$$\frac{dT_m}{dx} = \frac{\pi D q_w}{\dot{W} C_p}$$

c. $$\int_{T_m=T_0}^{T_m=T_L} dT_m = \frac{\pi D q_w}{\dot{W} C_p} \int_{x=0}^{x=L} dx \qquad T_L = T_0 + \frac{\pi D q_w L}{\dot{W} C_p}$$

8.6 A rapidly solidified ribbon is annealed by continuously passing it through a tube with countercurrent hot gas, which flows at 30 m s^{-1} and 730 K. Estimate the heat transfer coefficient, and calculate the temperature of the ribbon leaving the tube. *Data and thermal properties*: Ribbon thickness, 1 mm; ribbon width, 150 mm; ribbon velocity, 0.17 m s^{-1}.

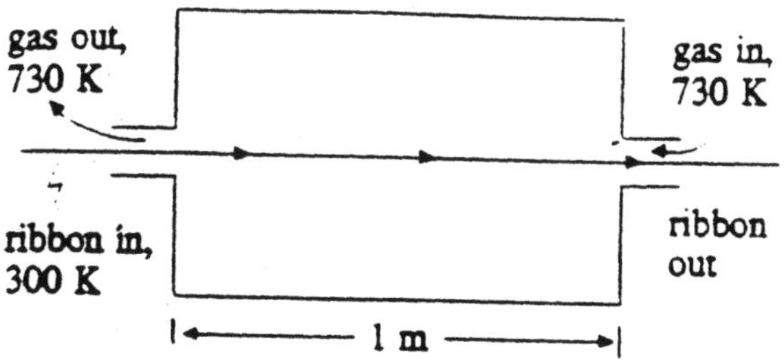

	ρ, kg m^{-3}	C_p, J kg^{-1} K^{-1}	k, W m^{-1} K^{-1}
Gas	0.64	2000	0.04
Ribbon	6400	500	30

$V_\infty = V_g(\text{gas}) + V_r(\text{ribbon}) = 30 + 0.17 = 30.17\ m\,s^{-1} \approx 30\ m\,s^{-1}$

Assume $Pr = 0.7$; $Pr = \frac{\nu}{\alpha} = \frac{\eta \rho C_p}{\rho k} = \frac{\eta C_p}{k}$

$$\eta = \frac{k Pr}{C_p} = \frac{0.04\ W}{m\,K}\,\Big|\,0.7\,\Big|\,\frac{kg\,K}{2000\ J}\,\Big|\,\frac{J}{s\,W}\,\Big|\,\frac{N\,s^2}{kg\,m} = 1.4\times10^{-5}\ N\,s\,m^{-2}$$

$$Re_L = \frac{L V_\infty \rho}{\eta} = \frac{(1)(30)(0.64)}{1.4\times10^{-5}} = 1.37\times10^{6}$$

Energy balance on ribbon

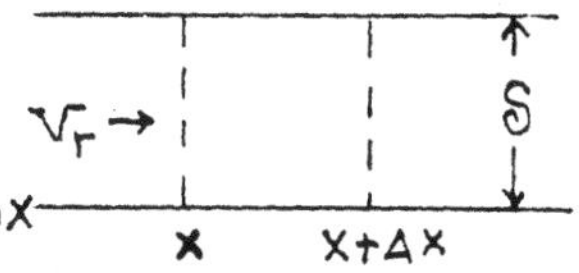

$$hW\Delta x(T_{gas} - T) + V_r W \delta \rho c_p T\big|_x = V_r W \delta \rho c_p T\big|_{x+\Delta x}$$

$$\lim_{\Delta x \to 0} \frac{dT}{dx} = -\frac{h}{V_r \rho C_p \delta}(T - T_{gas})$$

$$\int_{T_0}^{T_L} \frac{dT}{T - T_{gas}} = -\frac{h}{V_r \rho C_p \delta}\int_{x=0}^{L} dx\ ;\ \ln\frac{T_L - T_{gas}}{T_0 - T_{gas}} = -\frac{hL}{V_r \rho C_p \delta}$$

$$\frac{T_L - T}{T_0 - T_{gas}} = \exp\left(-\frac{hL}{V_r \rho C_p \delta}\right)$$

The following computer program solves the problem.

```
10 'Poirier and Geiger - Problem 8.6      SI units
20  L = 1 : W = .15 : THICK = .001 : VR = .17  'length, width, thickness and
      velocity of the ribbon
30  TO = 300 : TGAS = 730 : V = 30.17       'temperature of ribbon at entrance,
              temperature of the gas, and relative velocity
40 'Collect properties for gas
50   PR = .7 : VISC = .000014 : DENS = .64 : COND = .04 : CP = 2000
60 'Collect properties for ribbon
70   DENSR = 6400 : CONDR = 30 : CPR = 500
80 '
90   TOTAREA = 0 : DELX = .01    'TOTAREA and DELX are used to integrate hx
100  FOR X = DELX TO 1 STEP DELX
110     REX = X*V*DENS/VISC        'local Reynolds number
120     IF REX > 500000! THEN 150
130     NUX = .332 * PR^(.343) * SQR(REX)  'local Nusselt no. for laminar flow
140     GOTO 160
150     NUX = .0296 * REX^.8 * PR^(1/3)  'local Nusselt no. for turbulent flow
160     HX = NUX*COND/X              'local heat transfer coefficient
170     SUBAREA = HX*DELX : TOTAREA = TOTAREA + SUBAREA  'setup for averaging h
180  NEXT X
190    HAVG = TOTAREA/L            'average heat transfer coefficient
200    LPRINT "  The avg. heat transfer coefficient is";HAVG;"W/(m2 K)"
210    A = HAVG*L/( VR*DENSR*CPR*THICK )
220    RATIO = EXP(-A) : TL = (TO - TGAS)*RATIO + TGAS
230    LPRINT "  The temperature of the ribbon as it leaves is";TL;"K"
240    END
```

```
The avg. heat transfer coefficient is 74.44023 W/(m2 K)
The temperature of the ribbon as it leaves is 354.9923 K
```

8.7 In order to reduce the amount of dissolved hydrogen, dry helium is bubbled through molten copper. In order to prevent clogging of the submerged end of the lance, the helium must enter the melt at a temperature greater than the freezing point of copper (1356 K). What is the maximum mass flow rate of helium which can be employed if the supply of helium is at 300 K?

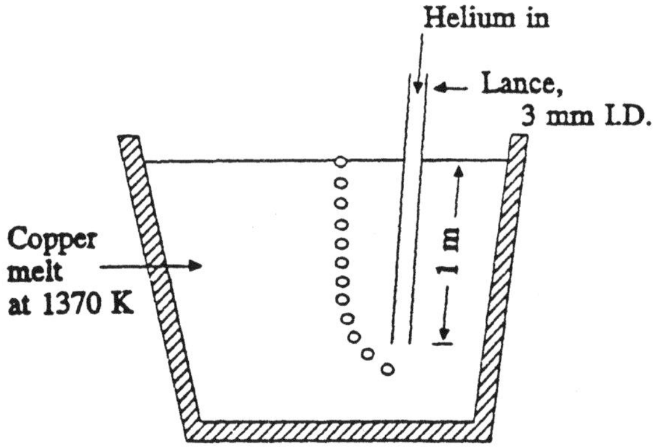

$$\text{Average temperature of He} = \frac{1357 + 300}{2} = 828.5\ \text{K}$$

The following program calculates the maximum flow rate of He.

```
10  Poirier and Geiger - Problem 8.7      SI units
20 'Collect properties for helium
30   PR = .654 : VISC = 3.591E-05: DENS = .0608 : COND = .285 : CP = 5193
40   L = 1 : D = .003 : PI = 3.1416      'lance dimensions
50   TMELT = 1370 : TM = 1356             'melt T and freezing point
60   DELTO = 1370 - 300                  'delta T at the entrance of the lance
70   DELTL = 1370 - 1356               'desired delta T at the exit of the lance
80  FOR I = 1 TO 100
90        W = I*.000001                    'mass flow rate
100       RE = 4*W/( PI*D*VISC )          'Reynolds number
110       Y = .074/SQR(RE)                'approximation of curve for L/D = 333
                                           and laminar flow in Fig. 8.2
120       NU = Y*RE*PR^(1/3)              'Nusselt no.
130       H = NU*COND/D                   'heat transfer coefficient
140       LEFT = - H*L*PI*D/(W*CP)
150       DELTT = DELTO*EXP(LEFT)         'delta T at exit for this flow rate
160       TL = TMELT - DELTT              'T of He at exit
170       IF TL < (TM + 2 )THEN 190
180 NEXT I
190 LPRINT " The mass flow rate is";W;"kg/s, and the gas leaving is at";TL;"K."
200 END
```

The mass flow rate is .000072 kg/s, and the gas leaving is at 1357.958 K.

8.8 Calculate the initial rate of losing energy (W) of an aluminum plate (1.2 m × 1.2 m × 10 mm) and heated uniformly to 370 K when it is a) cooled in a horizontal position by a stream of air at 290 K with a velocity of 2 m s^{-1}; b) suspended vertically in stagnant air at the same temperature.

a. $T_m = (T_s - T_f) = \frac{(370+290)\,K}{2} = 330\,K$

Thermal properties of air at 330 K

$k = 2.94\times10^{-2}\ W\,m^{-1}K^{-1};\ \eta = 1.96\times10^{-5}\ N\,s\,m^{-2};\ \rho = 1.11\ kg\,m^{-3};\ C_p = 1034\ J\,kg^{-1}K^{-1}$

$$Re_L = \frac{L\bar{V}\rho}{\eta} = \frac{1.2\,m}{} \left|\frac{2\,m}{s}\right| \frac{1.11\,kg}{m^3} \left| \frac{s\,m^2}{1.96\times10^{-5}\,N} \right| \frac{N\,s^2}{kg\,m} = 1.36\times10^5$$

$$Pr = \frac{\eta C_p}{k} = \frac{1.96\times10^{-5}\,N\,s}{m^2} \left| \frac{1034\,J}{kg\,K} \right| \frac{m\,K}{2.94\times10^{-2}\,W} \left| \frac{kg\,m}{N\,s^2} \right| \frac{W\,s}{J} = 0.690$$

Eq. (7.27) $Nu_L = 0.664\,Pr^{0.343}\,Re_L^{\frac{1}{2}} = (0.664)(0.690)^{0.343}(1.36\times10^5)^{\frac{1}{2}} = 216$

$$h = \frac{Nu\,k}{L} = \frac{216}{1.2\,m} \left| \frac{2.94\times10^{-2}\,W}{m\,K} \right. = 5.28\ W\,m^{-2}K^{-1}$$

$$Q = qA = hA\Delta T = \frac{5.28\,W}{m^2\,K} \left| (1.2)^2 m^2 \right| (370-290)\,K = 608\ W \text{ per side; } 1216\ W \text{ total}$$

b. $Gr_L = L^3 \rho^2 g \beta (T_c - T_\infty)(\eta^2)^{-1}$; $\beta = \frac{1}{T} = \frac{1}{330 K} = 3.03 \times 10^{-3} K^{-1}$

$$Gr_L = \frac{(1.2)^3 m^3}{} \left| \frac{(1.11)^2 kg^2}{m^6} \right| \frac{9.807 m}{s^2} \left| \frac{3.03 \times 10^{-3}}{K} \right| \frac{(370-290) K}{} \left| \frac{m^4}{(1.96 \times 10^{-5})^2 N^2 s^2} \right|$$

$$\frac{N^2 s^4}{kg^2 m^2} = 1.32 \times 10^{10}; \quad Gr_L Pr = (1.32 \times 10^{10})(0.690)$$

$\log_{10}(Gr_L Pr) = 9.96 \Rightarrow \log_{10} Nu_L = 2.15$ (Fig. 8.8); $Nu_L = 141$

$$h = \frac{Nu_L k}{L} = \frac{141}{1.2 m} \left| \frac{2.94 \times 10^{-2} W}{m K} \right. = 3.46 \; W m^{-2} K^{-1}$$

$$Q = h A \Delta T = \frac{3.46 W}{m^2 K} \left| (1.2)^2 m^2 \right| (370-290) K = 399 \text{ W per side}; \; Q = 798 \text{ W total}$$

8.9 Oil flows in a long horizontal 50 mm I.D. copper tube at an average velocity of 3 m s^{-1}. If the oil has a bulk temperature of 370 K and the air surrounding the tube is at 290 K, calculate a) the "liquid-side" heat transfer coefficient; b) the "vapor-side" heat transfer coefficient; c) the temperature of the copper tube; d) the rate of heat transfer to the air. *Data for oil*:

T (K)	ρ (kg m^{-3})	C_p (kJ kg^{-1} K^{-1})	$\eta \cdot 10^2$ (N s m^{-2})	$\nu \cdot 10^6$ (m^2 s^{-1})	$k \cdot 10^3$ (W m^{-1} K^{-1})	$\alpha \cdot 10^7$ (m^2 s^{-1})	Pr	$\beta \cdot 10^3$ (K^{-1})
290	890.0	1.868	99.9	1120	145	0.872	12 900	0.70
300	884.1	1.909	48.6	550	145	0.859	6400	0.70
320	871.8	1.993	14.1	161	143	0.823	1965	0.70
340	859.9	2.076	5.31	61.7	139	0.779	793	0.70
360	847.8	2.161	2.52	29.7	138	0.753	395	0.70
370	841.8	2.206	1.86	22.0	137	0.738	300	0.70

Since the properties of both the oil and air depend on temperature, we must estimate the temperature of the Cu. On the inside heat transfer is by forced convection of a liquid, whereas on the outside there is natural convection by a gas. Hence, h (inside) >> h (outside) So we can expect most of the temperature drop to be between the Cu tube and the air. To start the calculation, let's assume $T_{Cu} \simeq T_{oil} = 370 K$.

a. Properties of oil at 370 K: $\nu = 22.0 \times 10^{-6}\ m^2\ s^{-1}$; $Pr = 300$; $\eta = 1.86 \times 10^{-2}\ N\ s\ m^{-2}$;

$k = 137 \times 10^{-3}\ W\ m^{-1} K^{-1}$

$$Re_m = \frac{D\bar{V}}{\nu_m} = \frac{(0.050)(3)}{22.0 \times 10^{-6}} = 6820 \quad \text{Apply Fig. 8.2}$$

$$Nu_m = 0.004\ Re_m\ Pr_m^{1/3} \left(\frac{\eta_m}{\eta_o}\right)^{0.14} ; \quad \left(\frac{\eta_m}{\eta_o}\right) \simeq 1$$

$$Nu_m = (0.004)(6820)(300^{1/3})(1) = 182$$

$$h = 182 \frac{k}{D} = (182) \frac{(137 \times 10^{-3})}{(0.050)} = 499\ W\ m^{-2} K^{-1}$$

b. For the air side, the film temperature for property evaluation is:

$$T_f = \frac{1}{2}(370 + 290) = 330\ K$$

Properties of air at 330 K: $\nu = 19.05 \times 10^{-6}\ m^2 s^{-1}$; $\beta = \frac{1}{330} = 3.030 \times 10^{-3}\ K^{-1}$;

$Pr = 0.702$; $k = 24.0 \times 10^{-3}\ W\ m^{-1} K^{-1}$

we also need characteristic length.

Assume wall thickness of tubing is 3 mm; the outside diameter is

0.056 m. $L = \frac{\pi D}{2} = \frac{(\pi)(0.056)}{2} = 0.0880\ m$.

$$Gr_L = \frac{g\beta(T_s - T_\infty)L^3}{\nu^2} = \frac{(9.807)(3.030 \times 10^{-3})(370-290)(0.0880^3)}{(19.05 \times 10^{-6})^2} = 4.46 \times 10^6$$

$Gr_L Pr = 3.13 \times 10^6$ From Fig. 8.8: $\log_{10} Nu_L \simeq 1.23$; $Nu_L = 17.0$

$$h = 17.0 \frac{k}{L} = \frac{(17.0)(24.0 \times 10^{-3})}{(0.0880)} = 4.64\ W\ m^{-2} K^{-1}$$

c. Let i and o be subscript for inside and outside of the tube, respectively.

$$h_o \pi D_o (T_{cu} - T_o) = h_i \pi D_i (T_i - T_{cu}) : \quad T_{cu} = \frac{h_o D_o T_o + h_i D_i T_i}{h_o D_o + h_i D_i}$$

$$= \frac{(4.64)(0.056)(290) + (499)(0.050)(370)}{(4.64)(0.056) + (499)(0.050)} = 369.2\ K$$

The initial assumption that $T_{cu} \simeq 370\ K$ is valid, so that the estimates of the heat transfer coefficients are very good.

d. $Q = h_o A_o (T_s - T_o)$; ℓ = length of tube

$$\frac{Q}{\ell} = (4.64)\left(\frac{\pi}{4}\right)(0.056^2)(370 - 290) = 0.914\ W\ s^{-1} m^{-1}$$

8.10 Replace the oil with sodium and repeat Problem 8.9. Compare or contrast the results of the two problems.

Assume the copper temperature is 376 K. The heat transfer coefficient in the liquid Na is very high, so we need not estimate it in order to calculate the rate of heat loss. The solutions to parts b, c, and d are the same as given for Problem 8.9.

8.11 A heat treating furnace is 6 m long, 3 m wide and 6 m high. If a check with thermocouples indicates that the average wall temperature is 340 K and the top is 365 K, calculate the heat loss from the furnace in W. A quick estimate can be made by using the simplified equations given in Problem 8.14.

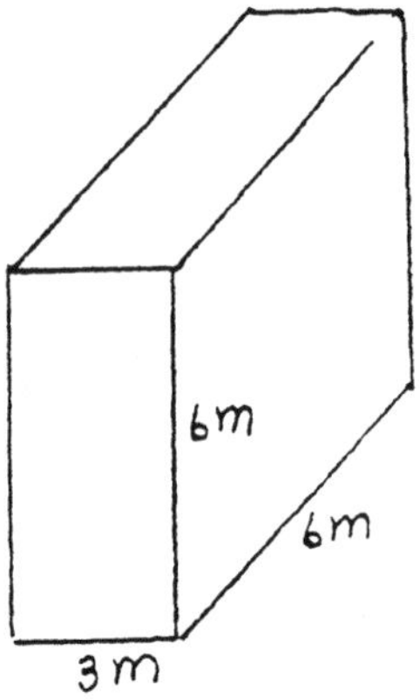

Assume $T_\infty = 300\,K$

Top $- L = \frac{A}{P} = \frac{18\,m^2}{18\,m} = 1\,m$

$L^3 \Delta T = (1)^3 m^3 (365-300) K = 65\ m^3 K$

$h = 1.52\,(65)^{1/3} = 6.11\ W\,m^{-2}K^{-1}$

$Q_{Top} = hA(T-T_\infty) = (6.11)(18)(365-300) = 7.15 \times 10^3\ W$

Sides $- L^3 \Delta T = 6^3(340-300) = 8.64 \times 10^3$

$h = 1.45\,(40)^{1/3} = 4.96\ W\,m^{-2}K^{-1}$

$Q_{sides} = (4.96)(18+18+36+36)(340-300)$

$Q_{sides} = 21.42 \times 10^3\ W$

$Q_{Total} = Q_{Top} + Q_{sides} = 28.6 \times 10^4\ W \quad (Q_{Bottom} \simeq 0)$

8.12 In flow past a flat plate, a laminar boundary layer exists over the forward portion between 0 and L_{tr}, and the turbulent boundary layer exists beyond L_{tr}. With this model, the average h over a plate of length L (with $L > L_{tr}$) can be determined as indicated

$$h = \frac{1}{L}\left[\int_0^{L_{tr}} h_{x\,(\mathrm{lam})}\,dx + \int_{L_{tr}}^{L} h_{x\,(\mathrm{turb})}\,dx\right]. \qquad (1)$$

Take $\mathrm{Re}_{tr} = 3.2 \times 10^5$ and show that

$$\frac{hL}{k} = 0.037\ \mathrm{Pr}^{1/3}\left(\mathrm{Re}^{0.8} - 15\ 500\right).$$

[*Hint*: $L_{tr}/L = \mathrm{Re}_{tr}/\mathrm{Re}_L$.]

Eq. (7.25) – $h_{x(\mathrm{lam})} = 0.332\, k\, Pr^{0.343}\left(\frac{V_\infty}{\nu}\right)^{1/2} x^{-1/2}$

The first integral becomes:

$$I_1 = 0.332\, k\, Pr^{0.343}\left(\frac{V_\infty}{\nu}\right)^{1/2}\int_0^{L_{tr}} x^{-1/2}dx = 0.664\, k\, Pr^{0.343}\left(\frac{V_\infty L_{tr}}{\nu}\right)^{1/2} \qquad (2)$$

Eq. (8.15) – $h_{x(\mathrm{turb})} = 0.0296\, k\, Re_x^{0.8}\, Pr^{1/3}\, x^{-1}$

$$= 0.0296\, k\, Pr^{1/3}\left(\frac{V_\infty}{\nu}\right) x^{-0.2}$$

The second integral becomes:

$$I_2 = 0.0296\, k\, Pr^{1/3}\left(\frac{V_\infty}{\nu}\right)^{0.8}\int_{L_{tr}}^{L} x^{-0.2}dx = 0.0370\, k\, Pr^{1/3}\left(\frac{V_\infty}{\nu}\right)^{0.8}\left(L^{0.8} - L_{tr}^{0.8}\right)$$

$$= 0.0370\, k\, Pr^{1/3}\left(\frac{L V_\infty}{\nu}\right)^{0.8}\left[1 - \left(\frac{L_{tr}}{L}\right)^{0.8}\right] \qquad (3)$$

Combine Eqs. (1), (2), (3) and approx. $Pr^{0.343} = Pr^{\frac{1}{3}}$

$$Nu_L = \frac{hL}{k} = Pr^{\frac{1}{3}}\left[0.664\, Re_{tr} + 0.0370\, Re_L^{0.8}\left[1 - \left(\frac{L_{tr}}{L}\right)^{0.8}\right]\right] ;\ \frac{L_{tr}}{L} = \frac{Re_{tr}}{Re_L}$$

$$Nu_L = Pr^{\frac{1}{3}}\left[0.037\left[Re_L^{0.8} - Re_{tr}^{0.8}\right] + 0.664\, Re_{tr}^{0.5}\right]$$

$Re_{tr} = 3.2 \times 10^5$

$$Nu_L = Pr^{\frac{1}{3}}\left[0.037\, Re_L^{0.8} + 0.664\,(3.2\times10^5)^{1/2} - 0.037\,(3.2\times10^5)^{0.8}\right]$$

$$Nu_L = 0.037\, Pr^{\frac{1}{3}}\left[Re_L^{0.8} + \frac{0.664}{0.037}(3.2\times10^5)^{1/2} - (3.2\times10^5)^{0.8}\right]$$

$$Nu_L = 0.037\, Pr^{1/3}\left[Re_L^{0.8} - 15\ 500\right]$$

8.13 A steel plate, 2.5 m by 2.5 m by 2.5 mm, is removed from an oven at 430 K and hung horizontally in a laboratory at 295 K. a) Calculate the initial heat loss (W) from the steel. b) Repeat if the plate is hung vertically.

a. Horizontal position (top and bottom surfaces)

$$T_f = \frac{1}{2}(430+295)\,K = 362\,K$$

Properties of air (362 K): $\rho_f = 0.981\ kg\ m^3$; $B_f \simeq 2.76\times10^{-3}K^{-1}$; $\eta_f = 2.13\times10^{-5}N\ s\ m^{-2}$;

$k_f = 3.1\times10^{-2}W\ m^{-1}K^{-1}$; $Pr = 0.696$

$$Gr_L = \frac{L^3\rho_f g B_f(T_o-T_\infty)}{\eta_f^2} = \frac{(2.5)^3(0.981)(9.807)(2.76\times10^{-3})(430-295)}{(2.13\times10^{-5})^2} = 1.23\times10^{11}$$

Assume Eq. (8.25) applies for the top surface and Eq. (8.27) for the bottom.

Top – $Nu_L = 0.14(Gr_L \cdot Pr)^{1/3} = 0.14\left[(1.23\times10^{11})(0.696)\right]^{1/3} = 618$

$$h_T = 618\frac{k_f}{L} = \frac{(618)(3.1\times10^{-2})}{2.5} = 7.66\ W\,m^{-2}K^{-1}$$

Bottom – $Nu_L = 0.27(Gr_L \cdot Pr)^{\frac{1}{4}} = 0.27\left[(1.23\times10^{11})(0.696)\right]^{\frac{1}{4}} = 146$

$$h_B = \frac{(146)(3.1\times10^{-2})}{2.5} = 1.81\ W\,m^{-2}K^{-1}$$

$$Q = (h_T+h_B)A(T_o-T_\infty) = (7.66+1.81)(2.5)^2(430-295) = 7.99\times10^3\,W$$

b. Vertical position – Eq. (8.20)

$$Nu_L = 0.0246\,Gr_L^{2/5}\,Pr^{7/15}(1+0.494\,Pr^{2/3})^{-2/5}$$

$$= 0.0246(1.23\times10^{11})^{2/5}(0.696)^{7/15}\left[1+(0.494)(0.696)^{2/3}\right]^{-2/5} = 497$$

$$h = \frac{(497)(3.1\times10^{-2})}{2.5} = 6.16\ W\,m^{-2}K^{-1}\ \text{per side}$$

$$Q = (6.16+6.16)(2.5)^2(430-295) = 10.40\times10^3\,W$$

8.14 Repeat Problem 8.13 using one of the following simplified equations, which apply reasonably well to air, CO, N_2, and O_2 in the range 310-1090 K. L and D are in m, ΔT in K, and h in W m^{-2} K^{-1}. $L = A/P$ with A = surface area and P = perimeter of the surface.

Vertical plates of length L:

$h = 1.42\,(\Delta T/L)^{1/4}$, $\quad 5 \times 10^{-6} < L^3\Delta T < 50$

$h = 1.45\,(\Delta T)^{1/3}$, $\quad 50 < L^3\Delta T < 5 \times 10^4$.

Horizontal pipes of diameter D:

$h = 1.22\,(\Delta T/D)^{1/4}$, $\quad 5 \times 10^{-6} < D^3\Delta T < 50$

$h = 1.24\,(\Delta T)^{1/3}$, $\quad 50 < D^3\Delta T < 5 \times 10^4$.

Horizontal plate,* hot surface up or cold surface down:

$h = 1.32\,(\Delta T/L)^{1/4}$, $\quad 5 \times 10^{-5} < L^3\Delta T < 0.30$

$h = 1.52\,(\Delta T)^{1/3}$, $\quad 0.30 < L^3\Delta T < 470$.

Horizontal plate,* hot surface down or cold surface up:

$h = 0.59\,(\Delta T/L)^{1/4}$, $\quad 0.005 < L^3\Delta T < 470$.

Are the results for the heat transfer coefficients within 20 pct of the results using the more complete correlations (as in Problem 8.13)?

$$L^3\Delta T = (2.5)^3\,m^3\,(430-295)\,K = 2.11\times10^3\ m^3\,K$$

a. $h_T = 1.52\,(430-295)^{\frac{1}{3}} = 7.80\ W\,m^{-2}K^{-1}$ (Top)

$h_B = 0.59\left[\dfrac{(430-295)}{2.5}\right]^{\frac{1}{4}} = 1.60\ W\,m^{-2}K^{-1}$ (Bottom)

$$Q = (h_T + h_B)A(T_0 - T_\infty) = (7.80+1.60)(2.5)^2(430-295) = 7.93\times10^3\ W$$

This is only 0.75 % less than the solution to Prob. 8.13

Yes, the results are within 20 % of the results of Prob. 8.13.

b. $h = 1.45\,(430-295)^{\frac{1}{3}} = 7.44\ W\,m^{-2}K^{-1}$

$Q = (7.44+7.44)(2.5)^2(430-295) = 12.55\times10^3\ W$

This is 20.7 % greater than the solution to Prob. 8.13.

No, the results are not within 20 % of the results of Prob. 8.13, but probably close enough for making rough estimates in many circumstances.

8.15 Two fluids are separated by a solid with a thickness of 10 mm and a thermal conductivity of 22 W m^{-1} K^{-1}. For each of the following scenarios, estimate the heat transfer coefficient on each side of the solid and the surface temperatures of the solid.

	Fluid	Free stream velocity, m s^{-1}	Free stream temperature, K
a)	A air	15.2	300
	B air	0	650
b)	A air	0	300
	B sodium	0	650
c)	A air	15.2	300
	B sodium	0	650
d)	A sodium	15.2	365
	B sodium	0	650

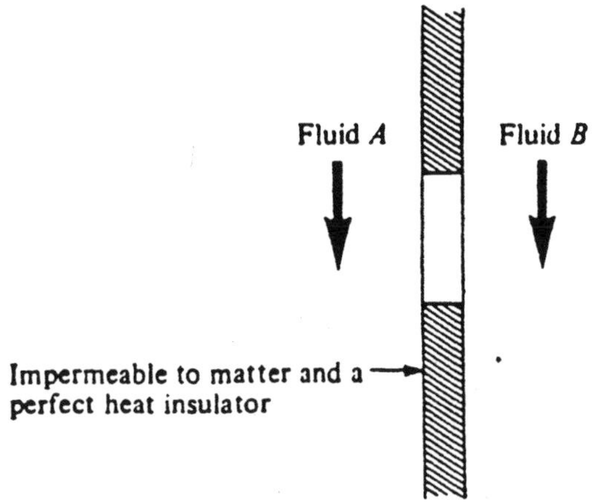

Properties of both air and sodium depend on temperature, so this is a difficult problem to solve exactly. The solid, itself, has a low thermal resistance. So for parts a)-c) that involve air as at least one of the fluids, we can assume that the temperature is uniform in the solid. With the aid of the following computer codes, the fluxes on both sides can be determined as a function of the temperature of the solid. When the fluxes are equal, the temperature of the solid is determined.

In part (d), however, the heat transfer coefficients on both sides of the solid are very high. Thus, we can assume that the surface temperatures are equal to the temperatures of the sodium.

At steady state, we can show that

$$T_B' - T_A' = (T_B - T_A)\left[\frac{k}{h_A\delta} + \frac{k}{h_B\delta} + 1\right]^{-1}$$

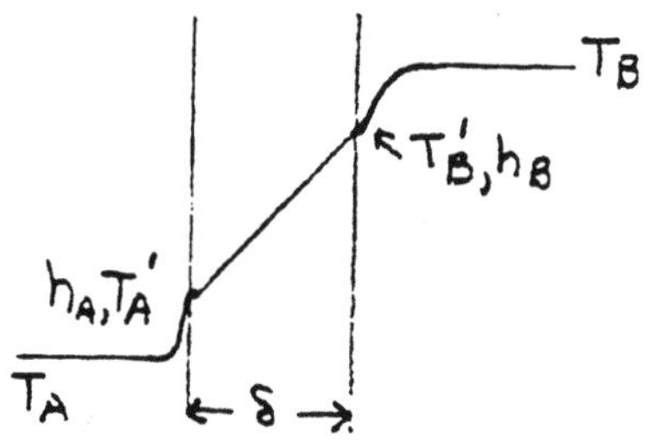

We make use of this equation for parts a)-d).

```
10 'Problem 8.15a
20   L = 1 : G = 9.807        'length of solid and gravitational acceleration
30   TINFF = 650                    'natural convection side
40   V = 15.2   :  TINF = 300      'forced convection side
50  LPRINT "  Solid   ------F L U X E S--------  -C O E F F I C I E N T S-"
60  LPRINT "  Temp.   Forced side  Natural side  Forced side  Natural side"
70  LPRINT "    K        W/m^2        W/m^2       W/(m^2 K)    W/(m^2 K)"
80  LPRINT "  ******  ***********  ************  ***********  ************"
90      FOR TS = 330 TO 340 STEP .5     'TS is temperature of the solid.
100       'forced convection side
110        TF = .5*( TS + TINF )          'film temp. for evaluating properties
120        T = TF : GOSUB 340             'retrieves properties of air
130        RE = L*V/NETA                  'Reynolds no.
140        IF RE < 500000! THEN 170
150        NU = .0347*RE^.8*PR^(1/3)      'Eq. (8.17) put on avg. basis for L
160        GOTO 180
170        NU = .664*PR^.343*SQR(RE)      'Eq. (7.27)
180        H = NU*K/L                     'H in W/(m2 K) for forced conv. side
190        DELT = ABS(TS - TINF) : FLUX = H*DELT    'flux on forced conv. side
200       'natural convection side
210        TFF = .5*( TS + TINFF )         'film temp. on natural convection side
220        DELT = ABS(TS - TINFF)          'delta T on natural convection side
230        T = TFF : GOSUB 340             'retrieves properties
240        GR = G*BETA*(DELT)*L^3/NETA^2 : RA = GR*PR  'Grashof and Rayleigh
250        IF RA < 1E+09 THEN 280
260        NU = .0246*GR^(2/5)*PR^(7/15)*(1+.494*PR^(2/3))^(-2/5)  'Nu turbul.
270        GOTO 290
280        NU = .56*(RA)^(1/4)             'Nusselt no. for laminar flow
290        HH = NU*K/L                     'H in W/(m2 K) for natural conv. side
300        FLUXX = HH*DELT                 'flux on natural conv. side
310        LPRINT USING "  ###.#    ##.##^^^^    ##.##^^^^    ##.##^^^^     ##.#
#^^^^";TS,FLUX,FLUXX,H,HH
320      NEXT TS
330   END
340 'SUBROUTINE for thermophysical properties of air, 300 < T < 700 K.
350    NETA = -6.40972 + .0499857*T + 8.07143E-05*T*T
360    NETA = NETA/1000000!                 'kinematic viscosity
370    K = -.35143 + .098871*T - 3.3571E-05*T*T
380    K = K/1000!                          'thermal conductivity
390    PR = .8069 - 4.6471E-04*T + 4.357141E-07*T*T    'Prandtl no.
400    BETA = 1/T
410    RETURN
```

Solid	------F L U X E S--------		-C O E F F I C I E N T S-	
Temp.	Forced side	Natural side	Forced side	Natural side
K	W/m^2	W/m^2	W/(m^2 K)	W/(m^2 K)
******	***********	************	***********	************
330.0	1.44E+03	1.69E+03	4.81E+01	5.29E+00
330.5	1.47E+03	1.69E+03	4.81E+01	5.29E+00
331.0	1.49E+03	1.68E+03	4.81E+01	5.28E+00
331.5	1.51E+03	1.68E+03	4.81E+01	5.28E+00
332.0	1.54E+03	1.68E+03	4.80E+01	5.27E+00
332.5	1.56E+03	1.67E+03	4.80E+01	5.26E+00
333.0	1.58E+03	1.67E+03	4.80E+01	5.26E+00
333.5	1.61E+03	1.66E+03	4.80E+01	5.25E+00
334.0	1.63E+03	1.66E+03	4.79E+01	5.25E+00
334.5	1.65E+03	1.65E+03	4.79E+01	5.24E+00 ←
335.0	1.68E+03	1.65E+03	4.79E+01	5.23E+00
335.5	1.70E+03	1.64E+03	4.79E+01	5.23E+00
336.0	1.72E+03	1.64E+03	4.78E+01	5.22E+00
336.5	1.75E+03	1.64E+03	4.78E+01	5.22E+00
337.0	1.77E+03	1.63E+03	4.78E+01	5.21E+00
337.5	1.79E+03	1.63E+03	4.78E+01	5.20E+00
338.0	1.81E+03	1.62E+03	4.78E+01	5.20E+00
338.5	1.84E+03	1.62E+03	4.77E+01	5.19E+00
339.0	1.86E+03	1.61E+03	4.77E+01	5.19E+00
339.5	1.88E+03	1.61E+03	4.77E+01	5.18E+00
340.0	1.91E+03	1.60E+03	4.77E+01	5.18E+00

$h_A = 47.9\ \mathrm{W\ m^{-2}\ K^{-1}}$; $h_B = 52.4\ \mathrm{W\ m^{-2}\ K^{-1}}$

$$T_B' - T_A' = (650-300)\left[\frac{22}{(47.9)(0.010)} + \frac{22}{(5.24)(0.010)} + 1\right]^{-1}$$

$T_B' - T_A' = 0.75\ \mathrm{K}$

Hardly any difference between T_B' and T_A'

$T_A' \cong T_B' = 334.5\ \mathrm{K}$

```
10 'Problem 8.15b
20  L = 1 : G = 9.807        'length of solid and gravitational acceleration
30  LPRINT "  Solid   ----F L U X E S----    --C O E F I C I E N T S--"
40  LPRINT "  Temp.   Air side   Na side       Air side    Na side"
50  LPRINT "    K      W/m^2      W/m^2        W/(m^2 K)  W/(m^2 K)"
60  LPRINT "  ******  ********  *********      *********  *********"
70  TINF = 650                     'sodium convection side
80  TINFF = 300                     'air convection side
90    FOR TS = 648 TO 650 STEP .1    'TS is temperature of the solid.
100      'air convection side
110      TFF = .5*( TS + TINFF )      'film temp. on air side
120      DELT = ABS(TS - TINFF)       'delta T on air side
130      T = TFF : GOSUB 350          'retrieves properties
140      GR = G*BETA*(DELT)*L^3/NETA^2 : RA = GR*PR  'Grashof and Rayleigh
150      IF RA < 1E+09 THEN 180
160      NU = .0246*GR^(2/5)*PR^(7/15)*(1+.494*PR^(2/3))^(-2/5)  'Nu turbul.
170      GOTO 190
180      NU = .56*(RA)^(1/4)          'Nusselt no. for laminar flow
190      HH = NU*K/L                  'H in W/(m2 K) for air side
200      FLUXX = HH*DELT              'flux on air side
210      'sodium convection side
220      TF = .5*( TS + TINF )       'film temp. on sodium side
230      DELT = ABS(TS - TINF)        'delta T on sodium side
240      T = TF  : GOSUB 430          'retrieves properties
250      GR = G*BETA*(DELT)*L^3/NETA^2 : RA = GR*PR  'Grashof and Rayleigh
260      IF RA < 1E+09 THEN 290
270      NU = .0246*GR^(2/5)*PR^(7/15)*(1+.494*PR^(2/3))^(-2/5)  'Nu turbul.
280      GOTO 300
290      NU = .56*(RA)^(1/4)          'Nusselt no. for laminar flow
300      H = NU*K/L                   'H in W/(m2 K) on sodium side
310      FLUX = H*DELT               'flux on sodium side
320      LPRINT USING "  ###.#  ##.##^^^^  ##.##^^^^      ##.##^^^^  ##.##^^^^
^";TS,FLUXX,FLUX,HH,H
330    NEXT TS
340  END
350 'SUBROUTINE for thermophysical properties of air, 300 < T < 700 K.
360   NETA = -6.40972 + .0499857*T + 8.07143E-05*T*T
370   NETA = NETA/1000000!              'kinematic viscosity
380   K = -.35143 + .098871*T - 3.3571E-05*T*T
390   K = K/1000!                       'thermal conductivity
400   PR = .8069 - 4.6471E-04*T + 4.357141E-07*T*T  'Prandtl no.
410   BETA = 1/T
420   RETURN
430 'SUBROUTINE for thermophysical properties of sodium, 366 < T < 977. Taken
            from Incropera and DeWitt, p. A24.
440   NETA = 119.1055 - 42.108963#*T^.25 + 3.794418*T^.5
450   NETA = NETA*.0000001              'kinematic viscosity
460   K = 109.2 - .0701041*T + 1.99052E-05*T^2   'thermal conductivity
470   PR = .165032 - .0580746*T^.25 + 5.22601E-03*T^.5  'Prandtl no.
480   BETA = 7.21E-08*T + .0002406
490   RETURN
```

Solid Temp. K	----F L U X E S---- Air side W/m^2	Na side W/m^2	--C O E F F I C I E N T S-- Air side W/(m^2 K)	Na side W/(m^2 K)
******	********	*********	*********	*********
648.0	1.97E+03	1.04E+04	5.65E+00	5.19E+03
648.1	1.97E+03	9.73E+03	5.65E+00	5.12E+03
648.2	1.97E+03	9.09E+03	5.65E+00	5.05E+03
648.3	1.97E+03	8.46E+03	5.65E+00	4.98E+03
648.4	1.97E+03	7.85E+03	5.65E+00	4.90E+03
648.5	1.97E+03	7.24E+03	5.65E+00	4.83E+03
648.6	1.97E+03	6.64E+03	5.65E+00	4.74E+03
648.7	1.97E+03	6.05E+03	5.65E+00	4.66E+03
648.8	1.97E+03	5.48E+03	5.65E+00	4.56E+03
648.9	1.97E+03	4.91E+03	5.65E+00	4.47E+03
649.0	1.97E+03	4.36E+03	5.65E+00	4.36E+03
649.1	1.97E+03	3.82E+03	5.65E+00	4.25E+03
649.2	1.97E+03	3.30E+03	5.65E+00	4.12E+03
649.3	1.97E+03	2.79E+03	5.65E+00	3.99E+03
649.4	1.97E+03	2.30E+03	5.65E+00	3.84E+03
649.5	1.97E+03	1.84E+03	5.65E+00	3.67E+03 ←
649.6	1.98E+03	1.39E+03	5.65E+00	3.47E+03
649.7	1.98E+03	9.70E+02	5.65E+00	3.23E+03
649.8	1.98E+03	5.85E+02	5.65E+00	2.92E+03
649.9	1.98E+03	2.47E+02	5.65E+00	2.45E+03
650.0	1.98E+03	3.17E-01	5.65E+00	6.48E+02

$h_A = 5.65 \ W\,m^{-2}K^{-1}$; $h_B = 3.67\times10^3 \ W\,m^{2}K^{-1}$

$$T_B' - T_A' = (650-300)\left[\frac{22}{(5.65)(0.010)} + \frac{22}{(3670)(0.010)} + 1\right]^{-1} = 0.90\,K$$

Again hardly any difference between the surface temperatures.

$T_B' \doteq 650\,K$; $T_A' \doteq 649\,K$

```
10 'Problem 8.15c
20   L = 1 : G = 9.807        'length of solid and gravitational acceleration
30   TINFF = 650                    'natural convection side, sodium
40   V = 15.2   :  TINF = 300       'forced convection side, air
50  LPRINT "  Solid    -----F L U X E S-----   -C O E F F I C I E N T S-"
60  LPRINT "  Temp.    Air side      Na side     Air side   Na side"
70  LPRINT "    K        W/m^2        W/m^2      W/(m^2 K)  W/(m^2 K)"
80  LPRINT "  #####    #########     #########  ########## ########## "
90      FOR TS = 648 TO 650 STEP .1  'TS is temperature of the solid.
100        'forced convection side, air
110         TF = .5*( TS + TINF )       'film temp. for evaluating properties
120         T = TF : GOSUB 340          'retrieves properties of air
130        RE = L*V/NETA               'Reynolds no.
140        IF RE < 500000! THEN 170
150        NU = .0347*RE^.8*PR^(1/3)   'Eq. (8.17) put on avg. basis for L
160        GOTO 180
170        NU = .664*PR^.343*SQR(RE)   'Eq. (7.27)
180        H = NU*K/L                  'H in W/(m2 K) for forced conv. side
190        DELT = ABS(TS - TINF) : FLUX = H*DELT    'flux on forced conv. side
200       'natural convection side, sodium
210        TFF = .5*( TS + TINFF )      'film temp. on natural convection side
220        DELT = ABS(TS - TINFF)       'delta T on natural convection side
230        T = TFF : GOSUB 420          'retrieves properties
240        GR = G*BETA*(DELT)*L^3/NETA^2 : RA = GR*PR  'Grashof and Rayleigh
250        IF RA < 1E+09 THEN 280
260        NU = .0246*GR^(2/5)*PR^(7/15)*(1+.494*PR^(2/3))^(-2/5)  'Nu turbul.
270        GOTO 290
280        NU = .56*(RA)^(1/4)          'Nusselt no. for laminar flow
290        HH = NU*K/L                  'H in W/(m2 K) for natural conv. side
300        FLUXX = HH*DELT              'flux on natural conv. side
310        LPRINT USING "  ###.#    ##.##^^^^     ##.##^^^^  ##.##^^^^  ##.##^^^^
^";TS,FLUX,FLUXX,H,HH
320      NEXT TS
330   END
340 'SUBROUTINE for thermophysical properties of air, 300 < T < 700 K.
350     NETA = -6.40972 + .0499857*T + 8.07143E-05*T*T
360     NETA = NETA/1000000!                 'kinematic viscosity
370     K = -.35143 + .0988714*T - 3.3571E-05*T*T
380     K = K/1000!                          'thermal conductivity
390     PR = .8069 - 4.6471E-04*T + 4.357141E-07*T*T   'Prandtl no.
400     BETA = 1/T
410     RETURN
420 'SUBROUTINE for thermophysical properties of sodium, 366 < T < 977. Taken
             from Incropera and DeWitt, p. A24.
430     NETA = 119.1055 - 42.10896*T^.25 + 3.794418*T^.5
440     NETA = NETA * .0000001              'kinematic viscosity
450     K = 109.2 - .0701041*T + 1.99052E-05*T^2    'thermal conductivity
460     PR = .165032 - .0580746*T^.25 + 5.22601E-03*T^.5  'Prandtl no.
470     BETA = 7.21E-08*T + .0002406
480     RETURN
```

Solid Temp. K	-----F L U X E S----- Air side W/m^2	Na side W/m^2	-C O E F I C I E N T S- Air side W/(m^2 K)	Na side W/(m^2 K)
648.0	5.18E+03	1.04E+04	1.49E+01	5.19E+03
648.1	5.18E+03	9.73E+03	1.49E+01	5.12E+03
648.2	5.18E+03	9.09E+03	1.49E+01	5.05E+03
648.3	5.18E+03	8.46E+03	1.49E+01	4.98E+03
648.4	5.19E+03	7.85E+03	1.49E+01	4.90E+03
648.5	5.19E+03	7.24E+03	1.49E+01	4.83E+03
648.6	5.19E+03	6.64E+03	1.49E+01	4.74E+03
648.7	5.19E+03	6.05E+03	1.49E+01	4.66E+03
648.8	5.19E+03	5.48E+03	1.49E+01	4.56E+03 ←
648.9	5.19E+03	4.91E+03	1.49E+01	4.47E+03
649.0	5.19E+03	4.36E+03	1.49E+01	4.36E+03
649.1	5.20E+03	3.82E+03	1.49E+01	4.25E+03
649.2	5.20E+03	3.30E+03	1.49E+01	4.12E+03
649.3	5.20E+03	2.79E+03	1.49E+01	3.99E+03
649.4	5.20E+03	2.30E+03	1.49E+01	3.84E+03
649.5	5.20E+03	1.84E+03	1.49E+01	3.67E+03
649.6	5.20E+03	1.39E+03	1.49E+01	3.47E+03
649.7	5.20E+03	9.70E+02	1.49E+01	3.23E+03
649.8	5.21E+03	5.85E+02	1.49E+01	2.92E+03
649.9	5.21E+03	2.47E+02	1.49E+01	2.45E+03
650.0	5.21E+03	3.17E-01	1.49E+01	6.48E+02

$h_A = 14.9\ W m^{2} K^{-1}; h_B = 4.51 \times 10^{3}\ W m^{-2} K^{-1}$

$$T_B' - T_A' = (650-300)\left[\frac{22}{(14.9)(0.010)} + \frac{22}{(4510)(0.010)} + 1\right]^{-1} = 2.35\ K$$

The difference is still small but noticeable because the air is subjett to forced convection.

$T_B' \simeq 650\ K;\ T_A' \simeq 647.8\ K$

```
110 'Problem 8.15d
20   L = 1 : G = 9.807          'length of solid and gravitational acceleration
30   TINFF = 650                     'natural convection side
40   V = 15.2   :  TINF = 365        'forced convection side
50  LPRINT "  Solid  ------F L U X E S------    ---C O E F I C I E N T S--"
60  LPRINT "  Temp.  Forced side  Natural side  Forced side   Natural side"
70  LPRINT "    K       W/m^2         W/m^2      W/(m^2 K)     W/(m^2 K)"
80  LPRINT "  #####  ###########  ############  ###########   ###########"
90       FOR TS = 448 TO 452 STEP .2     'TS is temperature of the solid.
100        'forced convection side
110         TF = .5*( TS + TINF )          'film temp. for evaluating properties
120         T = TF : GOSUB 330             'retrieves properties of sodium
130         RE = L*V/NETA                  'Reynolds no.
140        'The following equation is valid for laminar flow but used for all
            Re because one for turbulent flow could not be found.
150         TERM1 = 1.128*SQR(RE*PR) : TERM2 = ( 1+.9*SQR(PR) )
160         NU = TERM1/TERM2               'Eq. (7.30) averaged for length L
170         H = NU*K/L                     'H in W/(m2 K) for forced conv. side
180         DELT = ABS(TS - TINF) : FLUX = H*DELT     'flux on forced conv. side
190        'natural convection side, sodium
200         TFF = .5*( TS + TINFF )        'film temp. on natural convection side
210         DELT = ABS(TS - TINFF)         'delta T on natural convection side
220         T = TFF : GOSUB 330            'retrieves properties
230         GR = G*BETA*(DELT)*L^3/NETA^2 : RA = GR*PR  'Grashof and Rayleigh
240         IF RA < 1E+09 THEN 270
250         NU = .0246*GR^(2/5)*PR^(7/15)*(1+.494*PR^(2/3))^(-2/5)  'Nu turbul.
260         GOTO 280
270         NU = .56*RA^(1/4)              'Nusselt no. for laminar flow
280         HH = NU*K/L                    'H in W/(m2 K) for natural conv. side
290         FLUXX = HH*DELT                'flux on natural conv. side
300         LPRINT USING "  ###.#   ##.##^^^^    ##.##^^^^    ##.##^^^^    ##.#^^^
#^^^^";TS,FLUX,FLUXX,H,HH
310      NEXT TS
320    END
330 'SUBROUTINE for thermophysical properties of sodium, 366 < T < 977. Taken
          from Incropera and DeWitt, p. A24.
340    NETA = 119.1055 - 42.10896*T^.25 + 3.794418*T^.5
350    NETA = NETA * .0000001               'kinematic viscosity
360    K = 109.2 - .0701041*T + 1.99052E-05*T^2    'thermal conductivity
370    PR = .165032 - .0580746*T^.25 + 5.22601E-03*T^.5  'Prandtl no.
380    BETA = 7.21E-08*T + .0002406
390    RETURN
```

Solid	------F L U X E S------		---C O E F I C I E N T S--	
Temp.	Forced side	Natural side	Forced side	Natural side
K	W/m^2	W/m^2	W/(m^2 K)	W/(m^2 K)
#####	###########	############	###########	###########
448.0	3.42E+06	3.58E+06	4.12E+04	1.77E+04
448.2	3.43E+06	3.58E+06	4.12E+04	1.77E+04
448.4	3.44E+06	3.57E+06	4.12E+04	1.77E+04
448.6	3.45E+06	3.57E+06	4.12E+04	1.77E+04
448.8	3.45E+06	3.56E+06	4.12E+04	1.77E+04
449.0	3.46E+06	3.56E+06	4.12E+04	1.77E+04
449.2	3.47E+06	3.55E+06	4.12E+04	1.77E+04
449.4	3.48E+06	3.55E+06	4.12E+04	1.77E+04
449.6	3.49E+06	3.54E+06	4.12E+04	1.77E+04
449.8	3.49E+06	3.54E+06	4.12E+04	1.77E+04
450.0	3.50E+06	3.53E+06	4.12E+04	1.77E+04
450.2	3.51E+06	3.53E+06	4.12E+04	1.77E+04
450.4	3.52E+06	3.52E+06	4.12E+04	1.77E+04
450.6	3.53E+06	3.52E+06	4.12E+04	1.77E+04
450.8	3.53E+06	3.52E+06	4.12E+04	1.76E+04
451.0	3.54E+06	3.51E+06	4.12E+04	1.76E+04
451.2	3.55E+06	3.51E+06	4.12E+04	1.76E+04
451.4	3.56E+06	3.50E+06	4.12E+04	1.76E+04
451.6	3.57E+06	3.50E+06	4.12E+04	1.76E+04
451.8	3.57E+06	3.49E+06	4.12E+04	1.76E+04

Both h_A and h_B are quite high.

Hence $T_A' = T_A = 365\ K$

$T_B' = T_B = 650\ K$

8.16 A wire with a diameter of 0.03 mm and 0.3 m in length is heated by an electrical current and placed in helium at 280 K. a) If the surface temperature of the wire is 600 K, calculate the electric power. b) Calculate the electric power, for the same wire temperature, if there is a cross flow of helium with a free stream velocity of 20 m s^{-1}.

a. Apply Eq. (8.22) $Nu_D = \left\{0.60 + \dfrac{0.387(Gr_D Pr)^{1/6}}{\left[1+\left(\dfrac{0.559}{Pr}\right)^{9/16}\right]^{8/27}}\right\}^2$

$$Gr_D = \frac{g\beta(T_s - T_\infty)D^3}{\nu^2} \; ; \; T_f = \frac{1}{2}(600+280) = 440\text{ K}$$

Properties of He at 440 K: $\beta = 2.27\times10^{-3}\text{ K}^{-1}$; $\nu = 235\times10^{-6}\text{ m}^2\text{ s}^{-1}$; $Pr = 0.672$;

$k = 200\times10^{-3}\text{ W m}^{-1}\text{ K}^{-1}$

$$Gr_D = \frac{(9.807)(2.27\times10^{-3})(600-280)(3\times10^{-5})^3}{(235\times10^{-6})^2} = 3.48\times10^{-6} \; ; \; Gr_D Pr = 2.34\times10^{-6}$$

$$Nu_D = \left\{0.60 + \frac{0.387(2.34\times10^{-6})^{1/6}}{\left[1+\left(\frac{0.559}{0.672}\right)^{9/16}\right]^{8/27}}\right\}^2 = 0.406 \; ; \; Nu_D = \frac{hD}{k}$$

$$h = (0.406)\left(\frac{200\times10^{-3}}{3\times10^{-5}}\right) = 2707\text{ W m}^{-2}\text{ K}^{-1}$$

$$Q\,(\text{power}) = h(\pi DL)(T_s - T_\infty) = (2707)(\pi)(3\times10^{-5})(0.3)(600-280) = 24.5\text{ W}$$

b. $Re_f = \dfrac{D\bar{V}}{\nu_f} = \dfrac{(3\times10^{-5})(20)}{235\times10^{-6}} = 2.553$

Apply Eq. (8.8). $Nu_f Pr_f^{-0.3} = 0.35 + 56\, Re_f^{0.52}$

$$Nu_f = \left[0.35 + (56)(2.553^{0.52})\right](0.672^{0.3}) = 81.2$$

$$h = 81.2\left(\frac{200\times10^{-3}}{3\times10^{-5}}\right) = 5.42\times10^5\text{ W m}^{-2}\text{ K}^{-1}$$

$$Q = \left(\frac{5.42\times10^5}{2707}\right)24.5 = 4.90\times10^3\text{ W}$$

With forced convection, quite a bit of power can be passed through the wire. It is possible, however, that the temperature at the center of the wire could be much greater than at the surface. If the temperature difference in the wire is great enough, then there could be melting. Hence, we should consider conduction heat transfer within the wire.

8.17 A copper mold is used to make a casting of a nickel-base alloy. During most of the solidification period, the temperature profile in the copper is linear, with the surface on the casting side maintained constant at 1000 K. The other side of the mold contacts the water at 40°C (313 K). What is the surface temperature on the water side of the mold?

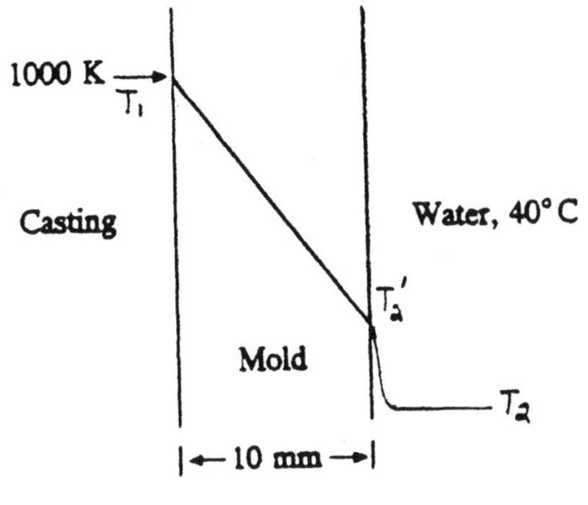

See the above figure for temperature nomenclature. Figure 8.12 is used to estimate the heat transfer coefficient on the water side.

Assume $T_2' = 200°C = 473$ K.

$h \simeq 2.8 \times 10^4 \ \text{W m}^{-2}\text{K}^{-1}$

k (Cu, 736 K) $= 368 \ \text{W m}^{-1}\text{K}^{-1}$

at steady state, $\frac{k}{L}(T_1 - T_2') = h(T_2' - T_2)$

$$T_2' = \frac{hT_2 + \frac{k}{L}T_1}{h + \frac{k}{L}} = \frac{(2.8\times10^4)(313) + \left(\frac{368}{0.01}\right)1000}{2.8\times10^4 + \left(\frac{368}{0.01}\right)}$$

$T_2' = 568$ K

Second iteration: $T_2' = 900$ K ; $h \simeq 4.1\times10^3 \ \text{W m}^{-2}\text{K}^{-1}$

k (Cu, 950 K) $= 360 \ \text{W m}^{-1}\text{K}^{-1}$

$$T_2' = \frac{(4.1\times10^3)(313) + \left(\frac{360}{0.01}\right)1000}{4.1\times10^3 + \left(\frac{360}{0.01}\right)} = 930 \text{ K}$$

Third iteration: $T_2' = 950$ K; $h \simeq 3.3\times10^3 \ \text{W m}^{-2}\text{K}^{-1}$

k (Cu, 975 K) $= 358 \ \text{W m}^{-1}\text{K}^{-1}$

$$T_2' = \frac{(3.3\times10^3)(313) + \left(\frac{358}{0.01}\right)1000}{3.3\times10^3 + \left(\frac{358}{0.01}\right)} = 942 \text{ K}$$

Answer : $T_2' = 940$ K

8.18 Glass plate can be made stronger by inducing compressive residual surface stresses. This is done by a process called thermal tempering. The glass is heated to a temperature between the softening point (~750 K) and the glass transition temperature (~600 K). It is then cooled to room temperature in a stream of air, or in some cases immersed in an oil bath. Compare the heat transfer coefficients, as the glass cools from 700 K to 300 K, in air (V_∞ = 30 m s^{-1}) and in "fast" oil.

```
20  ' Air cooling, use results of Problem 8.12; oil. use Fig. 8.13
30  FOR I = 1 TO 5                       'set up properties of air and h(oil)
40    READ TF(I), DENS(I), VISC(I), COND(I), PR(I), HOIL(I)
50  NEXT I
60  TINF = 300  : L = 2 : VINF = 30 : J = 0
70      LPRINT "    Temp.     Temp.                   h(air)    h(oil)"
80      LPRINT "   plate, K  fluid, K     Re(air)  W/m2 K   W/m2 K"
90  FOR TO = 300 TO 700 STEP 100       'TO = temp. of glass
100      J = J + 1 : HOIL = HOIL(J)
110      TF = (TO + TINF)/2                'film temp.
120      GOSUB 220                         'properties of air at TF
130      RE = L*VINF*DENS/VISC : IF RE^.8 < 15200 THEN 170
140      NU = .037 * PR^(1/3) * ( RE^.8 - 15200 )
150      IF NU < 0 THEN 180
160      H = NU*COND/L  : GOTO 190
170      PRINT RE;"less than transition"
180      PRINT "something  wrong"
190      LPRINT USING "     ###       ###       ##.##^^^^    ###.#    ####.";TO,TI
NF,RE,H,HOIL
200  NEXT TO
210  END
220 'SUBROUTINE - Interpolates properties of air
230   IF TF > TF(2) THEN  290
240   FRAC = ( TF - TF(1) )/( TF(2) - TF(1) )
250   DENS = ( DENS(2) - DENS(1) )*FRAC + DENS(1)
260   VISC = ( VISC(2) - VISC(1) )*FRAC + VISC(1)
270   COND = ( COND(2) - COND(1) )*FRAC + COND(1)
280   PR   = ( PR(2) - PR(1) )*FRAC + PR(1) : GOTO 460
290    IF TF > TF(3) THEN  350
300    FRAC = ( TF - TF(2) )/( TF(3) - TF(2) )
310    DENS = ( DENS(3) - DENS(2) )*FRAC + DENS(2)
320    VISC = ( VISC(3) - VISC(2) )*FRAC + VISC(2)
330    COND = ( COND(3) - COND(2) )*FRAC + COND(2)
340    PR   = ( PR(3) - PR(2) )*FRAC + PR(2) : GOTO 460
350     IF TF > TF(4) THEN  410
360     FRAC = ( TF - TF(3) )/( TF(4) - TF(3) )
370     DENS = ( DENS(4) - DENS(3) )*FRAC + DENS(3)
380     VISC = ( VISC(4) - VISC(3) )*FRAC + VISC(3)
390     COND = ( COND(4) - COND(3) )*FRAC + COND(3)
400     PR   = ( PR(4) - PR(3) )*FRAC + PR(3) : GOTO 460
410      FRAC = ( TF - TF(4) )/( TF(5) - TF(4) )
420      DENS = ( DENS(5) - DENS(4) )*FRAC + DENS(4)
430      VISC = ( VISC(5) - VISC(4) )*FRAC + VISC(4)
440      COND = ( COND(5) - COND(4) )*FRAC + COND(4)
450      PR   = ( PR(5) - PR(4) )*FRAC + PR(4) : GOTO 460
460   RETURN
470 '
480 DATA  300,1.1614, 1.846e-5, 2.63e-2, .707, 85
490 DATA  400, .8711, 2.301e-5, 3.38e-2, .690, 227
500 DATA  500, .6964, 2.701e-5, 4.07e-2, .684, 284
510 DATA  600, .5804, 3.058e-5, 4.69e-2, .685, 568
520 DATA  700, .4975, 3.388e-5, 5.24e-2, .695, 5394
```

Temp. plate, K	Temp. fluid, K	Re(air)	h(air) W/m2 K	h(oil) W/m2 K
300	300	3.77E+06	72.6	85
400	300	2.94E+06	66.3	227
500	300	2.27E+06	58.8	284
600	300	1.88E+06	54.3	568
700	300	1.55E+06	49.3	5394

8.19 Polymeric fibers are formed by heating the polymer to its viscous liquid state and then pumping it through small, round orifices. From each orifice, a single fiber is formed which solidifies almost immediately upon passing into air. Compare the heat transfer coefficients that can be obtained by forming the fibers (0.1 mm diameter) in air and in helium. The fibers are formed at 50 m s^{-1}.

Assume gas temperature = 300K

Properties of gasses:

Air – $\rho = 1.614\ \text{kg m}^{-3}$; $\nu = 15.69 \times 10^{-6}\ \text{m}^2\text{s}^{-1}$; $k = 26.3 \times 10^{-3}\ \text{W m}^{-1}\text{K}^{-1}$; $Pr = 0.707$

He – $\rho = 0.1625\ \text{kg m}^{-3}$; $\nu = 122.4 \times 10^{-6}\ \text{m}^2\text{s}^{-1}$; $k = 152 \times 10^{-3}\ \text{W m}^{-1}\text{K}^{-1}$; $Pr = 0.680$

The relative motion between the fibers and the gasses is parallel to the fibers' axes. Take $L = 1$ m of fiber.

Then $Re(\text{air}) = \dfrac{(1)(50)}{15.69 \times 10^{-6}} = 3.15 \times 10^6$

$Re(\text{He}) = \dfrac{(1)(50)}{122.4 \times 10^{-6}} = 4.09 \times 10^5$

For flow parallel to our surface we have no equation, but the equation for flow parallel to flat surfaces suggest that $Nu = C\,Re^{0.8}Pr^{1/3}$ where C is a constant. Based on this we can compare cooling in air to cooling in helium.

Set up ratio (1 ≑ air; 2 ≑ He)

$$\frac{Nu_2}{Nu_1} = \left(\frac{Re_2}{Re_1}\right)^{0.8}\left(\frac{Pr_2}{Pr_1}\right)^{1/3} \therefore \frac{h_2 k_1}{h_1 k_2} = \left(\frac{\nu_1}{\nu_2}\right)^{0.8}\left(\frac{Pr_2}{Pr_1}\right)^{1/3}$$

$$\frac{h_2}{h_1} = \left(\frac{k_2}{k_1}\right)\left(\frac{\nu_1}{\nu_2}\right)^{0.8}\left(\frac{Pr_2}{Pr_1}\right)^{1/3} = \left(\frac{152 \times 10^{-3}}{26.3 \times 10^{-3}}\right)\left(\frac{15.89 \times 10^{-6}}{122.4 \times 10^{-6}}\right)^{0.8}\left(\frac{0.680}{0.707}\right)^{1/3} = 1.11$$

This result is somewhat surprising because He is sometimes selected as a gas with "good cooling power". This selection is usually based only on its thermal conductivity, however, and does not appear to be justified in this case. An 11% increase in h would probably not justify its use.

8.20 A steel rod (25 mm diameter and 0.6 m length) is heated vertically in a large bath of molten salt at 920 K. Calculate the heat transfer coefficient when the bar is at 640 K.
Data for molten salt: k = 86 W m^{-1} K^{-1}; C_p = 870 J kg^{-1} K^{-1}; ρ = 3200 kg m^{-3}; β = 2 × 10^{-5} K^{-1}; η(920 K) = 1.03 × 10^{-3} N s m^{-2}; η(780 K) = 1.24 × 10^{-3} N s m^{-2}; η(640 K) = 1.65 × 10^{-3} N s m^{-2}.

$$Gr_L = \frac{L^3 \rho_f^2 \beta_f (T_0 - T_\infty) g}{\eta_f^2} = \frac{(0.6)^3 (3200)^2 (2\times10^{-5})(920-640)(9.81)}{(1.24\times10^{-3})^2}$$

$Gr_L = 79.02 \times 10^9$ Use $\eta_f = 1.24\times10^{-3}$ since $T_f = 780$ K.

$$\nu_f = \frac{\eta_f}{\rho} = \frac{1.24\times10^{-3}}{3200} = 3.88\times10^{-7} \text{ m}^2 \text{ s}^{-1}$$

$$\alpha_f = \frac{k}{\rho C_p} = \frac{86}{(3200)(870)} = 3.089\times10^{-5} \text{ m}^2 \text{ s}^{-1}$$

$$Pr = \frac{3.88\times10^{-7}}{3.089\times10^{-5}} = 1.26\times10^{-2}$$

$$\log_{10}(Gr_L Pr) = \log_{10}(9.96\times10^8) = 9.00$$

Fig. 8.8 gives $\log_{10} Nu_L = 1.88$; $Nu_L = 75.9$

$$h = Nu_L \frac{k}{L} = 75.9\left(\frac{86}{0.6}\right) = 1.09\times10^4 \text{ W m}^{-2} \text{ K}^{-1}$$

(Actually the diameter of the bar is too small to meet the criterion set forth as the foot note on p. 258. Hence, if this was an actual situation then one would be advised to seek another reference with more comprehensive correlations.)

8.21 Consider a vertical surface (1.5 m) at 600 K that loses heat by natural convection to nitrogen at 300 K. Using the simplified equations given in Problem 8.14, calculate and plot the heat transfer coefficient as the surface temperature decreases from 600 K to 300 K.

From Prob. 8.14 $h = 1.42(\Delta T/L)^{1/4}$, $\quad 5\times10^{-6} < L^3\Delta T < 50$

$h = 1.45(\Delta T)^{1/3}$, $\quad 50 < L^3\Delta T < 5\times10^4$

The following computer program can be used.

```
10 'Problem 8.21     S.I. units
20  TINF = 300 : L = 1.5
25  LPRINT "     Temp.,K          h, W/m2 K   "
30  FOR TO = 300 TO 600 STEP 50
32      DELT   = TO - TINF
40      PARAM = L^3 * DELT
50      IF PARAM > 50 THEN 70
60      H = 1.42* (DELT/L)^(1/4)  : GOTO 82
70      H = 1.45* DELT^(1/3)
82      LPRINT USING "      ###.              ##.## "; TO,H
90  NEXT TO
100  STOP
```

Temp.,K	h, W/m2 K
300	0.00
350	5.34
400	6.73
450	7.70
500	8.48
550	9.13
600	9.71

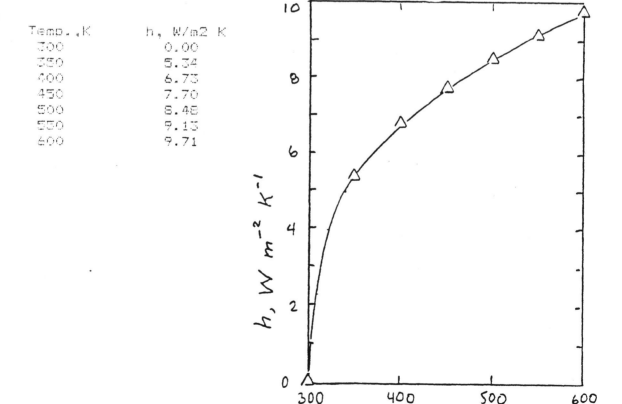

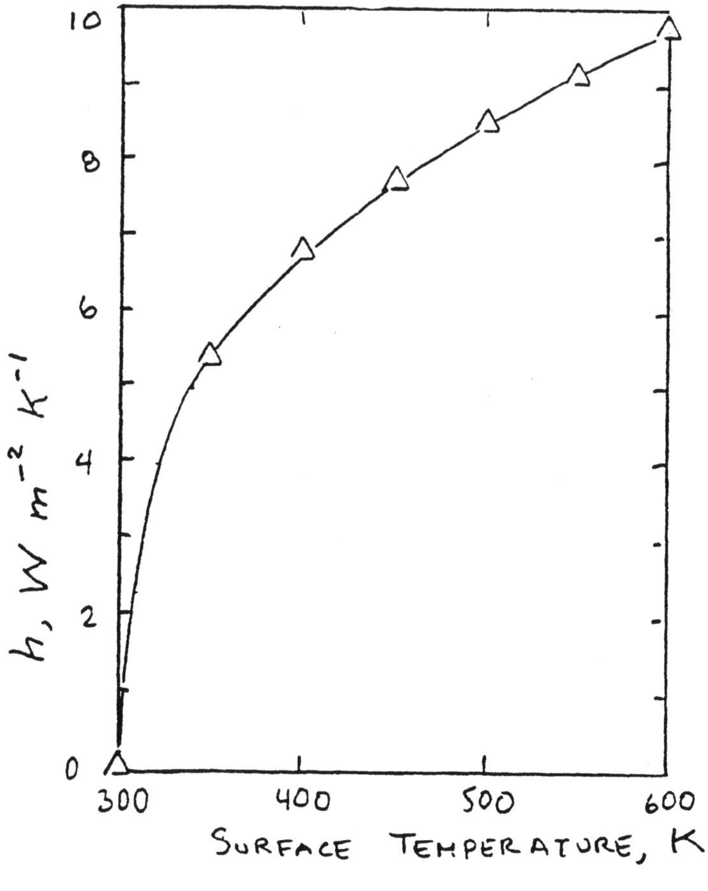

8.22 We can maximize the heat transfer coefficient in a fluidized bed by selecting the optimum superficial velocity. Start with Eq. (8.28), neglect the term for radiation, and assume that the thermophysical properties are constant. With these assumptions, the correlation is of the form:

$$\mathrm{Nu} = \mathrm{Nu}\,(\mathrm{Re},\omega)$$

Re and ω are interrelated, so Nu can be optimized with respect to Re. Carry out the optimization with

$$\omega^{4.7}\mathrm{Ar} = 18\ \mathrm{Re} + 2.70\ \mathrm{Re}^{1.687}.$$

(This equation is from C. Y. Wen and Y. H. Yu, *Fluid Particle Technology*, Chem. Eng. Prog. Symposium Series, No. 62, AIChE, New York, 1966.)

The results show that Nu and h decrease continually as Re and V_0 increase. Hence, the maximum heat transfer coefficient is obtained by operating the bed as closely as possible to minimum fluidization.

```
10 'Problem 8.22   Result shown with calculation using the data in Ex. 8.6.
20  G = 9.807
30  TS = 300 : DP = .001 : CP = 830 : DENSP = 3900   'properties of particles
40  TINF = 700 : DENSG = .696 : CG = 1030 : KG = .0407 :
    NETA = .000027 : PR = NETA/DENSG                'properties of the gas
50 'derived quantities and properties
60  AR = G*DP^3*DENSG*( DENSP - DENSG )/NETA^2   'Archimedes no., Eq. (8.29)
70  DENRATIO = (DENSP/DENSG)^.14 : CRATIO = (CP/CG)^.24
80  AA = .85*AR^.1*DENRATIO*CRATIO          '1st coefficient in Eq. (8.28)
90  LPRINT "  VO, m/s  Void frac.     Re           Nu     h, W/(m^2 K)
100 LPRINT "  *******  **********  *********  *********  ************"
110 'now we calculate Nu and h as a function of Vo
120  FOR VO =  .5 TO 20 STEP .5              'VO is superficial velocity
130     RE = DP* VO*DENSG/NETA               'RE is Reynolds no.
140     BB = .046*RE*PR                      '2nd coefficient in Eq. (8.28)
150     W = ( (18*RE + 2.7*RE^1.687 )/AR )^(1/4.7)    'W is void fract.
155     IF W >= 1 THEN 200
160     NU = AA*(1-W)^(2/3) + ( BB*(1-W)^(2/3) )/W
170     H = NU*KG/DP
180     LPRINT USING "  ###.#      #.###     ##.##^^^^  ##.##^^^^   ##.##^^^^";V
0,W,RE,NU,H
190  NEXT VO
200 END
```

V0, m/s	Void frac.	Re	Nu	h, W/(m^2 K)
*******	**********	*********	*********	************
0.5	0.389	1.29E+01	5.56E+00	2.26E+02
1.0	0.476	2.58E+01	5.02E+00	2.04E+02
1.5	0.538	3.87E+01	4.62E+00	1.88E+02
2.0	0.586	5.16E+01	4.28E+00	1.74E+02
2.5	0.631	6.44E+01	3.97E+00	1.62E+02
3.0	0.669	7.73E+01	3.69E+00	1.50E+02
3.5	0.704	9.02E+01	3.43E+00	1.40E+02
4.0	0.735	1.03E+02	3.19E+00	1.30E+02
4.5	0.764	1.16E+02	2.95E+00	1.20E+02
5.0	0.791	1.29E+02	2.72E+00	1.11E+02
5.5	0.816	1.42E+02	2.50E+00	1.02E+02
6.0	0.840	1.55E+02	2.27E+00	9.26E+01
6.5	0.863	1.68E+02	2.05E+00	8.35E+01
7.0	0.885	1.80E+02	1.83E+00	7.44E+01
7.5	0.906	1.93E+02	1.60E+00	6.52E+01
8.0	0.925	2.06E+02	1.37E+00	5.57E+01
8.5	0.945	2.19E+02	1.12E+00	4.57E+01
9.0	0.963	2.32E+02	8.56E-01	3.48E+01
9.5	0.981	2.45E+02	5.51E-01	2.24E+01
10.0	0.998	2.58E+02	1.15E-01	4.67E+00

8.23 Bars (50 mm diameter) of steel, on cooling beds, cool from a rolling temperature (1150 K) to a shearing temperature (650 K) in a horizontal position with all surfaces exposed to ambient air. The air temperature can vary from 250 K to 310 K. Calculate the convective heat transfer coefficients that apply to the two extremes of ambient conditions.

$$\text{Eq.}(8.22)\quad Nu_D = \left\{ 0.60 + \frac{0.387\,(Gr_D Pr)^{1/6}}{\left[1 + \left(\frac{0.559}{Pr}\right)^{9/16}\right]^{8/27}} \right\}^2$$

Case 1 - Steel at 1150 K, Air at 250 K

Case 2 - Steel at 1150 K, Air at 310 K

Case 3 - Steel at 650 K, Air at 250 K

Case 4 - Steel at 650 K, Air at 310 K

$$Gr_D \equiv \frac{D^3 \rho_f^2 \beta_f (T_o - T_\infty) g}{\eta_f^2}\ ;\ \text{Assume ideal gas } \beta_f = \frac{1}{T}\ ;\ T_f = \frac{1}{2}(T_o + T_\infty)$$

Properties of Air:

	Case 1, $T_f = 700$ K	Case 2, $T_f = 730$ K	Case 3, $T_f = 450$ K	Case 4, $T_f = 480$ K
β	1.429×10^{-3} K^{-1}	1.370×10^{-3} K^{-1}	2.22×10^{-3} K^{-1}	2.08×10^{-3}
ρ	0.4975 kg m^{-3}	0.4789	0.7838	0.7313
η	338.8×10^{-3} N s m^{-2}	348.1×10^{-7}	250.1×10^{-7}	262.1×10^{-7}
k	52.4×10^{-3} W m^{-1} K^{-1}	53.9×10^{-3}	37.25×10^{-3}	39.32×10^{-3}
Pr	0.695	0.699	0.637	0.685

Results:

	Case 1	Case 2	Case 3	Case 4
Gr_D	3.401×10^5	2.671×10^5	2.406×10^6	1.668×10^6
Nu_D	9.747	9.156	8.844	8.019

$$h = \frac{Nu_D k_f}{D} = \frac{Nu_D k_f}{0.05}$$

	Case 1	Case 2	Case 3	Case 4
h, W m^{-2} K^{-1}	10.21	9.87	6.59	6.31

9.1 A furnace wall is constructed of 7 in. of fire brick (k = 0.60 Btu h^{-1} ft^{-1} °F^{-1}), 4 in. of red brick (k = 0.40), 1 in. of glass-wool insulation (k = 0.04), and $\frac{1}{8}$ in. steel plate (k = 26) on the outside. The heat transfer coefficients on the inside and outside surfaces are 9 and 3 Btu h^{-1} ft^{-2} °F^{-1}, respectively. The gas temperature inside the furnace is 2500°F, and the outside air temperature is 90°F.

a) Calculate the heat-transfer rate through the wall (Btu h^{-1} ft^{-2}).

b) Determine the temperatures at all interfaces.

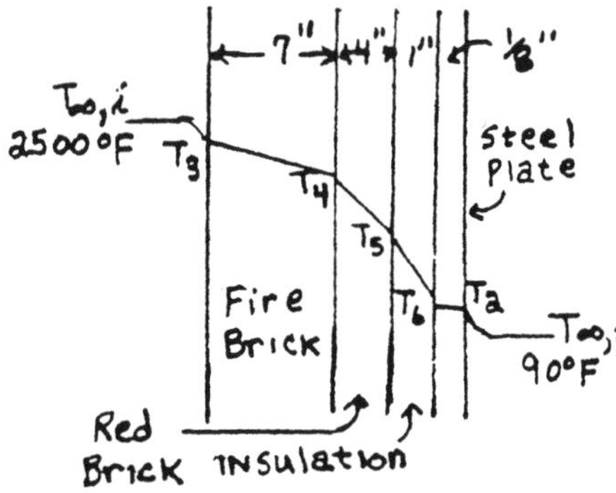

$$a.\ q = \frac{T_{\infty 1} - T_{\infty 2}}{\frac{1}{h_i} + \frac{L_3}{k_3} + \frac{L_4}{k_4} + \frac{L_5}{k_5} + \frac{L_6}{k_6} + \frac{1}{h_2}}$$

$$= \frac{2500-90}{\frac{1}{9} + \left(\frac{7}{12}\right)\left(\frac{1}{.6}\right) + \left(\frac{4}{12}\right)\left(\frac{1}{.4}\right) + \left(\frac{1}{12}\right)\left(\frac{1}{.04}\right) + \left(\frac{1}{8}\right)\left(\frac{1}{12}\right)\left(\frac{1}{26}\right) + \left(\frac{1}{3}\right)}$$

$$= 556 \text{ Btu h}^{-1}\text{ft}^{-2}$$

$$b.\ q = h_i (T_{\infty,i} - T_3);\ 556 = 9(2500 - T_3)$$

$$T_3 = 2438\,°F$$

$$q = \frac{k_3}{L_3}(T_3 - T_4);\ 556 = \frac{0.6}{\left(\frac{7}{12}\right)}(2438 - T_4)$$

$$T_4 = 1898\,°F$$

Similarly, $T_5 = 1434\,°F$, $T_6 = 269.4\,°F$, $T_2 = 269.2\,°F$

9.2 Consider the flow of heat through a spherical shell. For steady-state conditions, the inside surface ($r = R_1$) is at temperature T_1, and the outside surface ($r = R_2$) is at T_2.

a) Write the pertinent differential energy equation that applies.

b) Write the boundary conditions and develop an expression for the temperature distribution in the shell.

c) Develop an expression for the heat flow (Q, W) through the shell.

d) Determine the thermal resistance of the spherical shell.

$$a.\ \frac{1}{r^2}\frac{d}{dr}\left(r^2 \frac{dT}{dr}\right) = 0;\ \frac{d}{dr}\left(r^2 \frac{dT}{dr}\right) = 0$$

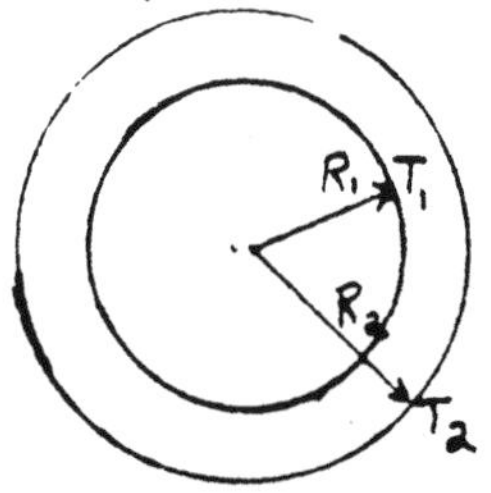

b. Boundary Conditions: at $r = R_1$, $T = T_1$

at $r = R_2$, $T = T_2$

$$r^2 \frac{dT}{dr} = C_1 \; ; \; \frac{dT}{dr} = \frac{C_1}{r^2} \; ; \; \therefore T = -\frac{C_1}{r} + C_2$$

$$T_1 = -\frac{C_1}{R_1} + C_2 \; ; \; \therefore C_2 = T_1 + \frac{C_1}{R_1} \; ; \; \therefore T = T_1 + C_1\left(\frac{1}{R_1} - \frac{1}{r}\right)$$

$$T_2 = T_1 + C_1\left(\frac{R_2 - R_1}{R_1 R_2}\right) ; \; \therefore C_1 = (T_2 - T_1)\left(\frac{R_1 R_2}{R_2 - R_1}\right)$$

$$\therefore T = T_1 + (T_2 - T_1)\left(\frac{R_2 R_1}{R_2 - R_1}\right)\left(\frac{1}{R_1} - \frac{1}{r}\right)$$

c. $$q_r = -k\frac{dT}{dr} = k\frac{1}{r^2}(T_1 - T_2)\left(\frac{R_1 R_2}{R_2 - R_1}\right)$$

$$Q = A q_r = 4\pi r^2\left[k\frac{1}{r^2}(T_1 - T_2)\frac{R_2 R_1}{(R_2 - R_1)}\right] = 4\pi k (T_1 - T_2)\left(\frac{R_1 R_2}{R_1 - R_2}\right)$$

d. Compare to Ohm's law

$$R_T = \left(\frac{R_1 - R_2}{R_1 R_2}\right)\frac{1}{4\pi k}$$

9.3 In order to reduce the heat loss through a large furnace wall, the decision has been made to add external insulation. Calculate the thickness of insulation required to reduce the heat loss by 75%. Before the change is made, no outer steel shell is used.

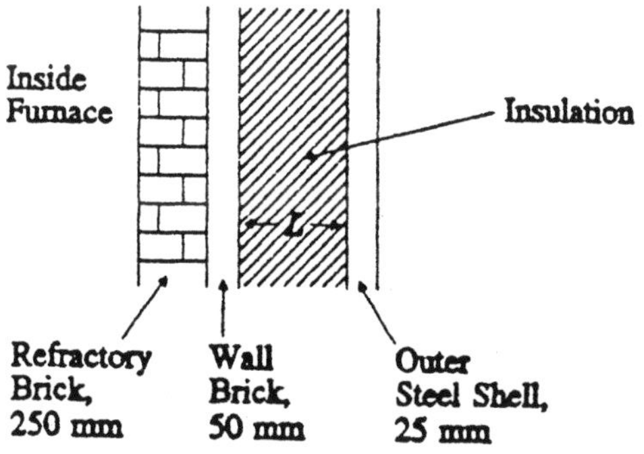

Data: Refractory brick and wall brick: $k = 0.87$ W m^{-1} K^{-1}.
Insulation: $k = 0.090$ W m^{-1} K^{-1}.
Steel: $k = 43$ W m^{-1} K^{-1}.
$h = 55$ W m^{-1} K^{-1} (inside furnace).
$h = 11$ W m^{-1} K^{-1} (outside furnace).

Let T_i = inside furnace temp., T_o = outside furnace temp.

Without added Insulation: $$q_1 = \frac{(T_i - T_o)}{\frac{1}{h_i} + \frac{L_1}{k_1} + \frac{L_2}{k_2} + \frac{1}{h_o}}$$

where h_i and h_o = inside and outside heat transfer coef., resp.

L_1 and L_2 = refractory and wall thickness, resp.

k_1 and k_2 = refractory and wall thermal cond., resp.

With added insulation: $q_2 = \dfrac{(T_i - T_o)}{\dfrac{1}{h_1} + \dfrac{L_1}{k_1} + \dfrac{L_2}{k_2} + \dfrac{1}{h_o} + \dfrac{L}{k} + \dfrac{L_S}{k_S}}$

where L and L_S = insulation and steel sheet thickness, resp.

k and k_S = insulation and steel thermal cond., resp.

To reduce heat loss by 75%

$$\frac{q_1}{q_2} = 4 = \frac{\left(\frac{1}{h_i} + \frac{L_1}{k_1} + \frac{L_2}{k_2} + \frac{1}{h_o}\right) + \left(\frac{L}{k} + \frac{L_S}{k_S}\right)}{\left(\frac{1}{h_i} + \frac{L_1}{k_1} + \frac{L_2}{k_2} + \frac{1}{h_o}\right)} = \frac{R_1 + R}{R_1}$$

$$\therefore R = 3R_1; \; R_1 = \frac{1}{55} + \frac{0.25}{0.87} + \frac{0.05}{0.87} + \frac{1}{11} = 0.454 \; m^2\,K\,W^{-1}$$

$$R = \frac{L}{k} + \frac{L_S}{k_S} = (3)(0.454)\; m^2\,K\,W^{-1}; \; L = \left[(3)(0.454) - \frac{(0.025)}{43}\right] 0.090 \qquad L = 0.122 \; m$$

9.4 The wall of a blast furnace is water-cooled. Given the inside- and outside-surface temperatures (2400°F and 180°F), what is the heat transfer coefficient for the water? The water, itself, is at 80°F. Assume steady-state conditions.

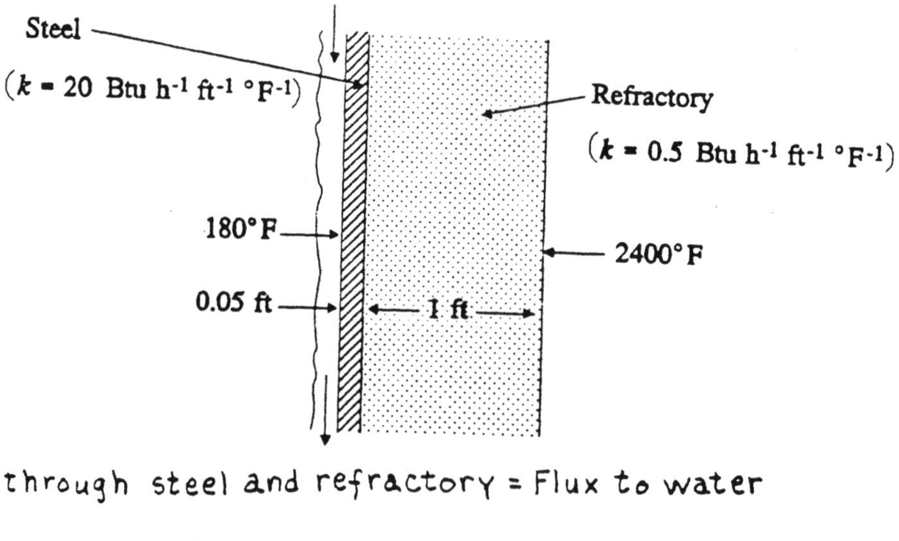

Flux through steel and refractory = Flux to water

$$\frac{T_i - T_o}{\left(\frac{L}{k}\right)_R + \left(\frac{L}{k}\right)_S} = h\,(T_o - T_f)$$

$$h = \frac{T_i - T_o}{T_o - T_f} \cdot \frac{1}{\left(\frac{L}{k}\right)_R + \left(\frac{L}{k}\right)_L} = \frac{(2400-180)}{(180-80)} \cdot \frac{1}{\left[\left(\frac{1}{0.5}\right) + \left(\frac{0.05}{20}\right)\right]} = 11.1 \text{ Btu } h^{-1}\,ft^{-2}\,°F^{-1}$$

9.5 Consider steady-state heat conduction through a cylindrical wall. The fluid on the inside is at 590 K with a heat transfer coefficient of 23 W m^{-2} K^{-1}. The temperature on the outside surface of the wall is known and maintained at 420 K. The heat flow rate through the cylindrical wall is 200 W per 1 m length of the cylinder. If the wall has a thermal conductivity of 0.17 W m^{-1} K^{-1}, what are the inside and outside radii of the cylindrical wall? The ratio of the outside radius to inside radius is 2.

use Eq (9.18) with $h_o = \infty$, $T_o = T_2$

$$\frac{Q}{L} = \frac{2\pi (T_i - T_2)}{\frac{1}{h_i r_1} + \frac{1}{k} \ln \frac{r_2}{r_1}}$$

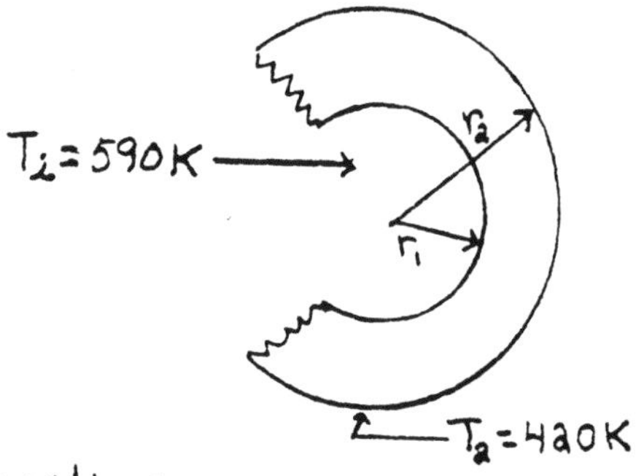

$$\frac{1}{r_1} = \frac{2\pi h_i (T_i - T_2)}{\frac{Q}{L}} - \frac{h_i}{k} \ln\left(\frac{r_2}{r_1}\right)$$

$$\frac{1}{r_1} = \frac{2\pi \mid 23\ W \mid (590-420)K \mid m}{\mid m^2 K \mid \quad \mid 200W} - \frac{23\ W \mid m\ K \mid \ln 2}{m^2 K \mid 0.17 W \mid}$$

$$= 2.906 \times 10^1 m^{-1}$$

$$\therefore r_1 = 3.44 \times 10^{-2} m \; ; \; r_2 = 6.88 \times 10^{-2} m$$

9.6 Small droplets of a molten glass maintain their amorphicity if they cool at a rate of at least 10 K s^{-1}, measured at 1070 K. For a spherical droplet with 0.1 mm diameter, what is the required heat transfer coefficient to achieve the minimum cooling rate? The quench environment is maintained at 293 K. *Data for the glass*: ρ = 3000 kg m^{-3}; C_p = 840 J kg^{-1} K^{-1}; k = 17 W m^{-1} K^{-1}.

Assume Newtonian Cooling

$$\text{Eq. (9.20)} \quad h = -\frac{V \rho C_P}{A (T - T_f)} \frac{dT}{dt} \; , \; \frac{V}{A} = \frac{D}{6}$$

$$h = -\frac{1 \times 10^{-4} m \mid 3000 kg \mid 840 J \mid -10 K \mid \quad \mid W s}{6 \mid m^3 \mid kg\ K \mid s \mid (1070-293) K \mid J} = 0.540\ W\ m^{-2} K^{-1}$$

check Biot No.

$$Bi_R = \frac{hR}{k} = \frac{0.540\ W \mid 1 \times 10^{-4} m \mid m\ K}{m^2 K \mid 2 \mid 17\ W} = 1.59 \times 10^{-6}$$

$Bi_R << 0.1$ so that Newtonian cooling applies and h is valid.

9.7 A small sphere (diameter = 0.30 mm), initially at 1365 K, drops through a gas layer of 150 mm with an average velocity of 3 m s^{-1} and then through a liquid layer of 30 mm with an average velocity of 3 mm s^{-1}. *Data for the sphere*: ρ = 2560 kg m^{-3}; C_p = 840 J kg^{-1} K^{-1}; k = 0.86 W m^{-1} K^{-1}. *Heat transfer coefficients*: h (gas) = 40 W m^{-2} K^{-1}; h (liquid) = 280 W m^{-2} K^{-1}.

a) What is the temperature of the droplet just before it impacts the liquid?

b) What is the temperature of the droplet when it reaches the bottom of the liquid layer?

Let t_1, t_2 = time in the gas, liquid; h_1, h_2 = heat tranfer coef. in gas, liq.

a. Apply Newtonian cooling

Eq. (9.21) $\frac{T-T_f}{T_i-T_f} = \exp\left(-\frac{h_1 A t_1}{\rho C_p V}\right)$; $\frac{A}{V} = \frac{6}{D} = \frac{6}{0.3\times10^{-3}} = 2\times10^{4}\ m^{-1}$

$t_1 = \frac{0.15\ m}{} \Big| \frac{s}{3\ m} = 0.05\ s$

Assume $T_f = 300K$

$T = (1365-300)\exp\left(-\frac{(40)(2\times10^4)(0.05)}{(2560)(840)}\right) + 300$

$T = 1345\ K$

b. $t_2 = \frac{30\times10^{-3}}{3\times10^{-3}} = 10\ s$; Assume $T_f = 300\ K$; $T_i = 1345\ K$

$T = (1345-300)\exp\left(-\frac{(280)(2\times10^4)(10)}{(2560)(840)}\right) + 300$

$T = 300\ K$

Check Bi_R. $Bi_R = \frac{h_2 R}{k} = \frac{(280)(0.15\times10^{-3})}{0.86} = 0.049 < 0.1$

Newtonian cooling is vAlid.

9.8 A thin wire is extruded at a fixed velocity through dies, and the wire temperature at the die is a fixed value T_0. The wire then passes through the air for a long distance before it is rolled onto large spools. It is desired to investigate the relationship between wire velocity and the distance from the extrusion nozzle for specific values of T_0.

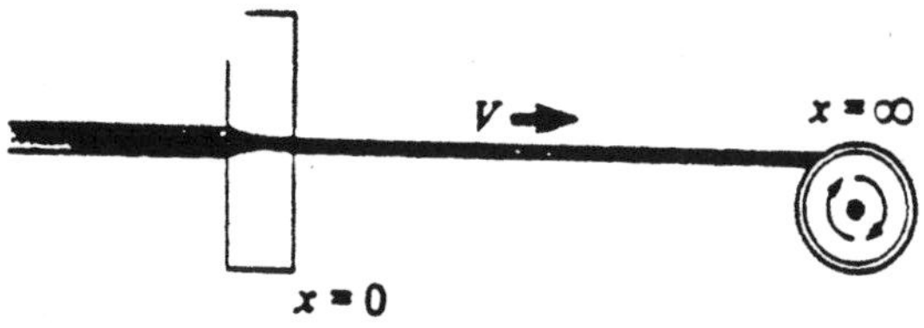

a) Derive the differential equation for determining wire temperature as a function of distance from the nozzle. [*Hint*: Since temperature gradients across the wire are certainly negligible, a slice between x and $x + \Delta x$ may be chosen that includes the wire surface. The heat balance then includes heat lost to the surroundings at T_∞.]

b) State boundary conditions and solve for the temperature in the extended wire.

a. Energy balance on length Δx:

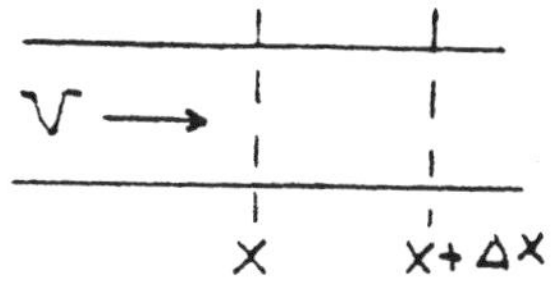

Rate of energy in $= VA\rho C_p T|_x - kA\frac{dT}{dx}\Big|_x$

Rate of energy out $= VA\rho C_p T|_{x+\Delta x} - kA\frac{dT}{dx}\Big|_{x+\Delta x} + hP\Delta x\,(T-T_\infty)$

$$\lim_{\Delta x \to 0} \quad 0 = VA\rho C_p\left(\frac{T|_{x+\Delta x} - T|_x}{\Delta x}\right) - kA\left(\frac{\frac{dT}{dx}\Big|_{x+\Delta x} - \frac{dT}{dx}\Big|_x}{\Delta x}\right) + hP(T-T_\infty)$$

$$\frac{d^2T}{dx^2} - \frac{V}{\alpha}\frac{dT}{dx} - \frac{hP}{kA}(T-T_\infty) = 0$$

b. B.C.: $T(0) = T_0$; $T(\infty) = T_\infty$

To make the differential Eq. homogeneous, let $\theta = T - T_\infty$. Then,

$$\frac{d^2\theta}{dx^2} - \frac{V}{\alpha}\frac{d\theta}{dx} - \frac{hP}{kA}\theta = 0; \quad \text{B.C.: } \theta(0) = T_0 - T_\infty = \theta_0; \ \theta(\infty) = 0$$

$\therefore \theta = c_1 \exp(r_1 x) + c_2 \exp(r_2 x)$ where $r_1 = \frac{V}{2\alpha} + \left(\frac{V^2}{4\alpha^2} + \frac{hP}{kA}\right)^{1/2}$, $r_2 = \frac{V}{2\alpha} - \left(\frac{V^2}{4\alpha^2} + \frac{hP}{kA}\right)^{1/2}$

$r_1 > 0$ and $r_2 < 0$ for finite h. $\therefore \theta(\infty) = 0$ requires $c_1 = 0$; $\theta(0) = \theta_0$ requires $\theta_0 = c_2$

$$\therefore \theta = \theta_0 \exp\left[\frac{V}{2\alpha} - \left(\frac{V^2}{4\alpha^2} + \frac{hP}{kA}\right)^{1/2}\right]x; \quad \frac{T - T_\infty}{T_0 - T_\infty} = \exp\left[\frac{V}{2\alpha} - \left(\frac{V^2}{4\alpha^2} + \frac{hP}{kA}\right)^{1/2}\right]x$$

9.9 A very long crystal (dia. = D) is slowly withdrawn with a velocity V from a melt maintained at only a few degrees above the freezing point. The diameter of the crystal is small enough so that radial temperature gradients can be ignored. There is heat loss from the crystal to the surroundings, maintained at T_∞; the heat transfer coefficient is h. Assume steady-state conditions.

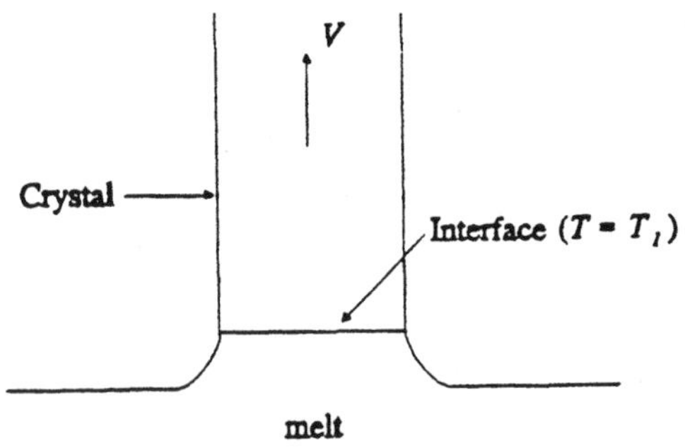

a) Distance in the crystal and measured from the interface is z. Derive the energy equation in terms of temperature.

b) Write appropriate boundary conditions and solve for temperature in the crystal.

c) If h = 110 W m^{-2} K^{-1}, what is the maximum diameter of the crystal so that the radial temperature gradient can be ignored?

[Note: If you have solved Problem 9.8, then you can write the equation for temperature by inspection.]

a. Same as Prob. 9.8 with x replaced by z.

$$\frac{d^2T}{dz^2} - \frac{V}{\alpha}\frac{dT}{dz} - \frac{hP}{kA}(T - T_\infty) = 0$$

b. Boundary Conditions: $T(0) = T_I$; $T(\infty) = T_\infty$

The solution is deduced from Prob. 9.8

$$\frac{T - T_\infty}{T_I - T_\infty} = \exp\left[\frac{V}{2\alpha} - \left(\frac{V^2}{4\alpha^2} + \frac{hP}{kA}\right)^{1/2}\right] z$$

c. $Bi_R = \frac{hR}{k} = \frac{hD}{2k}$; We want $Bi_R < 0.1$

$$\therefore D(\text{max.}) = \frac{2k\,Bi_R}{h} = \frac{(2)(k)(0.1)}{(110)}$$

$D(\text{max.}) = 1.82 \times 10^{-3}\,k$ with D in m and k in W m^{-1}K^{-1}

9.10 One end of a long thin rod is inserted into a furnace through a hole in the furnace door. Two thermocouples are inserted along the length of the bar and indicate steady-state temperatures of T_1 and T_2, respectively. Assume that the temperature in the bar only varies along its length and derive an equation which can be used to calculate h of the surrounding air as a function of T_1, T_2, P (perimeter), A (area), L, T_∞, and k (thermal conductivity of the bar).

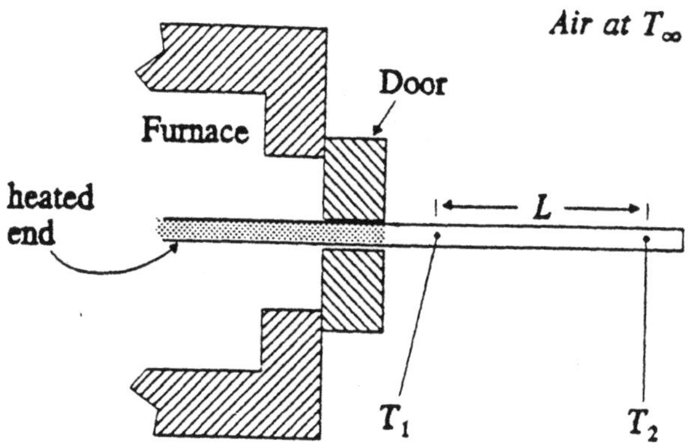

Given that $T=f(x)$, then an energy balance on a differential length of bar is: $\frac{d^2T}{dx^2} - \frac{hP}{kA}(T-T_\infty)=0$. This can be seen by taking $V=0$ in Prob. 9.8.

Let $\theta=T-T_\infty$; Then $\frac{d^2\theta}{dx^2} - \frac{hP}{kA}\theta=0$

$\therefore \theta = C_1 \exp(r_1 x) + C_2 \exp(r_2 x)$; where $r_1 = +\left(\frac{hP}{kA}\right)^{1/2}$ and $r_2 = -\left(\frac{hP}{kA}\right)^{1/2}$

The boundary conditions are: $\theta(0) = T_1 - T_\infty = \theta_1$; $\theta(L) = T_2 - T_\infty = \theta_2$

$$\therefore C_2 = \frac{(T_2-T_\infty)-(T_1-T_\infty)\exp(r_1 L)}{\exp(r_2 L)-\exp(r_1 L)}$$

$$C_1 = (T_1 - T_\infty) - C_2$$

9.11 One end of an amorphous rod is heated so that a portion crystallizes. Derive an equation that can be used for finding the position of the interface after steady state is achieved. The interface temperature is T^*. Assume one-directional conduction along the length of the rod and heat loss to the surroundings at T_∞ with a uniform heat transfer coefficient.

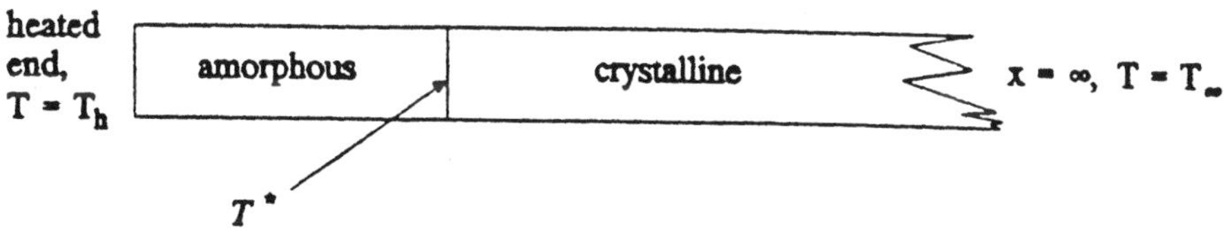

a) Assume equal thermal properties in the amorphous and crystalline states.
b) Assume k (crystal) $= 4k$ (amorphous).

a. Assume that $T=f(x)$. The rod is stationary so that we can use the energy eq. of Prob 9.10.

$$\frac{d^2T}{dx^2} - \frac{hP}{kA}(T-T_\infty)=0$$

$\therefore\ \theta \equiv T - T_\infty = C_1 \exp(r_1 x) + C_2 \exp(r_2 x)$; where $r_1 = +\left(\frac{hP}{kA}\right)^{1/2}$, $r_2 = -\left(\frac{hP}{kA}\right)^{1/2}$

Boundary Conditions: $\theta(0) = T_h - T_\infty = \theta_h$; $\theta(\infty) = 0$

$\theta(\infty) = 0$ requires that $C_1 = 0$; $\theta(0) = \theta_h$ gives $\theta_h = C_2$

$\therefore\ T - T_\infty = (T_h - T_\infty) \exp(r_2 x)$

Let x^* = position of interface at T^*

$$\frac{T^* - T_\infty}{T_h - T_\infty} = \exp(r_2 x^*) \quad \text{or} \quad x^* = \frac{1}{r_2} \ln\left(\frac{T^* - T_\infty}{T_h - T_\infty}\right)$$

b. Use "a" and "c" for subscript for amorphous and crystalline prop.

Part a - amorphous side: $T_a - T'_\infty = (T_h - T'_\infty) \exp(r_{2a} x)$, $\quad 0 \le x \le x^*$

crystalline side: $T_c - T_\infty = (T'_h - T_\infty) \exp(r_{2c} x)$, $\quad x^* \le x \le \infty$

T'_∞ and T'_h are integration constants.

The integration constants can be visualized by extrapolating the temperature in the two different regions. The broken curves are the extrapolations.

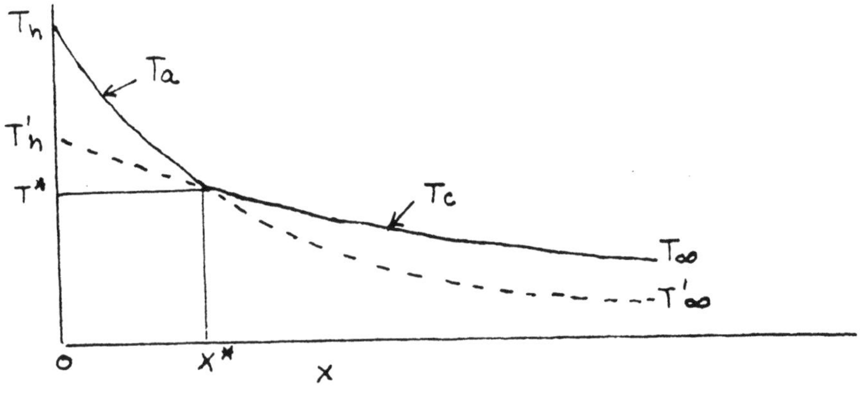

To solve for the two integration constants, we apply two conditions at the interface.

$$T^* = T_a = T_c \text{ at } x = x^* \quad \text{and} \quad k_a \frac{dT_a}{dx} = k_c \frac{dT_c}{dx} \text{ at } x = x^*$$

The first condition gives:

$$(T'_h - T_\infty) = (T_h - T'_\infty) \exp[(r_{2a} - r_{2c}) x^*] + (T'_\infty - T_\infty) \exp(-r_{2c} x^*) \qquad (i)$$

The second condition gives:

$$(T_h - T'_\infty) = \left(\frac{k_c}{k_a}\right)\left(\frac{r_{ac}}{r_{aa}}\right)\cdot \exp\left[(r_{ac} - r_{aa})x^*\right]\cdot (T'_h - T_\infty) \qquad (ii)$$

By combining the results of the two conditions, we get:

$$(T'_h - T_\infty) = \left(\frac{k_c}{k_a}\right)^{1/2}(T'_h - T_\infty) + (T'_\infty - T_\infty)\exp(-r_{ac}x^*)$$

or $$1 = \left(\frac{k_c}{k_a}\right)^{1/2} + \left(\frac{T'_\infty - T_\infty}{T'_h - T_\infty}\right)\exp(-r_{ac}x^*)$$

$$1 = \left(\frac{k_c}{k_a}\right)^{1/2} + \left(\frac{T'_\infty - T_\infty}{T^* - T_\infty}\right);$$ This gives $\frac{T'_\infty - T_\infty}{T^* - T_\infty} = 1 - \left(\frac{k_c}{k_a}\right)^{1/2}$, from which T'_∞ is determined.

Because T^* is specified, x^* is known and then Eq. (i) can be used to calculate T'_h and the problem is solved.

9.12 Consider the Newtonian cooling of a thin plate (15 mm thick) with a length and width of 1.2 m each. Initially the plate is at 530 K, and then it is cooled by natural convection as it is suspended vertically in air at 300 K. Use the simplified equations given in Problem 8.14 and answer the following questions.

a) Early in the cooling process, does h vary with $(\Delta T)^{1/4}$ or $(\Delta T)^{1/3}$?

b) Near the end of the cooling process, does h vary with $(\Delta T)^{1/4}$ or $(\Delta T)^{1/3}$? Determine the temperature at which the transition occurs.

c) Derive (an) equation(s) which give(s) the temperature of the bar consistent with the assumption of Newtonian cooling and h varying with $(\Delta T)^{1/3}$ and/or $(\Delta T)^{1/4}$.

$T_i = 530\text{ K}$, $T_\infty = 300\text{ K}$, $L = 1.2\text{ m}$: From Prob. (8.14)

a. $L^3\Delta T = (1.2)^3(530-300) = 397 \therefore h = 1.45\,\Delta T^{1/3}$; varies as $(\Delta T)^{1/3}$

b. Assume $\Delta T = 10\text{K}$; $L^3\Delta T = (1.2)^3(10) = 17 \therefore h = 1.42\left(\frac{\Delta T}{L}\right)^{1/4}$; varies as $(\Delta T)^{1/4}$

Transition occurs when $L^3\Delta T = 50$; $\Delta T = \frac{50}{(1.2)^3} = 29\text{ K}$

$\therefore$ Transition temperature $= 329\text{ K}$

c. $-V\rho c_p \frac{dT}{dt} = hA(T - T_f)$ where $T_f = 300\text{ K}$

Then with $h = h(T)$: $\frac{1}{(T-T_f)h}dT = -\frac{A}{V}\frac{1}{\rho c_p}dt$

(i) When $T \geq 329$, $h = 1.45(T - T_f)^{1/3}$

$$1.45^{-1}\int_{T_i}^{T}(T-T_f)^{-4/3}dT = -\frac{A}{V}\frac{1}{\rho c_p}\int_0^t dt$$

$$\frac{3}{1.45}\left[(T_i - T_f)^{-1/3} - (T - T_f)^{-1/3}\right] = -\frac{A}{V}\frac{1}{\rho c_p}t$$, where $T_i = 530\text{K}$ & $T_f = 300\text{K}$.

(ii) when $T < 329$, $h = 1.42 \left(\frac{\Delta T}{L}\right)^{1/4} = 1.36\ (\Delta T)^{1/4}$

$$(1.36)^{-1} \int_{T^*}^{T} (T-T_f)^{-5/4}\, dT = -\frac{A}{V}\frac{1}{\rho C_p}\int_{t^*}^{t} dt \quad \text{where } T^* = 329\text{K and } t^*$$

is determined from Part (i)

$$\left(\frac{4}{1.36}\right)\left[(T^*-T_f)^{-1/4} - (T-T_f)^{-1/4}\right] = -\frac{A}{V}\frac{1}{\rho C_p}(t-t^*)$$

9.13 In some alloys, grain refinement can be achieved by cycling the alloy above and below a transformation temperature.

Suppose an alloy sphere, initially at a uniform temperature T_0, is immersed in a bath of heated oil. The electric heaters are controlled so that the temperature of the oil (T_f) follows a cyclic variation given by

$$T_f - T_m = A \sin \omega t$$

where
T_m = time average mean oil temperature (constant),
A = amplitude of variation, and
ω = frequency.
Derive an expression for the temperature of the sphere as a function of time and the heat transfer coefficient. Assume that the temperature of the sphere is uniform.

$$-V\rho C_p \frac{dT}{dt} = hA'(T-T_f) \quad \text{where } A' = \text{area of the sphere}$$

Let $\theta = T - T_m$ $\therefore \frac{d\theta}{dt} + a\theta = b\sin\omega t$ where $a \equiv \left(\frac{A'}{V}\right)\left(\frac{h}{\rho C_p}\right) = \frac{3h}{R\rho C_p}$

and $b \equiv \left(\frac{A'}{V}\right)\left(\frac{h}{\rho C_p}\right)A = \frac{3hA}{R\rho C_p}$

Let $\theta = \theta_H + \theta_P$ where θ_H is the solution to $\frac{d\theta_H}{dt} + a\theta_H = 0$ and θ_P is the solution to $\frac{d\theta_P}{dt} + a\theta_P = b\sin\omega t$

$\therefore \theta_H = C_1 e^{-at}$ and $\theta_P = C_2 \sin\omega t + C_3\cos\omega t$ where C_1, C_2, C_3 are constants.

$$C_2 = \frac{ab}{a^2+\omega^2}\ ; \quad C_3 = -\frac{b\omega}{a^2+\omega^2}$$

with the initial condition $T = T_i$ at $t=0$; $\theta = \theta_i = T_i - T_m$ at $t=0$

$$\therefore \theta_i = C_1 - \frac{b\omega}{a^2+\omega^2} \quad \text{or} \quad C_1 = \theta_i + \frac{b\omega}{a^2+\omega^2}$$

Solution: $\theta = \theta_i e^{-at} + \frac{1}{a^2+\omega^2}\left[b\omega\left(e^{-at} - \cos\omega t\right) + ab\sin\omega t\right]$

This can be written as: $\theta = \left[\theta_i + \frac{b\omega}{a^2+\omega^2}\right]e^{-at} + \frac{b}{a^2+\omega^2}\left[a\sin\omega t - \omega\cos\omega t\right]$

$$\therefore\ \theta = \left[\theta_i + \frac{aA\omega}{a^2+\omega^2}\right] e^{-at} + \frac{aA}{a^2+\omega^2}\left[a \sin\omega t - \omega \cos\omega t\right] \quad \text{since } b = aA$$

1st term contains the initial condition and decays - transient portion.

2nd term is periodic - sustained portion.

Rewrite the 2nd term as: $\theta = \frac{aA}{(a^2+\omega^2)^{1/2}}\left[\frac{a}{(a^2+\omega^2)^{1/2}} \sin\omega t - \frac{\omega}{(a^2+\omega^2)^{1/2}} \cos\omega t\right]$

Let $\beta = \frac{1}{\tan\left(\frac{\omega}{a}\right)}$ then: $\sin\beta = \frac{\omega}{(a^2+\omega^2)^{1/2}}$; $\cos\beta = \frac{a}{(a^2+\omega^2)^{1/2}}$

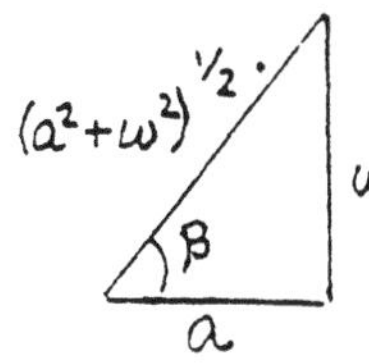

$$\therefore\ \theta = \frac{aA}{(a^2+\omega^a)^{1/2}}\left[\cos\beta \sin\omega t - \sin\beta \cos\omega t\right] = \frac{aA}{(a^2+\omega^2)^{1/2}} \sin(\omega t - \beta)$$

as $\omega \to 0$ then $\beta \to 0$ and $(a^2+\omega^2)^{1/2} = a$

$\theta = A \sin\omega t$; for small frequencies, the object can follow right along in phase and with the same amplitude.

as ω becomes infinity ; $\omega \to \infty$ then $\beta \to \frac{\pi}{2}$ and $(a^2+\omega^2)^{1/2} \to \omega$

$\theta = \frac{aA}{\omega} \sin\left(\omega t - \frac{\pi}{2}\right)$; the temperature of the object is 90° out of phase and it amplitude becomes smaller.

For grain refinement we would want low frequencies during the nucleation stage.

9.14 Ball bearings (12 mm diameter spheres) are austenitized at 1145 K and then quenched into a large tank of oil at 310 K. Calculate:

a) The time to cool the center of a bearing to 480 K.
b) The surface temperature when the center is at 480 K.
c) The space-mean temperature when the center is at 480 K.
d) If 10,000 balls are quenched per hour, calculate the rate of heat removal from the oil that is needed to maintain its temperature at 310 K.

Data for ball bearings: h = 1700 W m^{-2} K^{-1}; ρ = 7210 kg m^{-3}; C_p = 630 J kg^{-1} K^{-1}; k = 43 W m^{-1} K^{-1}.

a. $Bi = \frac{hR}{k} = \frac{(1700)(0.006)}{43} = 0.24$; $\alpha = \frac{43}{(7210)(630)} = 9.47 \times 10^{-6}\ m^2\ s^{-1}$

$$\frac{T - T_f}{T_i - T_f} = \frac{480-310}{1145-310} = 0.204$$

To estimate the time, we do a horizontal interpolation between Bi = 0.1 and Bi = 0.4.

Bi	Fo
0.1	5.4
0.24	4.0 (interpolated)
0.40	1.4

$\therefore\ t = \frac{\alpha}{R^2} F_o = \frac{9.47 \times 10^{-6}}{(0.006)^2}(4.0) = 1.05\ s$

Parts b) and c) are best done by plotting $\frac{T-T_f}{T_i-T_f}$ versus $\frac{r}{R}$ for Bi = 0.1 and 0.4

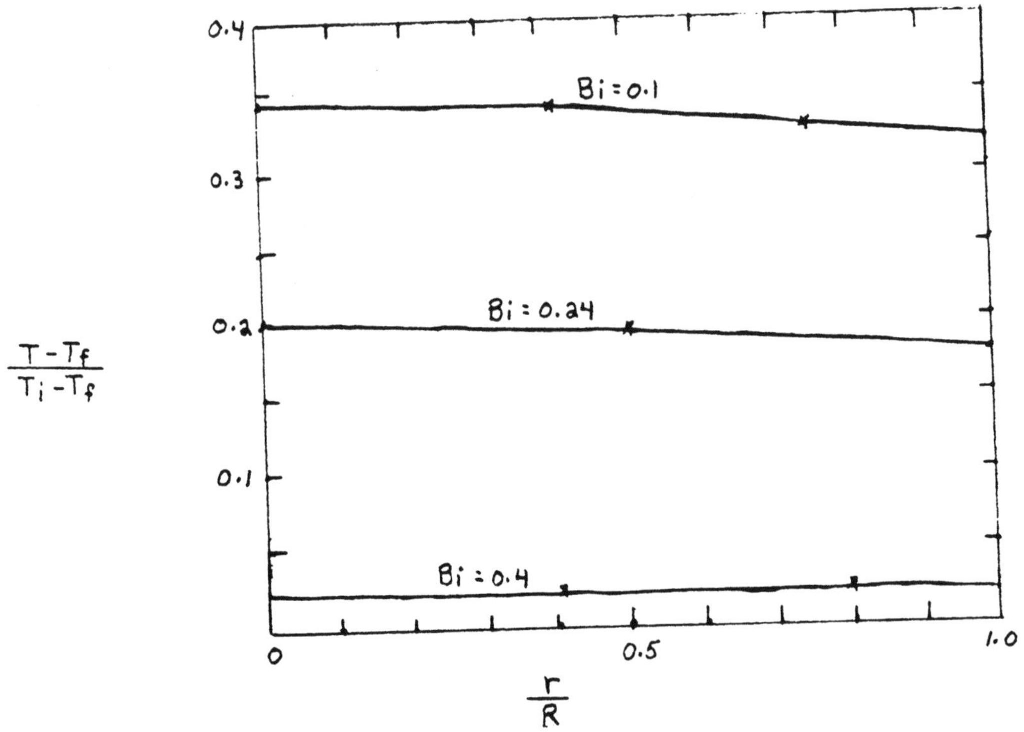

b. $\frac{r}{R} = 1;\ \frac{T-T_f}{T_i-T_f} = 0.18 \quad \therefore T = 0.18(1145-310)+310 = 460\ K$

c. From the graph: $\bar{\theta} = \frac{\bar{T}-T_f}{T_i-T_f} \simeq 0.185 \quad \therefore\ \bar{T} = 0.185(1145-310)+310 = 464\ K$

(If done precisely, it would be: $\bar{\theta} = \frac{3}{R^3}\int_0^r \theta r^2 dr$.)

d. Let ΔH = change in enthalpy per ball, J, and n = number of balls.

$\Delta H = \rho c_p (T_i - \bar{T}) V$, J ball^{-1}

Then $n\Delta H = \frac{10\,000\ \text{ball}}{3600\ \text{s}} \left| \frac{7210\ \text{kg}}{\text{m}^3} \right| \frac{630\ \text{J}}{\text{K kg}} \left| (1145-464)\ \text{K} \right| \frac{4\pi(0.006)^3\ \text{m}^3}{3\ \text{ball}} = 7.77 \times 10^3\ \text{J s}^{-1}$

Heat removal = 7.77×10^3 J s^{-1} = 2.80×10^7 J h^{-1}

9.15 Rapid solidification of Cu is effected by dropping molten droplets into water at 310 K. The droplets may be approximated as spheres with a diameter of 5 mm. Calculate the time for the droplets to cool to 365 K if they enter the water at 1450 K.

Data for Cu in S.I. units: Freezing point = 1358, C_p (solid) = 377; C_p (liquid) = 502; heat of fusion = 2.07×10^5 J kg^{-1}; ρ (solid) = 8970; ρ (liquid) = 8490; k (solid) = 346; k (liquid) = 311.

Data for water quench:

Temperature range	h, W m^{-2} K^{-1}
1450-920 K	450
920-360 K	2270

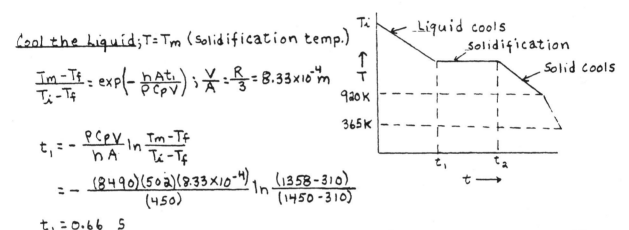

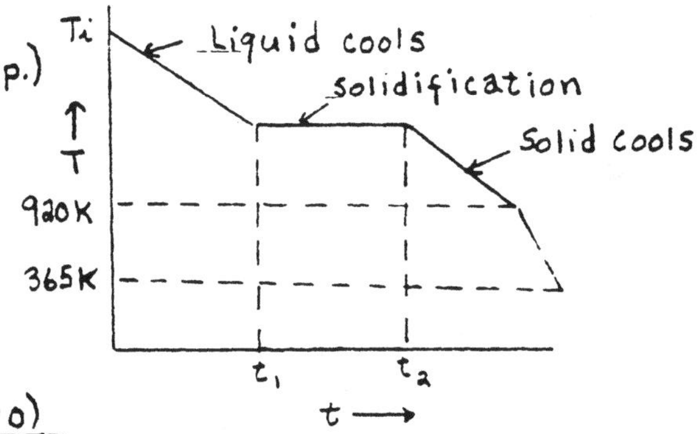

Cool the Liquid; T = T_m (solidification temp.)

$$\frac{T_m - T_f}{T_i - T_f} = \exp\left(-\frac{hAt_1}{\rho C_p V}\right); \quad \frac{V}{A} = \frac{R}{3} = 8.33\times10^{-4}\ \text{m}$$

$$t_1 = -\frac{\rho C_p V}{hA}\ln\frac{T_m - T_f}{T_i - T_f}$$

$$= -\frac{(8490)(502)(8.33\times10^{-4})}{(450)}\ln\frac{(1358-310)}{(1450-310)}$$

$t_1 = 0.66$ s

Solidification: $m = \text{mass} = (8490)\left(\frac{4}{3}\pi\right)(2.5\times10^{-3})^3 = 5.56\times10^{-4}$ kg

$$mH_f = hA(T_m - T_f)t_s$$

$$t_s(\text{solid. time}) = \frac{mH_f}{hA(T_m - T_f)} = \frac{(5.56\times10^{-4})(2.07\times10^{5})}{(450)(7.85\times10^{-5})(1358-310)} = 3.11\ \text{s}$$

$\therefore t_2 = 3.11 + 0.66 = 3.77$ s

Cool solid from 1358 to 920 K:

$$t = -\frac{\rho C_p V}{hA}\ln\frac{T - T_f}{T_m - T_f} = -\frac{(8970)(377)(8.33\times10^{-4})}{(450)}\ln\frac{(920-310)}{(1358-310)} = 3.39\ \text{s}$$

$\therefore t_3 = 3.39 + 3.77 = 7.16$ s

Cool solid from 920 to 365 K

$$t = -\frac{(8970)(377)(8.33\times10^{-4})}{(2270)}\ln\frac{(365-310)}{(920-310)} = 2.99\ \text{s}$$

$\therefore$ Total time = 7.16 + 2.99 = 10.15 s

9.16 Steel ball bearings (60 mm in diameter) are austenitized at 1089 K and then quenched in fluid *X* at 310 K. It is known by utilizing a thermocouple that a continuous vapor film surrounds the bearings for 72 s until the surface temperature drops to 530 K and at the same time the center temperature is 645 K. Knowing these results, determine the time it takes for the center of smaller bearings (6 mm in diameter) of the same steel to reach 920 K when quenched from 1089 K into fluid *X* at 310 K.

With $R = 0.03$ m

at surface with $T = 530$ K, $\dfrac{T - T_f}{T_i - T_f} = \dfrac{530 - 310}{1089 - 310} = 0.282$

at center with $T = 645$ K, $\dfrac{T - T_f}{T_i - T_f} = \dfrac{645 - 310}{1089 - 310} = 0.430$

By trial and error, determine values of Bi and Fo which give the relative temperatures of the surface and center using Figs. 9.10a and 9.10d.

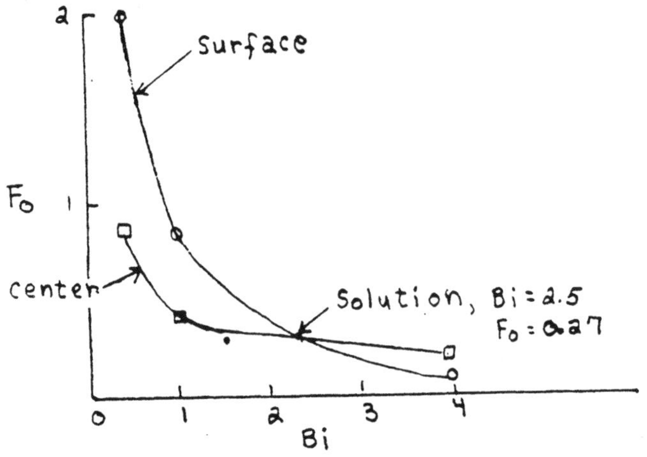

with Fo = 0.27, then $\alpha = \dfrac{(0.27)(0.030)^2}{(72)} = 3.375 \times 10^{-6}\ m^2\ s^{-1}$

With $R = 0.003$ m

Bi = 0.25

$$\left.\frac{T - T_f}{T_i - T_f}\right|_{center} = \frac{920 - 310}{1089 - 310} = 0.783$$

Fo ≃ 0.55 Fig. 9.10a.

$$\therefore t = \frac{(0.55)(0.003)^2}{(3.375 \times 10^{-6})} = 1.47\ s$$

9.17 An open-ended cylindrical section of a steel pressure vessel 10 ft in diameter with 8-in. thick walls is being heat-treated. The wall temperature is brought to a uniform value of 1750°F. Then the vessel is quenched into slow oil at 70°F. Gather the data and work the problem in English units.

a) How long does it take for the surface to reach 1000°F?

b) What is the temperature at the center of the wall at that time?

a. Treat as a flat wall

$$\frac{T-T_f}{T_i-T_f} = \frac{1000-70}{1750-70} = 0.554; \quad \rho_{steel} = 7850 \text{ kg m}^{-3} = 490 \text{ lb}_m \text{ ft}^{-3}$$

$$C_{p\,steel} = 712 \text{ J kg K} = 0.17 \text{ BTU lb}_m^{-1} \text{ °F}^{-1}$$

$h \simeq 670 \text{ W m}^{-2}\text{K}^{-1} \simeq 85 \text{ Btu h}^{-1}\text{ft}^{-2}\text{°F}^{-1}$ (slow oil, Fig. 8.14)

$k \simeq 27.5 \text{ W m}^{-1}\text{K}^{-1} \simeq 16 \text{ Btu hr}^{-1}\text{ft}^{-1}\text{°F}^{-1}$ (alloy 3, Fig. 6.8)

$L = 4 \text{ in} = \frac{1}{3} \text{ ft.}$

$Bi = \frac{(85)(1/3)}{17} = 1.77; \quad F_o = 0.16$ (Fig. 9.8d)

$$\therefore\ t = \frac{F_o L^2}{\alpha} = \frac{F_o L^2 \rho c_p}{k} = \frac{(0.16)(1/3)^2(490)(0.17)}{(16)} = 0.093 \text{ h} = 333 \text{ s}$$

b. $Bi = 1.77; \quad F_o = 0.16$

$\frac{T-T_f}{T_i-T_f} \simeq 0.95$ (Fig. 9.8a) $\qquad T = (0.95)(1750-70) + 70 = 1666\text{°F}$

9.18 A cylindrical piece of steel, 50 mm in diameter and initially at 1145 K, is quenched in water at 295 K (H = 59 m^{-1}). Calculate the temperature at the surface of the piece after 60 s, 120 s, and 300 s. Compare your results with the temperature at the same location if the piece had been quenched in oil (H = 20 m^{-1}). *Data*: $\alpha = 6.4 \times 10^{-6}$ m^2 s^{-1}.

Steel Properties: $\rho = 7850 \text{ kg m}^{-3}; \ C_p = 690 \text{ J kg}^{-1}\text{K}^{-1}; \ k = 34.6 \text{ W m}^{-1}\text{K}^{-1};$

$\alpha = 6.4 \times 10^{-6} \text{ m}^2 \text{ s}^{-1}$

<u>Quenched in water</u>

$Bi = HR = (59)(0.025) = 1.475 > 0.1 \quad \therefore$ Non-Newtonian Cooling

time	Fo	θ sur.	Tsurface
60 s	0.614	0.22	482 K
120 s	1.229	0.09	372 K
300 s	3.072	0.005	299 K

where $\theta_{sur} = \frac{T_s - T_f}{T_i - T_f}$; use Fig. 9.9 b

$T_s = \theta(T_i - T_f) + T_f = \theta(1145-295) + 295$

Quenched in oil

$Bi = HR = (20)(0.025) = 0.5 > 0.1$ ∴ Non-Newtonian Cooling

time	Fo	θsur	Tsurface
60 s	0.614	0.51	728 K
120 s	1.229	0.35	592 K
300 s	3.072	0.07	354 K

9.19 Compute the temperature, as a function of time, across a slab of steel 100 mm thick, cooled from 1145 K by water sprays from both sides. *Data*: $\alpha = 6.2 \times 10^{-6}$ m^2 s^{-1}.

Data for water sprays are provided in Fig. 8.18. Obviously the heat transfer coefficient changes significantly with surface temperature. Thus, by the methods discussed in this chapter, we can only make an approximate calculation. The average surface temperature is used to estimate the heat transfer coefficient. Furthermore, cooling to about 750 K is often considered, so we assume that is the case here.

Avg. T: $T = \frac{1}{2}(1145 + 750) = 948$ K; $h(948\text{ K}) = 2000$ W m^{-2}K^{-1} (extrapolate the curve for 13.7 L m^{-2}s^{-1}); $\rho(948\text{ K}) = 7640$ kg m^{-3} (Table B.1); $C_p(948\text{ K}) = 800$ J kg^{-1}K^{-1}; $L = 50$ mm $= 0.050$ m; $T_i = 1145$ K; $T_f = 300$ K; $\alpha = \frac{k}{\rho C_p} = 6.2 \times 10^{-6}$ m^2 s^{-1} (given); $k = (6.2 \times 10^{-6})(7640)(800)$ $= 38$ W m^{-1}K^{-1}; $Bi = \frac{hL}{k} = \frac{(2000)(0.050)}{38} = 2.63$

Take values from Figs. 9.8a – 9.8d for $1 < \frac{T - T_f}{T_i - T_f} < 0.53$

For the time scale, consult Fig. 9.8a, with $\theta = \frac{T - T_f}{T_i - T_f} = 0.50$ and $Bi = 2.6$. We get $F_o = 0.62$, so let's select $Fo = 0.1$, 0.3 and 0.60.

F_o	$\theta(\frac{x}{L}=0)$	$\theta(\frac{x}{L}=0.3)$	$\theta(\frac{x}{L}=0.6)$	$\theta(\frac{x}{L}=1.0)$
0.1	0.98(1128 K)	0.96(1111 K)	0.85(1018 K)	0.48(706 K)
0.3	0.78(959 K)	0.75(934 K)	0.61(815 K)	0.31(562 K)
0.6	0.51 (731 K)	0.49(714 K)	0.40(638 K)	0.21(477 K)

The corresponding times are $t = \frac{L^2}{\alpha} F_o$:

$$t = \frac{(0.050)^2}{6.2 \times 10^{-6}} F_o = 403\, F_o$$

$t = 40.3$, 121, and 242 s

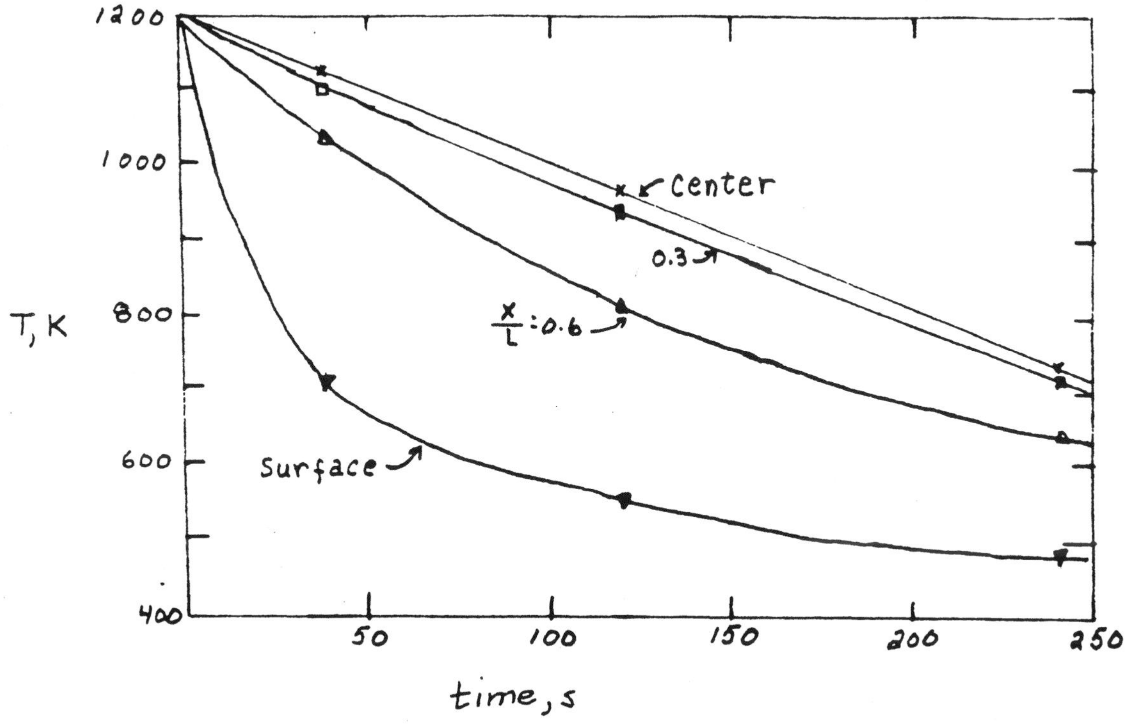

9.20 Consider a short cylinder 150 mm high and with a diameter of 150 mm. The cylinder is initially at a uniform temperature of 530 K and cools in ambient air at 300 K. Assume steel with $\alpha = 6.2 \times 10^{-6}$ m^2 s^{-1} and $k = 35$ W m^{-1} K^{-1}.

a) Write the partial differential equation that describes the temperature within the cylinder.
b) Calculate the temperature at the geometric center after 3600 s of cooling.
c) Calculate the temperature on the cylindrical surface midway between the end faces after 3600 s of cooling. Estimate heat transfer coefficients by consulting Problem 8.14.
d) In answering parts b) and c), show why your calculation procedure was justified, that is, demonstrate that the differential equation in part a) is satisfied.

a. The partial differential equation is deduced from Table 7.5, Eq. (B)

$\frac{\partial T}{\partial t} = \alpha\left[\frac{1}{r}\frac{\partial}{\partial r}\left(r\frac{\partial T}{\partial r}\right) + \frac{\partial^2 T}{\partial z^2}\right]$. However, if we assume Newtonian cooling,

Eq. (9.20): $-V\rho c_p \frac{dT}{dt} = hA(T - T_f)$. Use the simplified equations of Problem 8.14 for the heat transfer coefficents - assume the cylinder is horizontal.

For the ends, $L^3\Delta T = (0.15)^3(530-300) = 0.776$; $h = 1.42\left(\frac{230}{0.15}\right)^{1/4} = 8.89$ W m^{-2} K^{-1}

For the sides, $D^3\Delta T = (0.15)^3(530-300) = 0.776$; $h = 1.22\left(\frac{230}{0.15}\right)^{1/4} = 7.63$ W m^{-2} K^{-1}

An average, weighted according to area, is:

$$h = \left(\frac{2}{3}\right)(7.63) + \left(\frac{1}{3}\right)(8.89) = 8.05 \text{ W m}^{-2}\text{ K}^{-1}$$

With $k = 35$ W m^{-1} K^{-1}; $Bi_L = Bi_R = \frac{(8.05)(0.075)}{35} = 0.017 < .1$ ∴ Newtonian cooling is valid.

b. $\frac{T - T_f}{T_i - T_f} = \exp\left(-\frac{hAt}{\rho c_p V}\right) = \exp\left(-\frac{h\alpha}{k}\frac{At}{V}\right)$; $\frac{A}{V} = \frac{6}{0.15} = 40$ m^{-1}

$$T = (530 - 300)\exp\left(-\frac{(8.05)(6.02\times10^{-6})(40)(3600)}{35}\right) + 300 = 488 \text{ K}$$

c. Since Newtonian cooling applies, $T_{center} \simeq T_{surface}$. ∴ $T = 488$ K

d. The partial differential equation for Newtonian cooling is justified since Biot No. is less than 0.1.

9.21 A steel blank, 300 mm in diameter and 600 mm long, is heated in a preheating furnace maintained at 1410 K as the first step in a forging operation.

a) Calculate the temperature in the center of the blank after the blank has been heated for 5400 s from an initial temperature of 295 K.

b) Calculate the time required to heat a smaller blank, 150 mm in diameter and 300 mm long, to the same center temperature as the larger blank in part a).

Data: h = 110 W m^{-2} K^{-1}; k = 35 W m^{-1} K^{-1}; ρ = 7690 kg m^{-3}; C_p = 500 J kg^{-1} K^{-1}.

a. Application of Fig. 9.8a and 9.9a.

∞ slab: $L = 0.300\ m$; $\alpha = \dfrac{k}{\rho C_p} = \dfrac{35\ W}{m\ K}\cdot\dfrac{m^3}{7690\ kg}\cdot\dfrac{kg\ K}{500\ J}\cdot\dfrac{J}{W s} = 9.1\times10^{-6}\ m^2 s^{-1}$

$F_o = \dfrac{\alpha t}{L^2} = \dfrac{9.1\times10^{-6}\ m^2}{s}\cdot\dfrac{5400\ s}{(0.3)^2\ m^2} = 0.546$

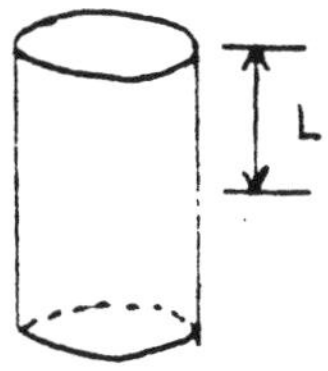

$B_i = \dfrac{hL}{k} = \dfrac{110\ W}{m^2\ K}\cdot\dfrac{0.300\ m}{}\cdot\dfrac{m\ K}{35\ W} = 0.943$

$\dfrac{T - T_f}{T_i - T_f} \simeq 0.74$ Fig. 9.8a.

∞ cylinder: $R = 0.150\ m$

$F_o = \dfrac{(9.1\times10^{-6})(5400)}{(0.15)^2} = 2.184$

$B_i = \dfrac{(110)(0.15)}{(35)} = 0.47$

$\dfrac{T - T_f}{T_i - T_f} \simeq 0.25$ Fig. 9.9a

$\therefore \dfrac{T - T_f}{T_i - T_f} = (0.74)(0.25) = 0.185$; $T = 0.185(295 - 1410) + 1410 = 1204\ K$

b. $\dfrac{\alpha_1 t_1}{L_1^2} = \dfrac{\alpha_2 t_2}{L_2^2}$ or $\dfrac{\alpha_1 t_1}{R_1^2} = \dfrac{\alpha_2 t_2}{R_2^2}$; geometrically similar

Since $\alpha_1 = \alpha_2$: $t_2 = \left(\dfrac{L_2}{L_1}\right)^2 t_1 = \left(\dfrac{1}{2}\right)^2 (5400) = 1350\ s$

$t_2 = 1350\ s$

9.22 The temperature field $T(x,y,t)$ in an infinitely long rectangular ($2L \times 2l$) bar must satisfy the partial differential equation

$$\frac{\partial^2 T}{\partial x^2} + \frac{\partial^2 T}{\partial y^2} = \frac{1}{\alpha}\frac{\partial T}{\partial t}.$$

Prove that $T(x,y,t)$ can be found by the product

$$T(x,y,t) = T_l(x,t) \cdot T_L(y,t),$$

where $T_l(x,t)$ is the solution for the temperature history in the semi-infinite plate bounded by $l < x < +l$, and $T_L(y,t)$ is the solution for the temperature history in the semi-infinite plate bounded by $-L < y < L$.

$$\frac{\partial^2 T}{\partial x^2} + \frac{\partial^2 T}{\partial Y^2} = \frac{1}{\alpha}\frac{\partial T}{\partial t} : \frac{\partial^2 T}{\partial x^2} = T_L \frac{\partial^2 T_l}{\partial x^2} ; \frac{\partial^2 T}{\partial Y} = T_l \frac{\partial^2 T_L}{\partial Y^2} ; \frac{\partial T}{\partial t} = T_l \frac{\partial T_L}{\partial t} + T_L \frac{\partial T_l}{\partial t}$$

$$\therefore T_L \frac{\partial^2 T_l}{\partial x^2} + T_l \frac{\partial^2 T_L}{\partial Y^2} = \frac{1}{\alpha}\left(T_l \frac{\partial T_L}{\partial t} + T_L \frac{\partial T_l}{\partial t}\right)$$

However: $\frac{\partial^2 T_l}{\partial x^2} = \frac{1}{\alpha}\frac{\partial T_l}{\partial t}$ and $\frac{\partial^2 T_L}{\partial Y^2} = \frac{1}{\alpha}\frac{\partial T_L}{\partial t}$

So that: $\frac{1}{\alpha}\left(T_L \frac{\partial T_l}{\partial t} + T_l \frac{\partial T_L}{\partial t}\right) = \frac{1}{\alpha}\left(T_L \frac{\partial T_l}{\partial t} + T_l \frac{\partial T_l}{\partial t}\right)$

9.23 A strip of spring steel (0.5 mm thick) is heated to 1090 K and quenched in "slow oil" maintained at 310 K. Using Fig. 8.14, calculate the cooling rate at 1090 K, 755 K, and 590 K. *Data*: k (1090 K) = 26 W m^{-1} K^{-1}; k (755 K) = 35 W m^{-1} K^{-1}; k (590 K) = 38 W m^{-1} K^{-1}; ρ = 7840 kg m^{-3}; C_p = 628 J kg^{-1} K^{-1}.

From Fig. 8.14: $h(1090\text{K}) = 550 \text{ W m}^{-2}\text{K}^{-1}$; $h(755\text{K}) = 600$; $h(590\text{K}) = 440$

$$Bi = \frac{hL}{k} = \frac{(6\times10^2)(2.5\times10^{-4})}{35} = 4.3\times10^{-3} << 0.1$$

$\therefore$ Newtonian cooling applies $\frac{dT}{dt} = -\frac{hA(T-T_f)}{V\rho C_p}$; $\frac{A}{V} = \frac{2}{5\times10^{-4}} = 4\times10^3 \text{ m}^{-1}$

$$\frac{dT}{dt} = -\frac{4\times10^3}{\text{m}}\left|\frac{\text{m}^3}{7.84\times10^3\text{ kg}}\right|\frac{\text{kg K}}{6.28\times10^2\text{ J}}\left|\frac{\text{J}}{\text{W s}}\right|h\,(T-310)$$

$$= -8.124\times10^{-4}\, h\,(T-310) \text{ m}^2\text{K W}^{-1}\text{s}^{-1}$$

$$T(1090\text{K}): \frac{dT}{dt} = (-8.124\times10^{-4})(550)(1090-310) = -349 \text{ K s}^{-1}$$

$T(755\,K)$: $\frac{dT}{dt} = (-8.124\times10^{-4})(6\times10^{2})(755-310) = -217\ K\ s^{-1}$

$T(590\,K)$: $\frac{dT}{dt} = (-8.124\times10^{-4})(440)(590-310) = -100\ K\ s^{-1}$

9.24 A thermoplastic (polypropylene) at 500 K is injected into a mold at 300 K to form a plate that is 4 mm thick. The plastic may not be ejected until the center-line temperature is 360 K. Estimate the time required so that the plastic can be ejected (this is called the "freeze-off" time).

$$\alpha = \frac{k}{\rho C_p} = \frac{0.12\ W}{m\ K}\left|\frac{m^3}{905\ kg}\right|\frac{kg\ K}{1.9\times10^3\ J}\left|\frac{J}{s\ W}\right. = 6.98\times10^{-8}\ m^2\ s^{-1}$$

$Bi = \frac{hL}{k} = 1000$ Assume excellent thermal contact between polymer and the mold.

$$\frac{T-T_f}{T_i-T_f} = \frac{360-300}{500-300} = 0.300\ ;\ \frac{x}{L}\ (\text{center}) = 0$$

Fig. 9.8a for ∞ plate gives: $F_o = \frac{\alpha t}{L^2} = 0.59$

$$t = (0.59)\frac{(2\times10^{-3})^2}{(6.98\times10^{-8})} = 33.8\ s$$

9.25 A very long cylinder is cooled in a fluid in which the heat transfer coefficient is constant but its value is unknown. At the center, the temperature does not noticeably decrease until the relative temperature at the surface cools to 0.5. Based on this information, deduce the heat transfer coefficient. *Data for solid*: radius = 80 mm; ρ = 2000 kg m^{-3}; C_p = 480 J kg^{-1} K^{-1}; k = 0.246 W m^{-1} K^{-1}.

From Fig. 9.9(a): $\frac{T-T_f}{T_i-T_f}$ starts to decrease when $F_o = \frac{\alpha t}{R^2} \simeq 0.05$

From Fig. 9.9(b) (surface), when $\frac{T-T_f}{T_i-T_f} = 0.5$ and $F_o = 0.05$ then $Bi \simeq 3.3$

$$\therefore h = \frac{k\,Bi}{R} = \frac{0.246\ W}{m\ K}\left|\,3.3\,\right|\frac{1}{0.08\ m} = 10.1\ W m^{-2} K^{-1}$$

9.26 The end ($z = 0$) of very long cylindrical bar is heated uniformly with a constant flux of q_0. The side of the bar loses heat to the surroundings at T_∞ with a uniform and constant value of h. Before heating, the bar is at a uniform temperature of T_i.

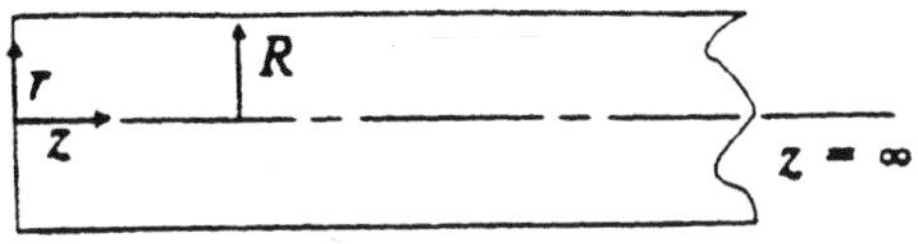

a) For constant thermal properties, write an appropriate form of the equation of energy for temperature within the bar during the transient period.
b) Give appropriate boundary conditions and an initial condition for part a).
c) Is it possible to reach a steady state? Give a reason for your answer.

a. $T = f(r, z, t)$ Table 7.5, Eq. (A) $C_v = C_p$

$$\frac{\partial T}{\partial t} = \alpha\left[\frac{1}{r}\frac{\partial}{\partial r}\left(r\frac{\partial T}{\partial r}\right) + \frac{\partial^2 T}{\partial z^2}\right]$$

b. $\frac{\partial T}{\partial r}(0, z, t) = 0$

$$\frac{\partial T}{\partial r}(R, z, t) = -\frac{h}{k}\left[T(R, z, t) - T_\infty\right]$$

$\frac{\partial T}{\partial z}(r, 0, t) = -\frac{q_0}{k}$ · $\frac{\partial T}{\partial z}(r, \infty, t) = 0$ I.C.: $T(r, z, 0) = T_i$

c. Yes. at steady-state and removed from $z = 0$, the bar will approach a temperature of T_∞. In addition the power added at $z = 0$ will equal the power lost from the side of the ∞-long cylinder.

9.27 A very large and thick slab of copper is initially at a uniform temperature of 600°F. The surface temperature is suddenly lowered to 100°F by a water spray.
a) What is the temperature at a depth of 3 in., 4 min. after the surface temperature has changed?
b) If it is necessary to predict the temperature in the slab for a period of 5 minutes, what must be the thickness of the slab so that it can be approximated as a semi-infinite solid?

Data for copper: ρ = 552 $lb_m\ ft^{-3}$; C_p = 0.100 Btu $lb_m^{-1}\ °F^{-1}$; k = 215 Btu $h^{-1}\ ft^{-1}\ °F^{-1}$. Work this out in English units.

a. Assume $h \to \infty$ for water spray.

$$\frac{T - T_s}{T_i - T_s} = \text{erf}\left(\frac{x}{2(\alpha t)^{1/2}}\right);\ x = \frac{3}{12} = 0.25\ ft.;\ \alpha = \frac{k}{\rho C_p} = \frac{(215)}{(552)(0.100)} = 3.89\ ft^2\ hr^{-1};$$

$$t = 4\ min. = 6.667 \times 10^{-2}\ h$$

$$\frac{x}{2(\alpha t)^{1/2}} = \frac{0.25}{2\left[(3.89)(6.667 \times 10^{-2})\right]^{1/2}} = 0.245$$

Table of error function gives

$$\frac{T-T_s}{T_i-T_s} = \text{erf}(0.245) = 0.271$$

$$T = (0.271)(600-100) + 100 = 235°F$$

b. $\dfrac{x}{2(\alpha t)^{1/2}} \simeq 2$

$$x = 4\left[(3.89)\left(\frac{5}{60}\right)\right]^{1/2} = 2.28 \text{ ft (distance from surface)}$$

$\therefore$ Thickness $= 2x = 4.56$ ft.

9.28 Initially the mold for a junction-shaped casting is at a uniform temperature, T_0. Then liquid metal at its freezing point, T_s, is poured into the mold. During the period of time it takes for the metal to solidify, the surface of the mold is maintained at T_s. Assume constant thermal properties of the mold. For the upper right quadrant of the mold which extends to ∞ for both x and y, write a solution that yields temperature as a function of coordinates in the mold and time.

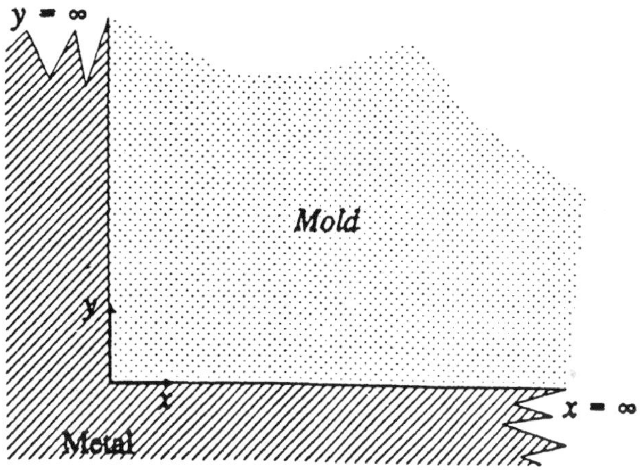

$$\frac{\partial T}{\partial t} = \alpha\left(\frac{\partial^2 T}{\partial x^2} + \frac{\partial^2 T}{\partial y^2}\right)$$

Boundary Conditions: $T(0,y,t) = T_s$; $T(x,0,t) = T_s$

$T(\infty,\infty,t) = T_0$

Initial Condition: $T(x,y,0) = T_0$

Product solution of "x-soln." and "y-soln."

"x-soln.": $\dfrac{T-T_s}{T_0-T_s} = \text{erf}\dfrac{x}{2(\alpha t)^{1/2}}$; "y-soln.": $\dfrac{T-T_s}{T_0-T_s} = \text{erf}\dfrac{y}{2(\alpha t)^{1/2}}$

$$\therefore \frac{T-T_s}{T_0-T_s} = \text{erf}\frac{x}{2(\alpha t)^{1/2}} \cdot \text{erf}\frac{y}{2(\alpha t)^{1/2}}$$

9.29 A ceramic brick, with dimensions 150 mm × 75 mm × 75 mm and initially at 310 K, is heated in a salt bath maintained at 865 K. The heat transfer coefficient is uniform for all faces and equals 280 W m^{-2} K^{-1}. Thermal properties of the brick are as follows: k = 1.7 W m^{-1} K^{-1}; ρ = 3200 kg m^{-3}; C_p = 840 J kg^{-1} K^{-1}. a) How long will it take for the center of the brick to reach 810 K? b) When the center is at 810 K, what is the maximum temperature difference in the brick?

$\frac{T-T_f}{T_i-T_f} = X(0,t) \cdot Y(0,t) \cdot Z(0,t)$ where X, Y, Z are solutions to $\frac{\infty}{2}$-plates.

With T specified, we need a trial and error solution to get t (time).

Let t = 600 s

$$\alpha = \frac{1.7}{(3200)(840)} = 6.32 \times 10^{-7}\ m^2 s^{-1}$$

$$Fo_x = \frac{\alpha t}{L^2} = \frac{(6.32\times10^{-7})(600)}{(75\times10^{-3})^2} = 6.74\times10^{-2}$$

$$Fo_Y = Fo_Z = \frac{(6.32\times10^{-7})(600)}{(37.5\times10^{-3})^2} = 2.70\times10^{-1}$$

$$Bi_x = \frac{hL}{k} = \frac{(280)(75\times10^{-3})}{(1.7)} = 12.4$$

$$Bi_Y = Bi_Z = \frac{(280)(37.5\times10^{-3})}{(1.7)} = 6.2$$

$$X(0,t) = \left.\frac{T-T_f}{T_i-T_f}\right)_x = 0.99 ;\ Y(0,t) = Z(0,t) = 0.70 \text{ (from Fig. 9.8 a)}$$

$$\therefore \frac{T-T_f}{T_i-T_f} = (0.99)(0.70)(0.70) = 0.485$$

We need $\frac{T-T_f}{T_i-T_f} = \frac{810-865}{310-865} = 0.099$ so a longer time is required.

Dimensional temps. for various times are calculated and plotted.

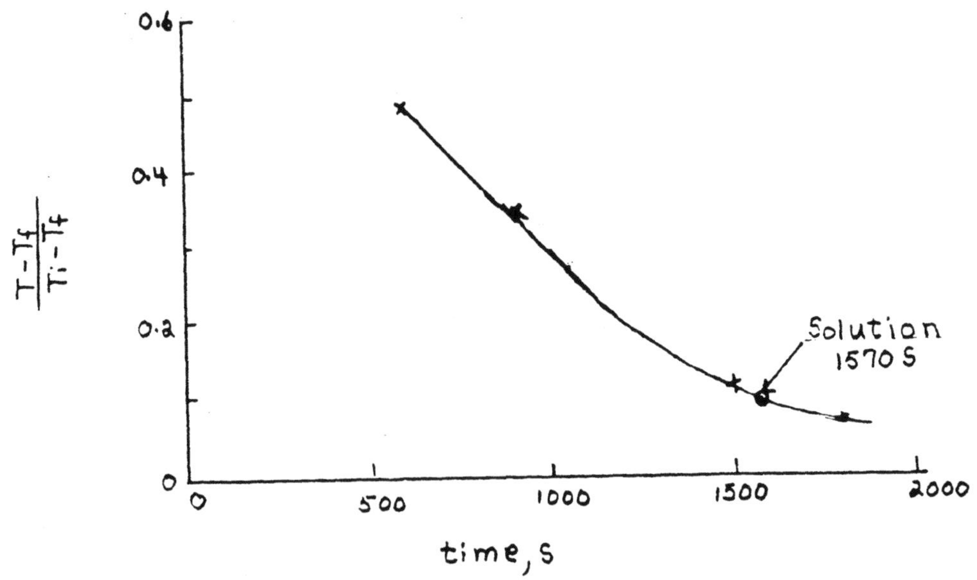

b. $Fo_x = \frac{(6.32\times10^{-7})(1570)}{(75\times10^{-3})^2} = 0.176$

$Bi_x = 12.4$

$X(1,t) = \left.\frac{T-T_f}{T_i-T_f}\right)_x = 0.11$ (surface)

$Fo_y = Fo_z = \frac{(6.32\times10^{-7})(1570)}{(37.5\times10^{-3})^2} = 0.706$

$Bi_y = Bi_z = 6.2$

$Y(1,t) = Z(1,t) = 0.04$ (surfaces)

$\therefore \left.\frac{T-T_f}{T_i-T_f}\right)_{corner} = (0.11)(0.04)(0.04) = 1.76\times10^{-4}$

$T_{corner} = (1.76\times10^{-4})(310-865) + 865 \cong 865$ K

Then ΔT (max.) = 865 − 810 = 55 K

9.30 A solid circular cylinder of steel, with a diameter of 240 mm and a length of 183 mm, is initially at 300 K. A treatment to transform retained austenite requires cooling in liquid nitrogen (78 K). For a cooling time of 2520 s, it is known (by measurement) that a point in the center of a circular face is at 100 K. What is the temperature at the geometric center of the cylinder. *Data:* ρ = 7690 kg m^{-3}; C_p = 795 J kg^{-1} K^{-1}; k = 35 W m^{-1} K^{-1}; h = 230 W m^{-2} K^{-1}.

With t = 2520 s

$$Fo_R = \frac{\alpha t}{R^2} = \frac{kt}{\rho C_p R^2} = \frac{35\,W}{m\,K}\Big|\frac{2520\,s}{}\Big|\frac{m^3}{7690\,kg}\Big|\frac{kg\,K}{795\,J}\Big|\frac{}{(0.12)^2 m^2}\Big|\frac{J}{W\,s} = 1.0$$

$$Bi_R = \frac{hR}{k} = \frac{230\,W}{m^2 K}\Big|\frac{0.12\,m}{}\Big|\frac{m\,K}{35\,W} = 0.788$$

$\frac{r}{R} = 0$; Fig. 9.9a.− $\frac{T-T_f}{T_i-T_f} \cong 0.33$ (infinite cylinder)

$Fo_L = \frac{\alpha t}{L^2} = 1\left(\frac{0.12}{0.092}\right)^2 = 1.70$

$Bi_L = \frac{hL}{k} = 0.788\left(\frac{0.092}{0.12}\right) = 0.604$

$\frac{x}{L} = 1$; Fig. 9.8d. $-\ \frac{T - T_f}{T_i - T_f} \cong 0.45$ (infinite plate)

$\therefore\ \frac{T - T_f}{T_i - T_f} = (0.45)(0.33) = 0.148$

$T = (0.148)(T_i - T_f) + T_f = (0.148)(300 - 78) + 78 = 111\ K$

9.31 A sheet of glass, 0.02 ft thick, is cooled from an initial temperature of 1800°F by air flowing over the top surface of the glass. The convective heat transfer coefficient for the air is 8 Btu h^{-1} ft^{-2} $°F^{-1}$, and the glass rests on a perfect insulator.

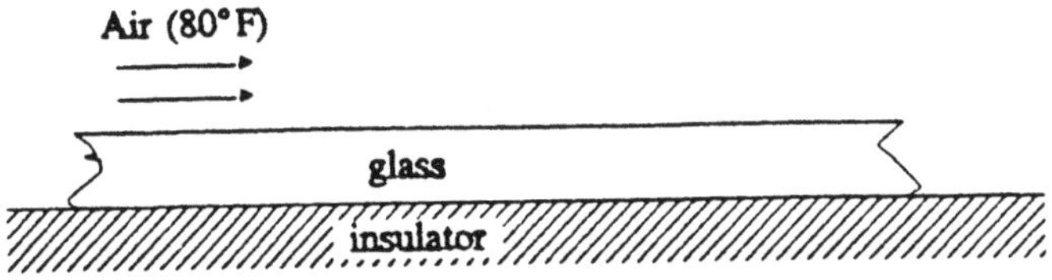

a) What time is required for the bottom surface of the glass to cool to 400°F?

b) When the bottom surface is at 400°F, what is the temperature of the top surface?

Do this problem in English units.

Thermal properties of the glass are k = 0.32 Btu h^{-1} ft^{-1} $°F^{-1}$, ρ = 200 lb_m ft^{-3} and C_p = 0.2 Btu lb_m^{-1} $°F^{-1}$.

a. $\frac{hL}{k} = \frac{8\ Btu}{h\ ft^2\ °F} \Big| \frac{0.02\ ft}{} \Big| \frac{h\ ft\ °F}{0.32\ Btu} = 0.5$; $Bi = 0.5$

$\frac{x}{L} = 1$

$\frac{T - T_f}{T_i - T_f} = \frac{400 - 80}{1800 - 80} = 0.186$

Fig. 9.8a. $F_o = \frac{\alpha t}{L^2} \cong 4.2$; $\alpha = \frac{k}{\rho C_p} = \frac{0.32\ Btu}{h\ ft\ °F} \Big| \frac{ft^3}{200\ lbm} \Big| \frac{lb_m °F}{0.2\ Btu} = 8 \times 10^{-3}\ ft^2\ h^{-1}$

$t = \frac{4.2}{} \Big| \frac{(0.02)^2\ ft^2}{} \Big| \frac{h}{8 \times 10^{-3}\ ft^2} = 0.21\ h = 756\ s$

b. $\frac{x}{L} = 1$; $F_o = 4.2$; $Bi = 0.5$

Fig. 9.8d. $-\ \frac{T - T_f}{T_i - T_f} = 0.16$; $\frac{T - 80}{1800 - 80} = 0.16$; $T = 0.16\ (1720) + 80 = 355\ °F$

$T = 355\ °F$

9.32 A cylinder, initially at 300°F, is plunged into a large melt of a low melting point metal maintained at 810 K; the heat transfer coefficient may be taken to be infinity. The dimensions of the cylinder are 230 mm diameter and 150 mm length.

a) After 100 s of heating in the melt, what is the temperature in the geometric center of the cylinder?

b) After 5 s, what is the temperature at the centers of the circular surfaces?

Data: $k = 5.2 \text{ W m}^{-1} \text{ K}^{-1}$; $\rho = 4810 \text{ kg m}^{-3}$; $C_p = 420 \text{ J kg}^{-1} \text{ K}^{-1}$.

a. $T_i = 80°F = 300 K$

$$\alpha = \frac{k}{\rho c_p} = \frac{5.2 \, W}{m \, K} \left| \frac{m^3}{4810 \, kg} \right| \frac{kg \, K}{420 \, J} \left| \frac{J}{s \, W} \right. = 2.57 \times 10^{-6} \, m^2 \, s^{-1}$$

$$F_o = \frac{\alpha t}{R^2} = \frac{2.57 \times 10^{-6} \, m^2}{s} \left| \frac{100 \, s}{(0.115)^2 m^2} \right. = 0.0194$$

$$B_i = \frac{hR}{k} = 1000 \simeq \infty$$

$$\frac{r}{R} = 0; \text{ Fig. 9.9a } \frac{T - T_f}{T_i - T_f} = 1 \text{ (infinite cylinder)}$$

$$F_o = \frac{\alpha t}{L^2} = \frac{(2.57 \times 10^{-6})(100)}{(0.075)^2} = 0.046$$

$$\frac{x}{L} = 0; \text{ Fig. 9.8a } \frac{T - T_f}{T_i - T_f} = 0.996 \text{ (infinite plate)}$$

$$\therefore \frac{T - T_f}{T_i - T_f} = (1)(0.996); \; T = 0.996(300 - 810) + 810 = 302 \text{ K}$$

b. Because $h = \infty$, the surface temperatures are 810 K.

9.33 A laser beam is used as a moving point source to harden the surface of a thick piece of steel by multiple passes (i.e., rapid scanning) across the surface; no melting occurs. To what depth is it possible to produce martensite with one pass if the critical cooling rate for the steel is 280 K s^{-1} at 810 K. For this steel it is known that austenite exists at $T \geq 1090$ K. *Data for steel*: $\alpha = 7.2 \times 10^{-6}$ m^2 s^{-1}; $k = 35$ W m^{-1} K^{-1}; initial temperature = 300 K. *Laser conditions*: Power = 4 kW at 50 percent efficiency; speed = 420 mm s^{-1}.

$\left|\frac{dT}{dt}\right| = 280$ K s^{-1} at $T = 810$ K (Critical cooling rate)

$T_i = 300$ K $\quad Q = 2000$ W $\quad V = 0.420$ m s^{-1}

Assume that $Pe_r > 5$ so that Eq. (9.81) can be used to calculate the peak temperature.

$$T_P - T_\infty = \left(\frac{QV}{2\pi k\alpha e}\right)\left(\frac{1}{2 + Pe_r^2}\right)$$

$$Pe_r^2 = \frac{QV}{2\pi k\alpha e(T_p - T_\infty)} - 2$$

$$= \frac{(2000)(0.420)}{(2\pi)(35)(7.2\times10^{-6})(e)(1090-300)} - 2 = 245$$

$Pe_r = 15.65$

Then

$$r = \frac{2\alpha Pe_r}{V} = \frac{(2)(7.2\times10^{-6})(15.65)}{0.420} = 5.39\times10^{-4}\text{ m}$$

$r = 0.539$ mm. This is distance where a peak temperature of 1090 K is achieved and austenite forms.

Now we check to see whether the critical cooling rate can be achieved.

Equation (9.78) gives T at the coordinates in a moving frame. Imagine that we attach a thermocouple to the steel and observe temperature versus time.

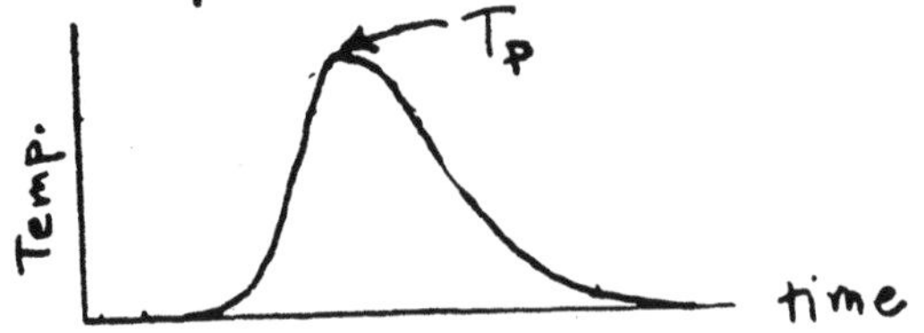

```
10 'Problem 9.33   Temperature versus time at the location where the peak
20 '               temperature is 1090 K. We want cooling rate at 810 K.
30 T = 810 : R = .000539 : V = .42 : ALPHA = .0000072 : TINF = 300
40 PER = V*R/(2*ALPHA) : Q = 2000  : K = 35 : PI = 3.1416
50 LPRINT "      z, m          t, s        T, K "
60 LPRINT "   *********     *********     *******"
70 'solve Eq.(9.78)
80 FOR Z = -.0005 TO .017 STEP .001
90     PEZ = V*Z/(2*ALPHA) : PARAM = PEZ*PEZ + PER*PER
100    FACTOR1 = Q*V/(4*PI*K*ALPHA) : FACTOR2 = 1/SQR(PARAM)
110    FACTOR3 = PEZ - SQR(PARAM) : FACTOR3 = EXP(FACTOR3)
120    DELTAT = FACTOR1*FACTOR2*FACTOR3
130    T = TINF + DELTAT : TIME = Z/V
140    LPRINT USING "   ##.##^^^^     ##.##^^^^     ####.# ";Z,TIME,T
150 NEXT Z
160 END
```

z, m	t, s	T, K
*********	*********	*******
-5.00E-04	-1.19E-03	300.0
5.00E-04	1.19E-03	313.0
1.50E-03	3.57E-03	668.9
2.50E-03	5.95E-03	965.8
3.50E-03	8.33E-03	1070.9
4.50E-03	1.07E-02	1085.3
5.50E-03	1.31E-02	1063.1
6.50E-03	1.55E-02	1027.4
7.50E-03	1.79E-02	988.0
8.50E-03	2.02E-02	949.0
9.50E-03	2.26E-02	912.1
1.05E-02	2.50E-02	877.9
1.15E-02	2.74E-02	846.6
1.25E-02	2.98E-02	818.0
1.35E-02	3.21E-02	791.9
1.45E-02	3.45E-02	768.0
1.55E-02	3.69E-02	746.2
1.65E-02	3.93E-02	726.2

On cooling through 810 K,

$$\frac{dT}{dt} \simeq \frac{(791.9-818)\,K}{2.381\times10^{-3}\,s} \simeq -10\,000\ K\ s^{-1}.$$

This certainly exceeds the critical cooling rate so all martensite will form wherever peak temperature is 1090 K or greater.

9.34 A laser beam is used to remelt silicon in a process to produce material for solar cells. The silicon is 100 mm wide and 2 mm thick. The molten pool passes across the width of the silicon, as shown in the diagram, with a velocity $V = 4.0 \text{ m s}^{-1}$. The bulk of the silicon is at 293 K. Assume that the pool can be treated as a moving source.

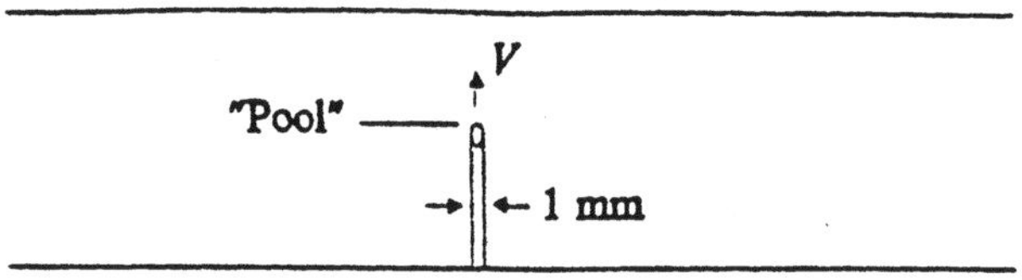

a) Calculate the maximum cooling rate in the silicon. At what location is this cooling rate achieved?

b) At what distance from the centerline of the path left by the laser beam, is a peak temperature of 1280 K achieved?

Data: Beam conditions, 5 kW at 50% efficiency. Silicon: melting temperature is 1700 K; heat of fusion is $1.41 \times 10^3 \text{ kJ kg}^{-1}$; thermal conductivity is $100 \text{ W m}^{-1} \text{ K}^{-1}$; thermal diffusivity is $5.2 \times 10^{-5} \text{ m}^2 \text{ s}^{-1}$.

a. The maximum cooling rate is achieved directly behind the beam where $r = 0$. For a line source (the Si is only 2mm thick):

$$T - T_\infty = \frac{Q}{\sqrt{8\pi}\, k\delta} \, \frac{\exp\left[(Pe_z) - (Pe_z^2 + Pe_r^2)^{1/2}\right]}{(Pe_z^2 + Pe_r^2)^{1/4}}$$

We want $\partial T'/\partial t = V(\partial T/\partial z)$ along the path $r = 0$. At the line source, itself, both the temperature and $\partial T'/\partial t$ are ∞, so let's calculate the cooling rate at the solid-liquid interface ($T = T_M$). The cooling rate along $r = 0$ is

$$\frac{\partial T'}{\partial t} = -2\pi k \rho c_p \left(\frac{V\delta}{Q}\right)^2 (T' - T_\infty)^3$$

$$k\rho c_p = \frac{k^2}{\alpha} \; \therefore \; \frac{\partial T'}{\partial t} = (2\pi)\left(\frac{100^2}{5.2 \times 10^{-5}}\right)\left[\frac{(4.0)(0.002)}{2500}\right]^2 (1700 - 293)^3 = 3.45 \times 10^7 \text{ K s}^{-1}$$

b. Apply Eq. (9.85): $T_p - T_\infty = \dfrac{Q}{\sqrt{8\pi e}\, k\delta Pe_r}$; $Pe_r = \dfrac{Q}{\sqrt{8\pi e}\, k\delta (T_p - T_\infty)}$

$$= \frac{2500}{\sqrt{8\pi e}\,(100)(0.002)(1280 - 293)} = 1.532$$

$$r = \frac{2\alpha}{V} Pe_r = \frac{(2)(5.2 \times 10^{-5})(1.532)}{400} = 3.98 \times 10^{-7} \text{ m} = 0.398 \ \mu\text{m}$$

Data for Problems 10.1-10.8 and Problems 10.27 and 10.28.

Mold material*	k, W m^{-1} K^{-1}	ρ, kg m^{-3}	C_p, J kg^{-1} K^{-1}
Silica sand	0.52	1600	1170
Mullite	0.38	1600	750
Plaster	0.35	1120	840
Zircon sand	1.0	2720	840
Ceramic shell	0.70	1800	1100
Copper	390	9000	380

Casting material	T_M, K	H_f, J kg^{-1}	ρ', kg m^{-3}	C_p', J kg^{-1} K^{-1}	k', W m^{-1} K^{-1}
Iron	1808	2.72×10^5	7210	750	40
Nickel	1728	2.91×10^5	7850	670	35
Aluminum	933	3.91×10^5	2400	1050	260

*Note: Only typical values can be given here. Actual properties depend on temperature, particle size, binders, porosity, etc.

10.1 Plot distance solidified versus the square root of time for the following metals (in each case the pure metal is poured at its melting point against a flat mold wall): a) Iron in a silica sand mold. b) Aluminum in a silica sand mold. c) Iron in a mullite mold heated to 1260 K.

Eq. (10.7) $M = \dfrac{2}{\pi^{1/2}}\left(\dfrac{T_M - T_0}{\rho' H_f}\right)(k\rho C_p)^{1/2}\, t^{1/2}$; Assume $T_0 = 300$K for a and b

a. $M = \dfrac{2}{\pi^{1/2}} \left| \dfrac{(1808-300)\text{K}}{} \right| \dfrac{\text{m}^3}{7210\,\text{kg}} \left| \dfrac{\text{kg}}{2.72\times10^5\,\text{J}} \right| \dfrac{(0.52)^{1/2}\,\text{W}^{1/2}}{\text{m}^{1/2}\,\text{K}^{1/2}} \left| \dfrac{(1600)^{1/2}\,\text{kg}^{1/2}}{\text{m}^{3/2}} \right|$

$\dfrac{(1170)^{1/2}\,\text{J}^{1/2}}{\text{kg}^{1/2}\,\text{K}^{1/2}} \left| \dfrac{\text{s}^{1/2}}{} \right| \dfrac{\text{J}^{1/2}}{\text{s}^{1/2}\,\text{W}^{1/2}} = 8.56\times10^{-4}\, t^{1/2}\, \text{m}$

b. $M = \dfrac{2\,(933-300)\,[(0.52)(1600)(1170)]^{1/2}}{\pi^{1/2}(2400)(3.91\times10^5)}\, t^{1/2} = 7.51\times10^{-4}\, t^{1/2}\, \text{m}$

c. $M = \dfrac{2\,(1808-1260)\,[(0.38)(1600)(750)]^{1/2}}{\pi^{1/2}\,(7210)(2.72\times10^5)}\, t^{1/2} = 2.13\times10^{-4}\, t^{1/2}\, \text{m}$

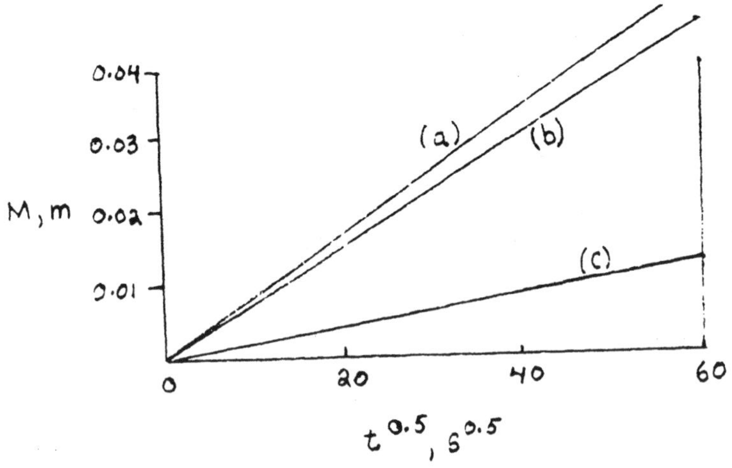

10.2 How long does it take to freeze a 100 mm diameter sphere of pure iron in a silica sand mold assuming: a) no superheat and neglecting the heat flow divergence? b) no superheat and realizing that a sphere is being cast? c) 110 K superheat and realizing that a sphere is being cast?

a. Eq. (10.10) $t = C\left(\frac{V}{A}\right)^2$

$$\left(\frac{V}{A}\right)^2 = \frac{R^2}{9} = \frac{(0.05)^2}{9} = 2.78\times10^{-4}\ m^2$$

$$C = \frac{\pi}{4}\left(\frac{\rho' H_f}{T_M - T_o}\right)^2 \frac{1}{K\rho C_p} = \frac{\pi}{4}\left[\frac{(7210)(2.72\times10^5)}{(1808-300)}\right]^2 \frac{1}{(0.52)(1600)(1170)} = 1.36\times10^6\ s\ m^{-2}$$

$$t = (1.36\times10^6)(2.78\times10^{-4}) = 378\ s \text{ or } 6.30\ min.$$

b. Eq. (10.11) $\beta = \gamma\left(\frac{2}{\sqrt{\pi}} + \frac{1}{3\beta}\right)$

$$\gamma = \left(\frac{T_M - T_o}{\rho' H_f}\right)\rho C_p = \frac{(1808-300)(1600)(1170)}{(7210)(2.72\times10^5)} = 1.439$$

$$\therefore \beta = 1.439\left(\frac{2}{\sqrt{\pi}} + \frac{1}{3\beta}\right) \text{ and } \beta = 1.88;\ \alpha = \frac{K}{\rho C_p} = \frac{0.52}{(1600)(1170)} = 2.78\times10^{-7}\ m^2 s^{-1}$$

$$t = \frac{\left(\frac{V}{A}\right)^2}{\alpha\beta^2} = \frac{2.78\times10^{-4}}{(2.78\times10^{-7})(1.88)^2} = 283\ s \text{ or } 4.72\ min.$$

c. $H_f' = H_f + C_{p,\ell}\,\Delta T_s = 2.72\times10^5 + (750)(110) = 3.545\times10^5$

$\gamma = 1.10;\ \beta = 1.485$

$$t = \frac{2.78\times10^{-4}}{(2.78\times10^{-7})(1.485)^2} = 453\ s \text{ or } 7.56\ min.$$

10.3 Plot distance solidified versus time for iron poured at its melting point into a heavy copper mold, assuming that a) there is a large flat mold wall and no resistance exists to heat flow at the mold-metal interface; b) interface resistance to heat flow is finite (h = 570 W m^{-2} K^{-1}).

a. $$\left[\frac{k'\rho'c_p'}{k\rho c_p}\right] = \left[\frac{(40)(7210)(750)}{(390)(9000)(380)}\right]^{1/2} = 0.403$$

$$(T_M - T_0)\frac{c_p'}{H_f} = (1808 - 300)\frac{750}{2.72\times10^5} = 4.16$$

From Fig. 10.8 $\frac{T_s - T_0}{T_M - T_0} = 0.32$; $T_s = (1808-300)0.32 + 300 = 783$ K

$$(T_M - T_s)\frac{c_p'}{H_f\sqrt{\pi}} = (1808-783)\frac{750}{2.72\times10^5\sqrt{\pi}} = 1.59; \quad \alpha' = \frac{40}{(7210)(750)} = 7.397\times10^{-6},$$

From Fig. 10.5, $\beta = 0.88$

$$M = 2\beta(\alpha' t)^{1/2} = 2(0.88)(7.397\times10^{-6})^{1/2}t^{1/2} = 4.79\times10^{-3}t^{1/2}$$

$$\therefore M = 4.79\times10^{-3}t^{1/2}$$

b. $$h_M = \left[1 + \sqrt{\frac{k\rho c_p}{k'\rho'c_p'}}\right]h = \left[1 + \sqrt{\frac{(390)(9000)(380)}{(40)(7210)(750)}}\right](570) = 1985 \text{ W m}^{-2}\text{ K}^{-1}$$

$$h_c = \left[1 + \sqrt{\frac{k'\rho'c_p'}{k\rho c_p}}\right]h = 800 \text{ W m}^{-2}\text{ K}^{-1}$$

Following the procedure on p. 347.

$T_s = 783$ K (from part a)

$$M = \frac{h_c(T_M - T_s)}{\rho' H_f a}t - \frac{h_c}{2k'}M^2 \quad \text{or} \quad t = \frac{\rho' H_f a}{h_c(T_M - T_s)}\left[M + \frac{h_c}{2k'}M^2\right]$$

$$a = \frac{1}{2} + \sqrt{\frac{1}{4} + \frac{c_p'(T_M - T_s)}{3H_f}} = \frac{1}{2} + \sqrt{\frac{1}{4} + \frac{750(1808-783)}{3(2.72\times10^5)}} = 1.592$$

$$\frac{h_c}{2k'} = \frac{800}{(2)(40)} = 10$$

$$\frac{\rho' H_f a}{h_c(T_M - T_s)} = \frac{(7210)(2.72\times10^5)(1.592)}{(800)(1808-783)} = 3807$$

$t = 3807 (M + 10 M^2)$

M, mm	t, s
10	41.9
20	91.4
30	148.5

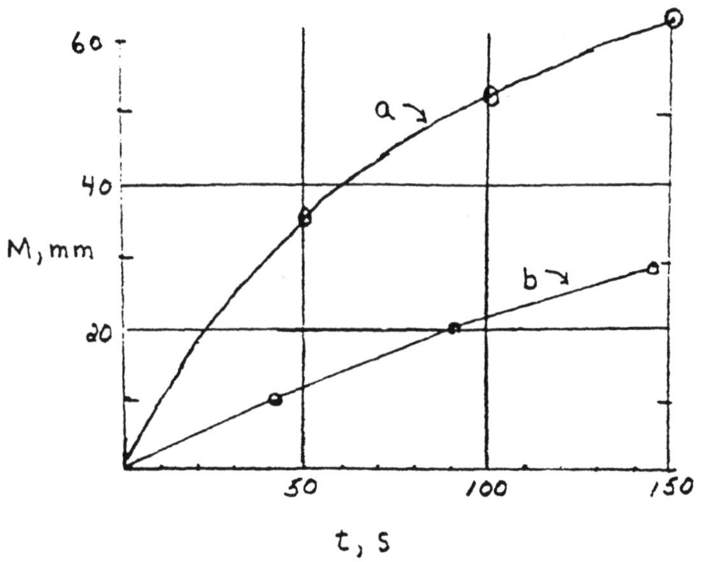

10.4 Show whether iron can be cast against a very thick aluminum mold wall without causing the aluminum to melt.

Eq. (10.25) $\dfrac{(T_M - T_0) c_p'}{H_f \sqrt{\pi}} = \beta e^{\beta^2} \left[\left(\dfrac{k' \rho' c_p'}{k \rho c_p} \right)^{1/2} + \operatorname{erf} \beta \right]$

$\left(\dfrac{k' \rho' c_p'}{k \rho c_p} \right)^{1/2} = \left[\dfrac{(45)(7210)(750)}{(260)(2400)(1050)} \right]^{1/2} = 0.575$

$\dfrac{(T_M - T_0) c_p'}{H_f} = \dfrac{(1808 - 300)(750)}{(2.72 \times 10^5)} = 4.16$

From Fig. 10.8 $\dfrac{T_s - T_0}{T_M - T_0} = 0.44$

$T_s = (1808 - 300)(0.44) + 300 = 964$ K

The aluminum is above the melting point. However, if interface resistance is taken into account, it is likely that $T_s < 933$ K and the aluminum would not melt.

10.5 Slab-shaped steel castings are prone to center-line porosity, which—for our purposes—is simply an alignment of defects along the plane of last solidification. The sketch below shows the solidification of a slab cast in silica sand and the location of the centerline porosity.

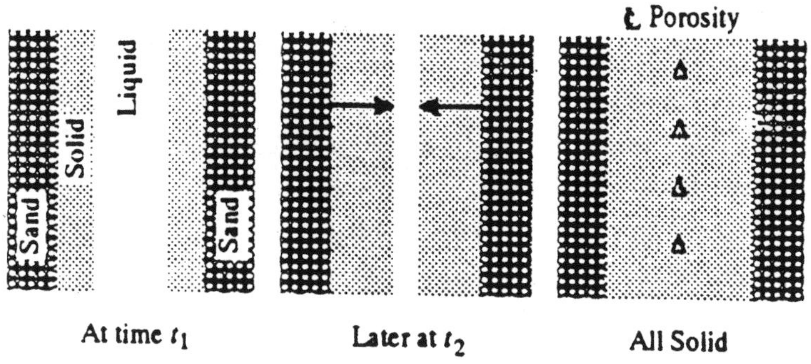

The solidification time for the 2-in. slab cast in sand is known to be 6 min; when cast in an insulating mullite mold, the time is 60 min. If a casting is made in the composite mold depicted to the right, determine the thickness T the casting should have to yield $1\frac{7}{8}$ in. of sound metal after machining.

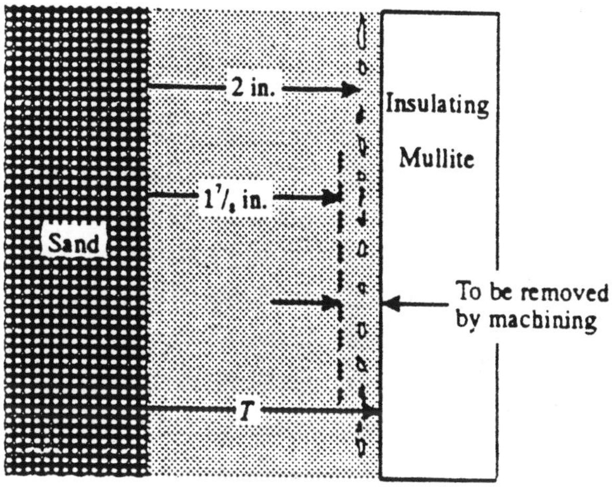

Eq. 10.7 $M = At^{1/2}$ where $A = \frac{2}{\pi^{1/2}}\left(\frac{T_M - T_0}{\rho' H_f}\right)(k\rho C_p)^{1/2} t^{1/2}$

$A_1 = \frac{M_1}{t^{1/2}} = \frac{1}{6^{1/2}}$ silica mold

$A_2 = \frac{M_2}{t^{1/2}} = \frac{1}{60^{1/2}}$ mullite mold

For a sound casting from the composite, M should be 2 in. from the sand side.

$\therefore t^{1/2} = 2(6)^{1/2}$ min$^{1/2}$

In this time, the thickness solidified (M_2) from the mullite side is

$M_2 = \frac{1}{(60)^{1/2}}(2\sqrt{6}) = 0.63$ in.

$\therefore T = 2 + 0.63 = 2.63$ in.

10.6 A 2-in. thick slab of aluminum is cast in a mold made of silica sand (forming one face) and a proprietary material (forming the other face). The aluminum is poured with no superheat, and the as-cast structure of the slab is examined after solidification and cooling. The examination shows that the plane of last solidification (i.e., the plane where the two solidification fronts meet) is located $1\frac{1}{2}$ in. from the sand side. Knowing this, calculate the *heat diffusivity* (not the thermal diffusivity) of the proprietary material.

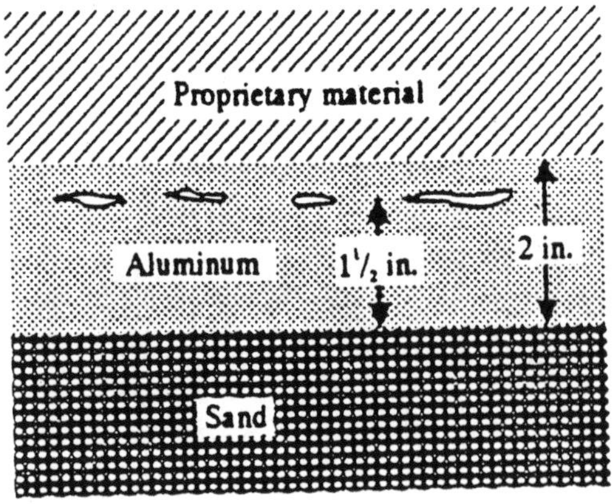

Let $M_1 = 0.5$ in. From proprietary material side.

$M_2 = 1.5$ in From silica sand side.

$k_1 \rho_1 c_{p1}$ proprietary material properties

$k_2 \rho_2 c_{p2}$ silica sand properties.

Eq. (10.7) $M = \dfrac{2}{\sqrt{\pi}}\left(\dfrac{T_M - T_0}{\rho' H_f}\right)(k\rho c_p)^{1/2}(t)^{1/2}$

$t_1 = t_2$

$$\therefore \frac{M_1}{M_2} = \frac{(k_1 \rho_1 c_{p1})^{1/2}}{(k_2 \rho_2 c_{p2})^{1/2}} = \frac{1}{3}$$

$$k_1 \rho_1 c_{p1} = (k_2 \rho_2 c_{p2})\left(\frac{M_1}{M_2}\right)^2 = (0.52)(1600)(1170)\left(\frac{1}{3}\right)^2 = 1.082 \times 10^5 \; W^2 s \, m^{-4} K^{-2}$$

10.7 Consider solidification in a flat ceramic shell mold with a thickness L. There is heat loss from the outside surface to the surroundings with a constant heat transfer coefficient (h = 150 W m^{-2} K^{-1}). Except for very early times, the temperature in the mold is at steady state. a) Derive an equation for thickness solidified versus time. b) A plate of nickel (38 mm thick) is cast in a ceramic shell mold (10 mm thick). Calculate the solidification time.

a. Heat evolved $= \rho' H_f \dfrac{dM}{dt}$

Heat conducted away $= \dfrac{k(T_M - T_s)}{L} = h(T_s - T_0)$

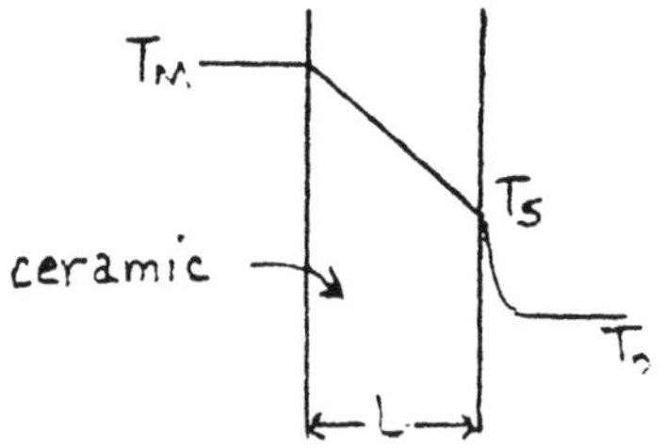

T_M ——/\/\/\/\—— T_s ——/\/\/\/\—— T_0

$R_1 = \dfrac{L}{k}$ $R_2 = \dfrac{1}{h}$

$$\rho' H_f \frac{dM}{dt} = \frac{T_M - T_0}{R_1 + R_2} = \frac{T_M - T_0}{\frac{1}{h} + \frac{L}{k}} = \frac{hk}{Lh + k}(T_M - T_0)$$

$$\left(\frac{Lh+k}{hk}\right)\frac{\rho' H_f}{(T_M - T_0)}\int_0^M dM = \int_0^t dt$$

$$\therefore \left(\frac{Lh+k}{hk}\right)\frac{\rho' H_f}{(T_M - T_0)} M = t \Rightarrow M = \left(\frac{T_M - T_0}{\rho' H_f}\right)\left(\frac{hk}{hL+k}\right)$$

$$\text{b. } t = \left[\frac{(0.010)(150)+(0.70)}{(150)(0.70)}\right]\frac{(7850)(2.91\times10^5)(0.019)}{(1728-300)} = 637 s$$

10.8 Repeat Problem 10.7, but replace the flat mold with a cylindrical shell of thickness L. A cylinder of nickel (38 mm diameter) is cast into the mold with a thickness of 10 mm.

a. Thermal resistances are from Eq. (9.17)

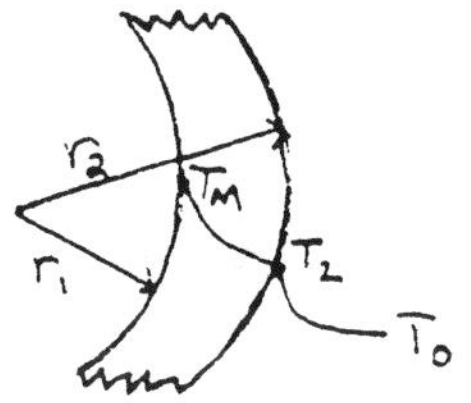

$$\rho' H_f \frac{dV}{dt} = \frac{T_M - T_0}{\dfrac{\ln(r_2/r_1)}{2\pi L k} + \dfrac{1}{2\pi r_2 L h}}$$

$$\therefore \int_0^V dV = \left(\frac{T_M - T_0}{\rho' H_f}\right)\left[\frac{\ln(r_2/r_1)}{2\pi L k} + \frac{1}{2\pi r_2 L h}\right]^{-1}\int_0^t dt \quad \text{where } V = \text{volume solidified.}$$

$$V = \left(\frac{T_M - T_0}{\rho' H_f}\right)\left[\frac{\ln(r_2/r_1)}{2\pi L k} + \frac{1}{2\pi r_2 L h}\right]^{-1} t$$

If we want thickness solidified: $V = \pi r_1^2 L - \pi(r_1 - \delta)^2 L$ where δ = thickness solidified.

b. When $V = \pi r_1^2 L$ then solidification is complete.

$$t = (\pi r_1^2 L)\left(\frac{\rho' H_f}{T_M - T_0}\right)\left[\frac{\ln(r_2/r_1)}{2\pi L k} + \frac{1}{2\pi r_2 L h}\right] = r_1^2\left(\frac{\rho' H_f}{T_M - T_0}\right)\left[\frac{\ln(r_2/r_1)}{2k} + \frac{1}{2 r_2 h}\right]$$

$$t = (0.019)^2\left[\frac{(7850)(2.91\times10^5)}{1728-300}\right]\left[\frac{\ln\left(\frac{0.029}{0.019}\right)}{2(0.70)} + \frac{1}{(2)(0.029)(150)}\right] = 240.8 \text{ s} = 4.01 \text{ min.}$$

10.9 Low density polyethylene is injected into a water-cooled copper mold. The temperature of the melt entering the mold is 465 K. The polyethylene is molded to form a plate that is 10 mm thick. Estimate the time required for all of the polyethylene reach less than 335 K, when it can be safely ejected from the mold. Assume that the heat transfer coefficient at the polyethylene-copper interface is relatively high (h = 4000 W m^{-2} K^{-1}) because cooling occurs while the molding is under pressure. The enthalpy of polyethylene is given below. Other properties of low density polyethylene are k = 0.26 W m^{-1} K^{-1} and ρ = 920 kg m^{-3}.

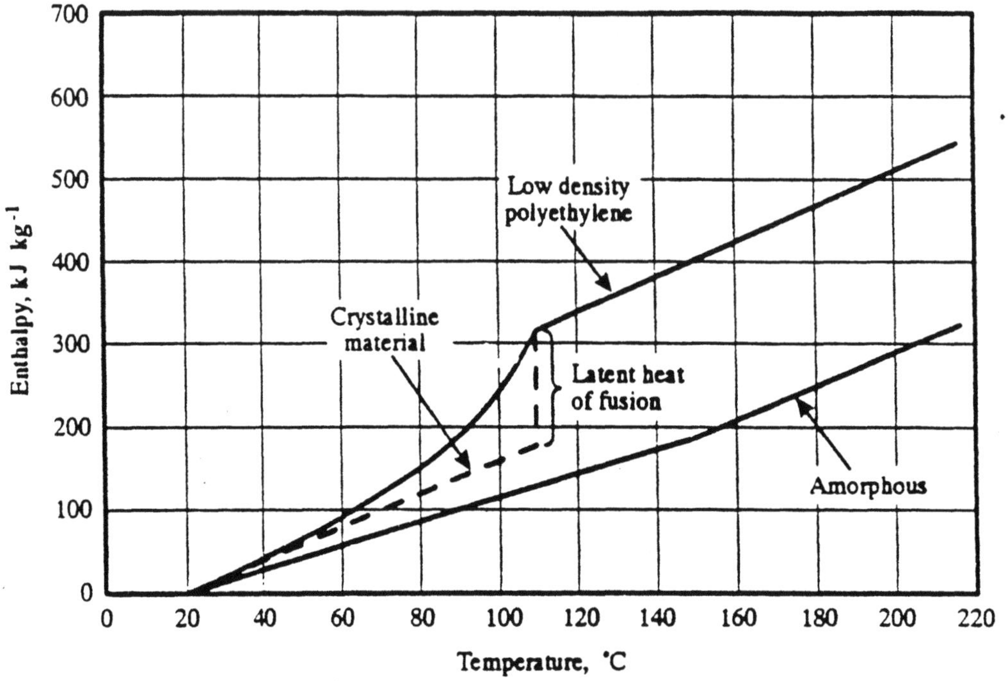

In this case the "casting" is a poor conductor whereas the mold is a good conductor. Hence we can assume that the major resistance is within the polymer and the mold is at a uniform and constant temperature.

Let H_2 = enthalpy at 465 K (T_2)

H_1 = enthalpy at 335 K (T_1)

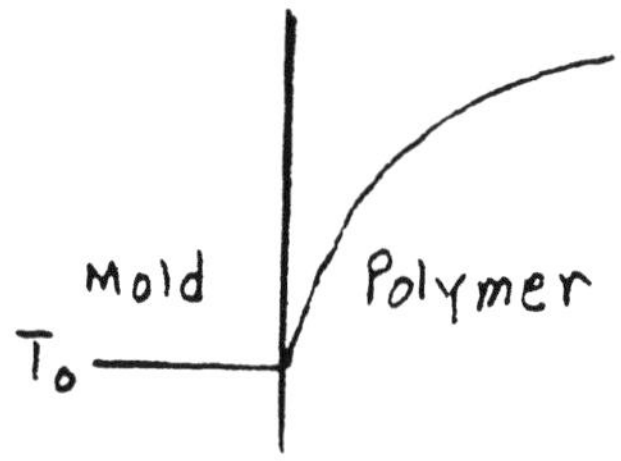

$$\bar{C}_p \simeq \frac{H_2 - H_1}{T_2 - T_1} = \frac{490 - 95}{465 - 335} = 3.039 \text{ kJ kg}^{-1}\text{K}^{-1}$$

$$= 3.039 \times 10^3 \text{ J kg}^{-1} \text{ K}^{-1}$$

$$\alpha = \frac{0.26}{(920)(3.039 \times 10^3)} = 9.30 \times 10^{-8} \text{ m}^2 \text{ s}^{-1}$$

Apply Fig. 9.8 a with Bi = 1000

$$\frac{T-T_f}{T_i-T_f} = \frac{335-300}{465-300} = 0.212\ ;\ F_o \simeq 0.69 = \frac{\alpha t}{L^2}\quad (L = 5mm)$$

$$t = 0.69\frac{L^2}{\alpha} = 0.69\frac{(0.005)^2}{9.30\times10^{-8}} = 185\ s$$

10.10 Repeat Problem 10.9 for amorphous polyethylene (ρ = 970 kg m^{-3}). This polymer can be safely ejected at 315 K. Compare the achievable production rates of the two forms of polyethylene.

$$\bar{C}_p = \frac{H_2-H_1}{T_2-T_1} = \frac{275-37}{465-315} = 1.587\ kJ\ kg^{-1}\ K^{-1} = 1.587\times10^3\ J\ kg^{-1}\ K^{-1}$$

$$\alpha = \frac{0.26}{(1.587\times10^3)(970)} = 1.69\times10^{-7}\ m^2\ s^{-1}$$

$$\frac{T-T_f}{T_i-T_f} = \frac{315-300}{465-300} = 0.0909\ ;\ F_o \simeq 1.05$$

$$t = 1.05\frac{L^2}{\alpha} = 1.05\frac{(0.005)^2}{1.69\times10^{-7}} = 155\ s$$

Production rate is proportional to inverse of time.

$$\frac{\text{Rate (amorphous)}}{\text{Rate (low density)}} = \frac{185\ s}{155\ s} = 1.19$$

10.11 Aluminum oxide is solidified in a water-cooled molybdenum mold to form continuous fibers with a diameter of 200 μm. Estimate the length of the mold required to completely solidify the aluminum oxide as it exits from the mold, as a function of the fiber velocity. Assume that h = 4000 W m^{-2} K^{-1} in the mold. *Data for* Al_2O_3: k = 11 W m^{-1} K^{-1}; C_p = 1230 J kg^{-1} K^{-1}; ρ = 3016 kg m^{-3}; T_M = 2327 K; H_f = 1.07 × 10^6 J kg^{-1}.

This is similar to continuous casting. To make use of Fig. 10.15, we must make an approximation because of the cylindrical symmetry.

In Fig. 10.15, M is the thickness solidified. Let's take M to V/A, where V is the volume solidified and A is the contact area with the mold. For complete solidification of the Al_2O_3 fibers:

$$\frac{V}{A} = \frac{\pi R^2 L}{2\pi R L} = \frac{R}{2}$$

Hence, the ordinate in Fig. 10.15 is $\frac{hR}{2k'} = \frac{(4000)(100\times10^{-6})}{(2)(11)} = 1.82\times10^{-2}$

Also $\frac{H_f'}{C_p'(T_M - T_0)} = \frac{1.07\times10^{6}}{(1230)(2327-300)} = 0.429$

Then with $y = L$; $\frac{h^2 L}{u k' \rho' C_p'} \simeq 0.04$; $L = 0.04 \frac{(11)(3016)(1230)}{4000^2} u$

$L = 0.102\,u$ with L in m and u in m s^{-1}

10.12 A continuous casting machine forms molten steel into a slab, 1.93 m wide and 229 mm thick, at a production rate of 52.5 kg s^{-1}. Assume that h = 1135 W m^{-2} K^{-1}. a) Determine the vertical length of the mold if the solid shell must be 12 mm thick at the mold exit. b) Calculate the cooling water requirement (kg s^{-1}) if its temperature rise is from 300 to 307 K. *Data for low carbon steel*: k = 35 W m^{-1} K^{-1}; $C_p = C_{p,l}$ = 670 J kg^{-1} K^{-1}; ρ = 7690 kg m^{-3}; H_f = 2.79 × 10^5 J kg^{-1}; T_M = 1790 K.

a. $u = \frac{52.5\text{ kg}}{\text{s}} \cdot \frac{\text{m}^3}{7690\text{ kg}} \cdot \frac{1}{(1.93)(0.229)\text{ m}^2} = 1.54\times10^{-2}$ m s^{-1}

$$\frac{hM}{k'} = \frac{1135\text{ W}}{\text{m}^2\text{ K}} \cdot 0.012\text{ m} \cdot \frac{\text{m K}}{35\text{ W}} = 0.389$$

$$\frac{H_f'}{C_p'(T_M - T_0)} = \frac{\left[2.79\times10^5 + (1840-1790)(670)\right]\text{ J}}{\text{kg}} \cdot \frac{\text{kg K}}{670\text{ J}} \cdot \frac{1}{(1790-335)\text{ K}} = 0.321$$

From Fig. 10.15 $\frac{h^2 Y}{u k' \rho' C_p'} = 0.21$

$$\therefore Y = \frac{(0.21)(1.54\times10^{-2})(35)(7690)(670)}{(1135)^2} = 0.45\text{ m}$$

b. From Fig. 10.17, with $Y = L$

$$Q = (0.38)L(T_M - T_0)(L u \rho' C_p' k')^{1/2} = (0.38)(0.45)(1790-335)\left[(0.45)(1.54\times10^{-2})(7690)(670)(35)\right]^{1/2} = 2.78\times10^5\text{ W}$$

In terms of the temperature rise of the water, $Q = \dot{M} C_p \Delta T$ where $\dot{M}$ is the mass flow rate. C_p = 4184 J kg^{-1} K^{-1} (water).

$$\therefore \dot{M} = \frac{2.78\times10^5}{(4184)(70)} = 0.95\text{ kg s}^{-1}$$

10.13 The dwell time in the mold of a continuous casting machine is defined as the period which the metal spends in the mold during solidification; that is, $t = L/u$, in which t = dwell time, L = length of mold over which solidification is occurring, and u = velocity of metal through the mold. Since the skin solidified in the mold is thin, a simple analysis might be expected to apply. a) Neglect conduction in the withdrawal direction and write an expression for thickness solidified versus time. b) Compare the results for dwell time in Problem 10.12 and the dwell time calculated from part a) for the same conditions.

a. Heat evolved at the interface $\rho' H_f' \frac{dM}{dt}$

Heat conducted away from interface through solid. $\frac{k'(T_M - T_s)}{M}$ Note that T_s varies.

$$\therefore \rho' H_f' \frac{dM}{dt} = k' \frac{(T_M - T_s)}{M}$$

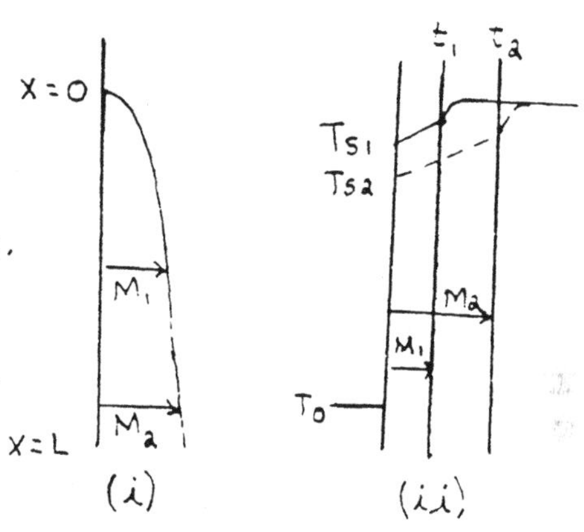

Assume the temperature profile in the solid is linear, then $k' \frac{(T_M - T_s)}{M} = h(T_s - T_0)$

This is a problem of heat flow through two thermal resistances.

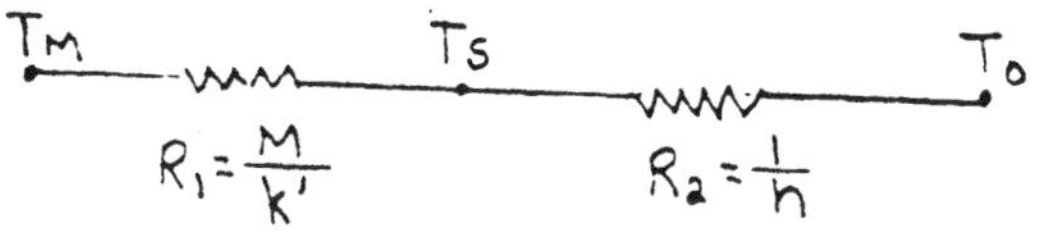

$$R_T = \frac{M}{k'} + \frac{1}{h}$$

The rate of heat flow expressed in terms of the total temperature drop $(T_M - T_0)$ is: $\rho' H_f' \frac{dM}{dt} = \left[\frac{1}{\frac{1}{h} + \frac{M}{k'}}\right](T_M - T_0)$

Integrate:

$$\int_0^M \left(\frac{1}{h} + \frac{M}{k'}\right) dM = \frac{T_M - T_0}{\rho' H_f} \int_0^t dt$$

$$\therefore M = \frac{h(T_M - T_0)}{\rho' H_f} t - \frac{h}{2k'} M^2$$

b. $$t = \frac{\rho' H_f M}{h(T_M - T_0)}\left(1 + \frac{hM}{2k'}\right) = \frac{(7690)\left[2.79 \times 10^5 + (670)(50)\right](0.012)}{(1135)(1840 - 300)}\left[1 + \frac{(1135)(0.012)}{(2)(35)}\right] = 19.7 \text{ s}$$

From problem 10.12 $t = \frac{L}{u} = \frac{0.45}{1.54 \times 10^{-2}} = 29.2$ s

Simple analysis underestimates the dwell time by 33% in this case.

10.14 A junction-shaped casting as depicted below is made in a sand mold. The junction may be considered infinitely long in the z-direction. For the upper right quadrant of sand:

a) Write a differential equation for temperature.
b) Write the boundary conditions (for time and space) that apply.
c) Write the solution yielding temperature as a function of position in the sand and time.
d) Derive an equation for the heat absorbed by the sand as a function of time and the lengths of the junction legs.

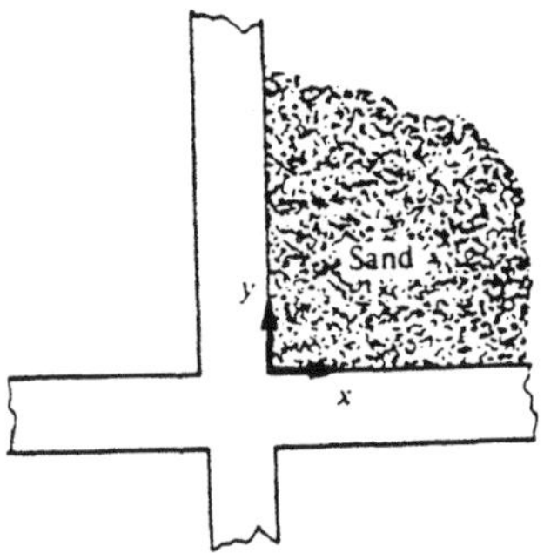

a. $\frac{\partial^2 T}{\partial X^2} + \frac{\partial^2 T}{\partial Y^2} = \frac{1}{\alpha}\frac{\partial T}{\partial t}$ where α = thermal diffusivity of the sand.

b. Initial Conditions: $T(X, Y, 0) = T_0$.

Boundary Conditions: $T(0, Y, t) = T_M$; $T(X, 0, t) = T_M$

$T(\infty, \ , t) = T_0$

Where T_M = freezing point of metal.

T_0 = initial mold temperature

c. $\frac{T - T_M}{T_0 - T_M} = \left(\text{erf}\frac{X}{2(\alpha t)^{1/2}}\right)\left(\text{erf}\frac{Y}{2(\alpha t)^{1/2}}\right)$.

d. Average T in the mold: $\frac{\bar{T} - T_M}{T_0 - T_M} = \frac{1}{L_1 L_2}\int_{Y=0}^{L_2}\int_{X=0}^{L_1} \text{erf}\frac{X}{2(\alpha t)^{1/2}}\,\text{erf}\frac{Y}{2(\alpha t)^{1/2}}\,dX\,dY$

where L_1 and L_2 define the junction legs.

See approximation b) under Table 9.3; $\text{erf}\frac{X}{2(\alpha t)^{1/2}} \simeq \frac{X}{(\pi \alpha t)^{1/2}}$

$$\therefore \frac{\bar{T} - T_M}{T_0 - T_M} \simeq \frac{1}{L_1 L_2}\int_{Y=0}^{L_2} \frac{Y}{(\pi\alpha t)^{1/2}}\left[\int_{X=0}^{L_1} \frac{X}{(\pi\alpha t)^{1/2}}\,dX\right] dY \simeq \frac{L_1 L_2}{4}\,\frac{1}{\pi \alpha t},$$

provided L_1 and L_2 are less than $0.4\sqrt{\alpha t}$.

Heat absorbed by mold ($0 \le Y \le L_2$, $0 \le X \le L_1$); $V = L_1 L_2 W$, $W \perp$ to L_1 and L_2

$$Q_M \simeq V \rho C_p (\bar{T} - T_0);\ \bar{T} = (T_0 - T_M)\frac{L_1 L_2}{4\pi\alpha t} + T_M;\ \bar{T} - T_0 = (T_0 - T_M)\frac{L_1 L_2}{4\pi \alpha t} + T_M - T_0$$

$$Q_M \simeq V\rho C_p\left[(T_0 - T_M)\frac{L_1 L_2}{4\pi\alpha t} + (T_M - T_0)\right] \simeq V\rho C_p\left[1 - \frac{L_1 L_2}{4\pi\alpha t}\right](T_M - T_0)$$

Further discussion (10.14d)

Carslaw and Jaeger* give an exact solution for infinitely long junction legs. The heat flux into a mold with a flat surface:

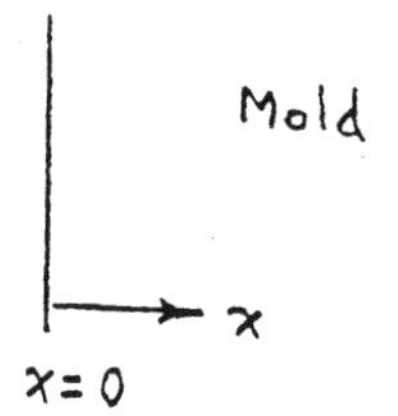

$$q_{x=0} = \frac{k\,(T_M - T_0)}{\sqrt{\pi \alpha t}}$$

Heat flux from surface $x = 0$ in a junction:

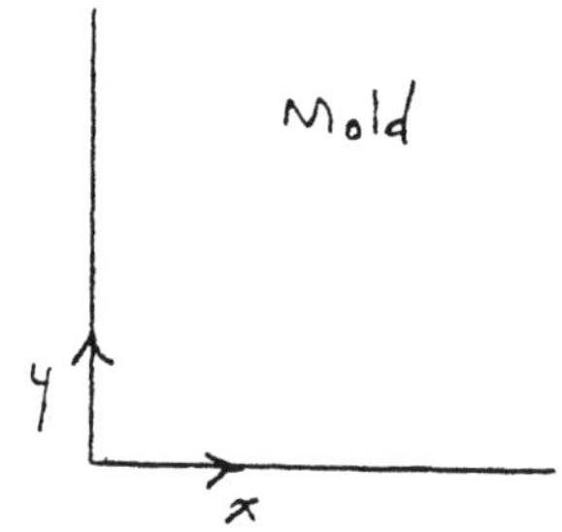

$$q'_{x=0} = \frac{k\,(T_M - T_0)}{\sqrt{\pi \alpha t}}\, \mathrm{erf}\left(\frac{y}{2\sqrt{\alpha t}}\right)$$

The difference between $q_{x=0}$ and $q'_{x=0}$ is

$$q_{x=0} - q'_{x=0} = \frac{k\,(T_m - T_0)}{\sqrt{\pi \alpha t}}\, \mathrm{erfc}\left(\frac{y}{2\sqrt{\alpha t}}\right).$$

Now suppose we determine the difference integrated along the surface $x = 0$. According to Carslaw and Jaeger it is

$$\int_{y=0}^{\infty} \left(q_{x=0} - q'_{x=0}\right) dy = \frac{2k\,(T_m - T_0)}{\pi}.$$

As a practical matter, we know that $q_{x=0} - q'_{x=0} \simeq 0$ provided $L_2/\sqrt{\alpha t} > 4$. With this proviso, we take L_2 as the length of the junction in y-direction. Then the average of the difference is

$$\frac{1}{L_2} \int_{y=0}^{L_2} \left(q_{x=0} - q'_{x=0}\right) dy = \frac{2k\,(T_m - T_0)}{\pi\, L_2}.$$

---

* H.S. Carslaw and J.C. Jaeger: *Heat Conduction in Solids*, 2nd ed., Oxford Science Public., pp. 171-172.

Similarly;

$$\frac{1}{L_1}\int_{x=0}^{L_1}\left(q_{y=c} - q'_{y=0}\right)dx = \frac{2k(T_M - T_0)}{\pi L_1}$$

The average of the difference in the heat (J) absorbed by the mold at $x = 0$ is

$$WL_2\left\{\frac{1}{L_2}\int_0^t\int_{y=0}^{L_2}\left(q_{x=0} - q'_{x=0}\right)dy\,dt\right\} = WL_2\,\frac{2k(T_M - T_0)}{\pi L_2}\,t.$$

where W is the dimension perpendicular to the dimensions L_1 and L_2.

Hence, for both surfaces the difference is

$$Q_M(\text{flat}) - Q_M(\text{junction}) = (4W/\pi)k\,(T_M - T_0)\,t.$$

But

$$Q_M(\text{flat}) = \frac{2(L_1 + L_2)W\,k\,(T_M - T_0)\sqrt{t}}{\sqrt{\pi\alpha}},$$

so that

$$Q_M(\text{junction}) = 2Wk\,(T_M - T_0)\left[\frac{(L_1 + L_2)\sqrt{t}}{\sqrt{\pi\alpha}} - \frac{2t}{\pi}\right].$$

10.15 Equations (10.32) and (10.33) are approximations for the case depicted in Fig. 10.9. The exact solution for the temperature in the solidified skin is

$$\frac{T - T_0}{T_M - T_0} = \left[\frac{1}{\operatorname{erf}\beta}\right]\operatorname{erf}\left[\beta\,\frac{M_0 + x}{M_0 + M}\right],$$

where β satisfies

$$\beta e^{\beta^2}\operatorname{erf}\beta = (T_M - T_0)\,\frac{C'_p}{H_f\sqrt{\pi}}$$

and M_0 satisfies

$$\frac{2k'\beta^2}{hM_0} = (T_M - T_0)\,\frac{C'_p}{H_f}.$$

a) Show that the above relationships are exact.

b) If t_0 is a time defined as

$$t_0 = \frac{M_0^2}{4\beta^2\alpha'},$$

show that the thickness solidified is

$$M = 2\beta\sqrt{\alpha'}\left[(t_0 + t)^{1/2} - t_0^{1/2}\right].$$

a.

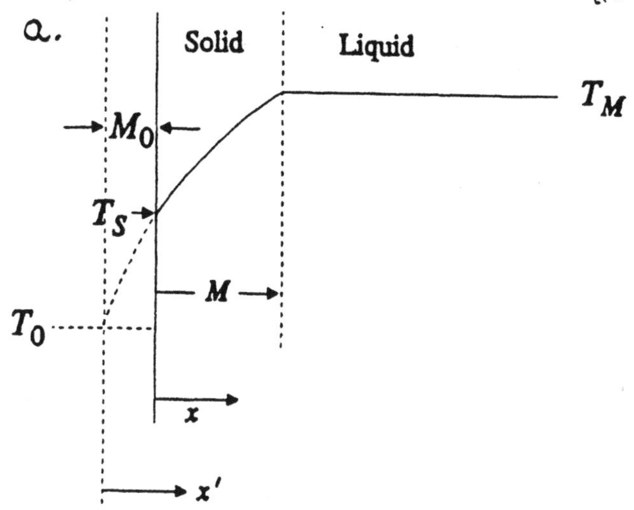

Notice that there are two coordinate systems: x and x'. The two systems are related by

$$x' = M_0 + x \qquad (1)$$

$$M' = M_0 + M \qquad (2)$$

$$t' = t_0 + t \qquad (3)$$

where t_0 is the time required for the thickness M_0 to be formed.

In the x'-system the problem is the same as the case of an infinite h, so the temperature profile is given by combining Eqs. (10.15) and (10.18).

$$\frac{T - T_0}{T_M - T_0} = \left[\frac{1}{erf\ \beta}\right] erf\ \frac{x'}{2\sqrt{\alpha' t}} \qquad (4)$$

in which

$$\beta = \frac{M'}{2\sqrt{\alpha' t}} \qquad (5)$$

With the aid of Eqs. (1), (2) and (5), it can be shown that

$$\frac{x'}{2\sqrt{\alpha' t}} = \beta_0\ \frac{M_0 + x}{M_0 + M} \qquad (6)$$

so the temperature profile is

$$\frac{T - T_0}{T_M - T_0} = \left[\frac{1}{erf\ \beta}\right] erf\left[\beta\ \frac{M_0 + x}{M_0 + M}\right] \qquad (7)$$

Notice that Eq. (7) contains two unknown constants, β and M_0. β can be calculated using Eq. (10.22). In order to determine M_0, we write a heat balance at $x = 0$ at the instant (t_0) M_0 has been established.

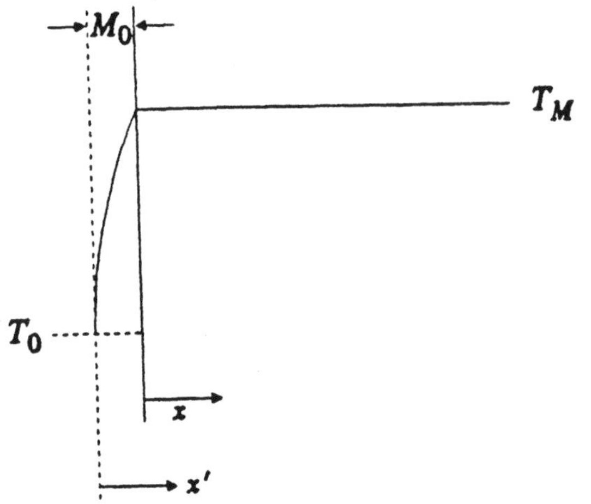

$$h(T_M - T_0) = k' \left[\frac{\partial T}{\partial x'}\right]_{x'=M_0} \tag{8}$$

Equation (4) gives

$$\left[\frac{\partial T}{\partial x'}\right]_{M_0} = \frac{2\beta(T_M - T_0)}{\sqrt{\pi}\, M_0\, erf\, \beta \exp \beta^2} \tag{9}$$

and by combining Eqs. (8) and (9) we get

$$\frac{2k'\beta^2}{hM_0} = \sqrt{\pi}\, \beta\, erf\, \beta \exp \beta^2. \tag{10}$$

Finally if we compare Eqs.(10.22) and (10), we see that

$$\frac{2k'\beta}{hM_0} = (T_M - T_0)\frac{C_p'}{H_f}. \tag{11}$$

After β has been evaluated, Eq. (11) is used to calculate M_0. Then the temperature profile is completely specified by Eq. (7).

b. To get the thickness solidified, Eqs. (2), (3) and (5) are combined with the following:

$$\beta = \frac{M_0}{2\sqrt{\alpha' t_0}}.$$

The result is

$$t_0 = \frac{M_0^2}{4\beta^2\alpha'} \tag{12}$$

and

$$M = 2\beta\sqrt{\alpha'(t_0 + t)} - M_0 \tag{13}$$

or

$$M = 2\beta\sqrt{\alpha'}\left[(t_0 + t)^{1/2} - t_0^{1/2}\right] \tag{14}$$

10.16 Usually when Eq. (10.48) is invoked, the solid-liquid interface is considered to be at the freezing temperature (i.e., at the equilibrium temperature). In *rapid solidification processing*, however, the interface can be significantly undercooled. If H_f is the latent heat at the equilibrium temperature: a) rewrite Eq. (10.48) in a more precise manner, taking into account that C_p of the solid and liquid phases are not necessarily equal; b) rewrite Eq. (10.48) in accordance with part a) and also account for the densities not being equal.

a. At the interface, let $T = T^*$ and $T^* < T_M$

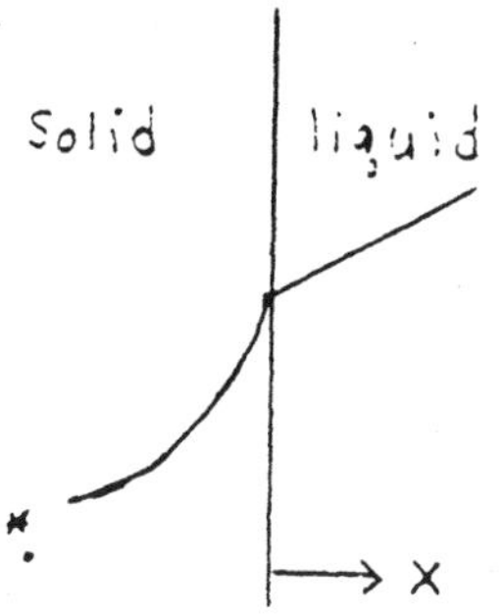

$$k_s \left.\frac{\partial T}{\partial x}\right|_S = k_L \left.\frac{\partial T}{\partial x}\right|_L + \rho U \Delta H \quad \text{where } \Delta H = H_L - H_S$$

and H_L, H_S = enthalpies of the liquid and solid at T^*.

$$H_S = H_S(T_M) + C_{p,s}(T^* - T_M)$$

$$H_L = H_L(T_M) + C_{p,L}(T^* - T_M)$$

$$\Delta H = H_L(T_M) - H_S(T_M) + (C_{p,L} - C_{p,s})(T^* - T_M) \quad \text{or}$$

$$\Delta H = H_f + (C_{p,L} - C_{p,s})(T^* - T_M)$$

Then with $G_S = \left.\frac{dT}{dx}\right)_S$ and $G_L = \left.\frac{dT}{dx}\right)_L$, we have

$$k_S G_S = k_L G_L + \rho U \left[H_f + (C_{p,L} - C_{p,s})(T^* - T_M) \right]$$

b. Suppose $\rho_S > \rho_L$. Then to satisfy continuity, there will be a velocity normal to the interface. Continuity requires $\frac{V}{U} = \frac{\rho_S - \rho_L}{\rho_L}$ where U is the velocity of the interface and V is the velocity of the liquid at the interface. Since liquid flows toward the interface, it brings with it (i.e., advects) some enthalpy. Thus, at the interface, we have:

$$k_S G_S = k_L G_L + U(\rho_L H_L - \rho_S H_S) + \rho_L V H_L$$

$$\text{or } k_S G_S = k_L G_L + U(\rho_L H_L - \rho_S H_S) + (\rho_S - \rho_L) U H_L$$

$$k_S G_S = k_L G_L + U \rho_L H_L - U \rho_S H_S + U \rho_S H_L - U \rho_L H_L$$

$$k_S G_S = k_L G_L - U(\rho_S H_S + \rho_L H_L)$$

where $H_S = H_S(T_M) + C_{p,s}(T^* - T_M)$

$$H_L = H_L(T_M) + C_{p,L}(T^* - T_M)$$

10.17 Is it possible that $G_L < 0$ during crystal growth? Can there be solidification? Do you think the planar interface will be stable?

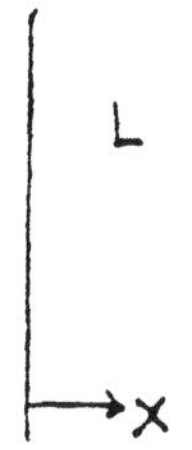

With growth $U > 0$: $k_S G_S = k_L G_L + \rho U H_f$

Even if $G_L < 0$, these energy balance at the interface can be satisfied. For example, solid could be growing into an undercooled liquid. The temperature profile would appear as

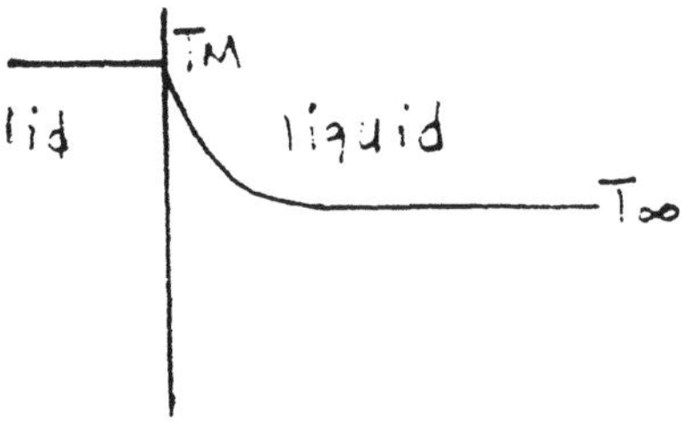

The interface would not be stable because the liquid is at a temperature below TM. Actually, a stable interface could support a small undercooling ($T_M - T_\infty$) because of interfacial energy that would play a role should the flat interface become perturbed.

10.18 A single crystal in the form of a long thin rod is grown by the Czochralski method (Fig. 10.18(a)). Assume that the heat conduction is one dimensional and that the heat transfer coefficient is uniform along the length of rod. a) Write the energy equation for temperature along the length of the crystal and obtain the solution for the temperature distribution. b) Use the properties given in Example 10.8 for silicon and assume that $h = 100\ \mathrm{W\ m^{-2}\ K^{-1}}$ (uniform). Calculate G_S for growth rates of 10^{-6}, 10^{-5} and $10^{-4}\ \mathrm{m\ s^{-1}}$. c) Calculate the corresponding values of G_L and briefly discuss your results.

a. Refer back to the solution for Problem 9.9

$$\frac{\partial^2 T}{\partial z^2} - \frac{V}{\alpha}\frac{\partial T}{\partial z} - \frac{hP}{kA}(T - T_\infty) = 0$$

$$\frac{T - T_\infty}{T_M - T_\infty} = \exp\left[\frac{V}{2\alpha} - \left(\frac{V^2}{4\alpha^2} + \frac{hP}{kA}\right)^{1/2}\right] z \quad \text{where } \frac{P}{A} = \frac{\pi D}{\frac{\pi D^2}{4}} = \frac{4}{D}$$

$$\text{b. } G_S = \left.\frac{dT}{dz}\right|_{z=0} = (T_M - T_\infty)\left[\frac{V}{2\alpha} - \left(\frac{V^2}{4\alpha^2} + \frac{4h}{Dk}\right)^{1/2}\right]$$

From Example 10.8: $T_M = 1683$ K; $T_\infty = 300$ K; $\alpha = 1.32 \times 10^{-5}$ $m^2 s^{-1}$;

$k = 31$ W $m^{-1} K^{-1}$; $h = 100$ W $m^{-2} K^{-1}$; $H_f = 1.80 \times 10^6$ J kg^{-1}

Assume maximum D for Newtonian cooling: Problem 9.9 c

$$D(max.) = \frac{2kBi_e}{h} = \frac{2(31)(0.1)}{100} = 0.062 \text{ m}$$

c. $G_L = \dfrac{kG_S + \rho V H_f}{k_L}$ where $k_L = 50$ W $m^{-1} K^{-1}$, $\rho = 2300$ kg m^{-3}, and $V > 0$ to be consistent with parts (a) and (b).

```
10 'Problem 10.18
20 ALPHA = .0000132 : K =31 : D =.062 : KL = 50 : RHO = 2300
30 TM = 1683 : TINF = 300 : H =100 : HF = 1800000!
40 FOR I = 4 TO 6
50    V = 10^(-I) : A = V/(2*ALPHA) : B = 4*H/(D*K)
60    BRACK = A - SQR(A*A + B)
70    GS = (TM-TINF)*BRACK
80    NUMER = K*GS + RHO*V*HF : DENOM = KL
90    GL = NUMER/DENOM
100   LPRINT V,GS,GL
110 NEXT I
120 END
```

V, m s^{-1}	G_s, K m^{-1}	G_L, K m^{-1}
.0001	-15389.15	-1261.273
.00001	-19434.51	-11221.4
.000001	-19899.18	-12254.69

Anything greater than $|G_L| = 5 \times 10^3$ K m^{-1} would be difficult to achieve in a process so a growth rate of 10^{-5} m s^{-1} is not practical.

10.19 The properties of aluminum near the melting point are given: $T_M = 933$ K; $k = 210$ W m^{-1} K^{-1}. Consult Chapter 11 for emissivity. Calculate the power loss to maintain a floating zone in aluminum. Compare your results to Fig. 10.19 and discuss.

From Fig. 11.8, $\varepsilon \simeq 0.1$

$$Q = 2\pi R \varepsilon \sigma T_M^4 \lambda_0 \left(\frac{6}{5} + \frac{L}{\lambda_0}\right)$$

$\sigma = 5.670 \times 10^{-8}$ W $m^{-2} K^{-4}$; $T_M = 933$ K; $L \simeq 1$ cm $= 0.01$ m; $R = 0.5$ cm $= 0.005$ m

(same as Fig. 10.19); $k = 210$ W m^{-1} K^{-1}

$$\lambda_0 = \left[\frac{(5)(210)(0.005)}{(9)(0.1)(5.67 \times 10^{-8})(933^3)}\right]^{1/2} = 0.356 \text{ m}$$

$$Q = (2)(\pi)(0.005)(0.1)(5.670\times10^{-8})(933^4)(0.356)\left[\frac{6}{5} + \frac{0.01}{0.356}\right]$$

$Q = 59.0$ W [This compares with Fig. 10.19]

10.20 Welding of a "thin plate" can be analyzed by starting with Eq. (9.85) for the peak temperature around a moving line source. With r as the cylindrical radius from the line source, we can set $T_p = T_M$ (the melting point of the base metal) at $r = r_M$; then

$$T_M - T_0 = \frac{Q}{\rho C_p \delta V r_M \sqrt{2\pi e}}.$$

Show that

$$\frac{1}{T_p - T_0} = \frac{V r' \delta C_p \sqrt{2\pi e}}{Q} + \frac{1}{T_M - T_0}.$$

where $r' = r - r_M$. For $r' \geq 0$ this gives the peak temperature in the base material next to the weld.

Let $T_\infty = T_0$; then,

$$\frac{1}{T_M - T_0} = \frac{\rho c_p \delta V r_M \sqrt{2\pi e}}{Q}$$

$$T_p - T_0 = \frac{Q}{\rho c_p \delta V r \sqrt{2\pi e}} \;;\; \frac{1}{T_p - T_0} = \frac{\rho c_p \delta V r \sqrt{2\pi e}}{Q}$$

$$\frac{1}{T_p - T_0} - \frac{1}{T_M - T_0} = \frac{\rho c_p \delta V \sqrt{2\pi e}}{Q}(r - r_M)$$

$$\therefore \frac{1}{T_p - T_0} = \frac{V r' \delta \rho c_p \sqrt{2\pi e}}{Q} + \frac{1}{T_M - T_0}$$

10.21 Welding of a "thick plate" can be analyzed by starting with Eq. (9.81) for the peak temperature around a moving point source. As in Problem 10.20, set $T_p = T_M$ at $r = r_M$ and show that

$$\frac{1}{T_p - T_0} = \frac{2\pi k \alpha e}{QV}\left[\frac{V r'}{2\alpha}\right]^2 + \frac{1}{T_M - T_0},$$

where $r' = (r^2 - r_M^2)^{1/2}$, provided $r' \geq 0$.

$$T_p - T_\infty \simeq \left(\frac{QV}{2\pi k\alpha e}\right)\left(\frac{1}{2+Pe_r^2}\right) \quad \text{where } Pe_r = \frac{Vr}{2\alpha} \quad \text{and } T_\infty = T_0$$

$$T_p - T_0 = \left(\frac{QV}{2\pi k\alpha e}\right)\left[\frac{1}{2+\left(\frac{Vr}{2\alpha}\right)^2}\right]$$

$$\frac{1}{T_p - T_0} = \left(\frac{2\pi k\alpha e}{QV}\right)\left[2+\left(\frac{Vr}{2\alpha}\right)^2\right]$$

$$T_M - T_0 \simeq \left(\frac{QV}{2\pi k\alpha e}\right)\left[\frac{1}{2+\left(\frac{Vr_M}{2\alpha}\right)^2}\right] \quad \text{where } T_p = T_M \text{ at } r = r_M$$

$$\frac{1}{T_M - T_0} = \left(\frac{2\pi k\alpha e}{QV}\right)\left[2+\left(\frac{Vr_M}{2\alpha}\right)^2\right]$$

$$\frac{1}{T_p - T_0} - \frac{1}{T_M - T_0} = \left(\frac{2\pi k\alpha e}{QV}\right)\left[\left(2+\left(\frac{Vr}{2\alpha}\right)^2\right) - \left(2+\left(\frac{Vr}{2\alpha}\right)^2\right)\right] = \left(\frac{2\pi k\alpha e}{QV}\right)\left(\frac{V}{2\alpha}\right)^2\left(r^2 - r_m^2\right)$$

$$\therefore \frac{1}{T_p - T_0} = \frac{2\pi k\alpha e}{QV}\left(\frac{Vr'}{2\alpha}\right)^2 + \frac{1}{T_M - T_0} \quad \text{where } r' = \left(r^2 - r_M^2\right)^{1/2}$$

Problems 10.22-10.26 should be attempted after Problems 10.20 and 10.21 have been solved.

10.22 Sketch temperature versus time at $r = 0$, $r = r_M$ and $r > r_M$ in a thin plate that is welded.

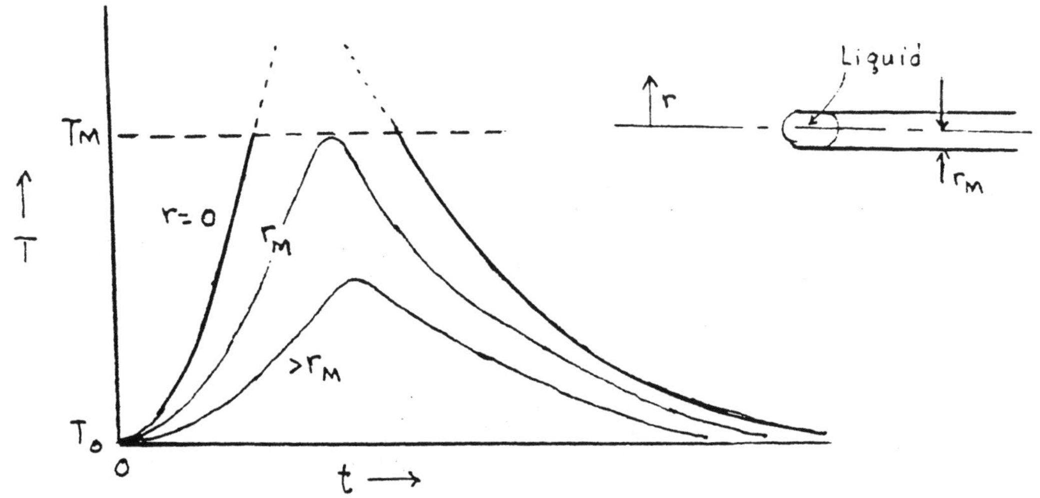

10.23 A very thick steel plate is welded with 5 kW and 75 pct. efficiency at a welding speed of 4.2 mm s^{-1}. Prior to welding the plate is preheated to 480 K. Assume that the fusion zone (i.e., the edge of the molten weld metal) corresponds to the 1644 K isotherm. a) Calculate the peak temperature at a distance of 1 mm from the fusion zone. b) What is the cooling rate at that location when the temperature is 1070 K? *Data for steel:* k = 35 W m^{-1} K^{-1}; ρ = 7690 kg m^{-3}; C_p = 754 J kg^{-1} K^{-1}.

a. From Problem 10.21; $\frac{1}{T_p - T_o} = \frac{2\pi k \alpha e}{QV}\left(\frac{Vr'}{2\alpha}\right)^2 + \frac{1}{T_M - T_o}$ where $T_o = 480$ K;

$\alpha = \frac{k}{\rho C_p} = \frac{35}{(7690)(754)} = 6.04 \times 10^{-6}\ m^2 s^{-1}$; $Q = (5000)(0.75) = 3750$ W;

$V = 4.2 \times 10^{-3}\ m\ s^{-1}$; $r' = 1 \times 10^{-3}\ m$; $T_M = 1644$ K; $k = 35\ W\ m^{-1} K^{-1}$

$$\therefore \frac{1}{T_p - T_o} = \frac{(2\pi)(35)(6.04\times10^{-6})e}{(3750)(4.2\times10^{-3})}\left[\frac{(4.2\times10^{-3})(1\times10^{-3})}{2(6.04\times10^{-6})}\right]^2 + \frac{1}{1644-480} = 8.87\times10^{-4}\ K^{-1}$$

$T_p - T_o = 1128$ K $\therefore T_p = 1608$

b. Close enough to the center so we can use the following:

at $r=0$, Eq. (9.78) gives: $T - T_\infty = \frac{QV}{4\pi k \alpha}\left(\frac{Vz}{2\alpha}\right)^{-1}$

To transform to a time basis, we use:

$$\frac{\partial T}{\partial t} = -V\frac{\partial T}{\partial z} = V\frac{QV}{4\pi k\alpha}\left(\frac{Vz}{2\alpha}\right)^{-2}\left(\frac{V}{2\alpha}\right)$$

$$\frac{\partial T}{\partial t} = V(T-T_o)\left(\frac{Vz}{2\alpha}\right)^{-1}\left(\frac{V}{2\alpha}\right) = V(T-T_o)^2\frac{4\pi k\alpha}{QV}\frac{V}{2\alpha} = \frac{2\pi Vk}{Q}(T-T_o)^2$$

$$\frac{\partial T}{\partial t} = \frac{2\pi k V}{Q}(T-T_o)^2 = \frac{(2\pi)(35)(4.2\times10^{-3})}{3750}(1070-480)^2 = 85.7\ K\ s^{-1}$$

10.24 Derive an equation for the thermal gradient, $\partial T/\partial r$, at the edge of a weld pool ($r = r_M$) of a material with a freezing point of T_M. The plate is steel and very thick. Thermal properties and welding conditions are given in Problem 10.23.

we want the gradient at this point.

Start with Eq. (9.78)

$$\frac{\partial T}{\partial Pe_r} = \frac{QV}{4\pi k\alpha}\left\{\left(Pe_z^2 + Pe_r^2\right)^{-1/2}\frac{\partial}{\partial Pe_r}\exp\left[Pe_z - \left(Pe_z^2 + Pe_r^2\right)^{1/2}\right] + \exp\left[Pe_z - \left(Pe_z^2 + Pe_r^2\right)^{1/2}\right]\right.$$

$$\left.\frac{\partial}{\partial Pe_z}\left(Pe_z^2 + Pe_r^2\right)^{-1/2}\right\}$$

$$\frac{\partial T}{\partial Pe_r} = \frac{QV}{4\pi k\alpha}\left\{(Pe_z^2+Pe_r^2)^{-1/2}\exp\left[Pe_z-(Pe_z^2+Pe_r^2)^{1/2}\right]\frac{\partial}{\partial Pe_r}\left[Pe_z-(Pe_z^2+Pe_r^2)^{1/2}\right]\right.$$

$$\left. + \exp\left[Pe_z-(Pe_z^2+Pe_r^2)^{1/2}\right]\left(-\frac{1}{2}\right)(Pe_z^2+Pe_r^2)^{-3/2}(2Pe_r)\right\}$$

$$\frac{\partial T}{\partial Pe_r} = (T-T_\infty)(-1)(Pe_z^2+Pe_r^2)^{-1/2}(Pe_r)+(T-T_\infty)\left(-\frac{1}{2}\right)(Pe_z^2+Pe_r^2)^{-1}(2Pe_r)$$

$$\frac{\partial T}{\partial Pe_r} = (T-T_\infty)(Pe_r)\left[-(Pe_z^2+Pe_r^2)^{-1/2} - (Pe_z^2+Pe_r^2)^{-1}\right]$$

We use approx. of Eq. (9.80) and set $r = r_M$:

$$\left.\frac{\partial T}{\partial Pe_r}\right)_{r=r_M} = (T-T_\infty)(Pe_r)\left[-\left(\frac{Pe_{rM}^4}{4}+Pe_{rM}^2\right)^{-1/2} - \left(\frac{Pe_{rM}^4}{4}+Pe_{rM}^2\right)^{-1}\right]$$

$$\left.\frac{\partial T}{\partial r}\right)_{r=r_M} = (T-T_\infty)\left(\frac{V}{2\alpha}\right)^2(r_M)\left[-\left(\frac{Pe_{rM}^4}{4}+Pe_{rM}^2\right)^{-1/2} - \left(\frac{Pe_{rM}^4}{4}+Pe_{rM}^2\right)^{-1}\right]$$

In order to find r_M, we set $T_P = T_M$ and $r = r_M$ and use Eq. (9.81)

$$(T_M - T_\infty) = \left(\frac{QV}{2\pi k\alpha e}\right)\left(\frac{1}{2+Pe_{rM}^2}\right)$$

$$1644-480 = \frac{(3750)(4.2\times10^{-3})}{(2\pi e)(35)(6.04\times10^{-6})}\left(\frac{1}{2+Pe_{rM}^2}\right)$$

$$\frac{1}{2+Pe_{rM}^2} = 0.2668 : Pe_{rM} = 1.322$$

$$r_M = \frac{2\alpha}{V}Pe_{rM} = \frac{(2)(6.04\times10^{-6})(1.322)}{4.2\times10^{-3}} = 3.803\times10^{-3}\ m$$

$$\left.\frac{\partial T}{\partial r}\right)_{r=r_M} = (1644-480)\left(\frac{0.0042}{2\times6.04\times10^{-6}}\right)^2(3.803\times10^{-3})\left[-\left(\frac{1.322^4}{4}+1.322^2\right)^{-1/2} - \left(\frac{1.322^4}{4}+1.322^2\right)^{-1}\right]$$

$$\left.\frac{\partial T}{\partial r}\right)_{r=r_M} = 5.351\times10^5\left[-0.631-0.398\right] = -5.506\times10^5\ K\ m^{-1}$$

$$\left.\frac{\partial T}{\partial r}\right)_{r=r_M} = 5.351\times10^5\left[0.631-0.398\right] = 1.247\times10^5\ K\ m^{-1}$$

10.25 The surface of damaged silicon can be annealed by passing a laser beam over the surface in order to effect melting to a depth below the damaged layer. When the silicon resolidifies, it does so epitaxially so that the crystal structure of the underlying single crystal is maintained.

Assume that the silicon is at 293 K and is "thick." The molten pool is 1 mm thick, and the laser beam moves at a velocity of 20 cm s^{-1} with a power of 8 kW at 75% efficiency. a) Calculate the distance from the edge of the molten pool where a peak temperature of 1273 K is achieved. b) Calculate the maximum cooling rate directly behind the molten pool. [Hint: Start with the steady-state temperature associated with a moving point source (Chapter 9) and make a transformation from moving to stationary coordinates.] Consult Example 10.8 for properties of silicon.

From example 10.8: $k = 31\ W\,m^{-1}K^{-1}$; $k_L = 50\ W\,m^{-1}K^{-1}$; $\alpha = 1.32\times10^{-5}\ m^2 s^{-1}$;

$\alpha_L = 1.94\times10^{-5}\ m^2 s^{-1}$; $T_M = 1683\ K$; $H_f = 1.80\ J\,kg^{-1}$; $\rho = 2300\ kg\,m^{-3}$

a. From Problem 10.21: $\frac{1}{T_p - T_o} = \left(\frac{2\pi k\alpha e}{QV}\right)\left(\frac{Vr'}{2\alpha}\right)^2 + \frac{1}{T_M - T_o}$

$Q = (8000)(0.75) = 6000\,W$; $T_p = 1273\,K$; $T_o = 293\,K$; $V = 0.2\ m\,s^{-1}$

Now solve for r':

$$\frac{QV}{2\pi k\alpha e}\left(\frac{1}{T_p - T_o} - \frac{1}{T_M - T_o}\right) = \left(\frac{Vr'}{2\alpha}\right)^2 \;:\; (T_p - T_o) = (1273 - 293) = 980\,K$$

$$(T_M - T_o) = (1683 - 293) = 1390\,K$$

$$\frac{QV}{2\pi k\alpha e} = \frac{(6000)(0.2)}{(2\pi e)(31)(1.32\times10^{-5})} = 1.717\times10^5\ K$$

$$\therefore \left(\frac{Vr'}{2\alpha}\right)^2 = 1.717\times10^5\left(\frac{1}{980} - \frac{1}{1390}\right) = 51.68$$

$$r' = \frac{(51.68)^{1/2}(2)(1.32\times10^{-5})}{(0.2)} = 9.49\times10^{-4}\,m$$

b. For derivation, see solution to Problem 10.23b

$$\frac{dT}{dt} = \frac{2\pi kV}{Q}(T' - T_o)^2 \;:\; T' = 1683\ K \text{ (directly behind molten pool)}$$

$$\therefore \frac{dT}{dt} = \frac{2\pi \cdot 31\,W \cdot 0.2\,m \cdot (1683-293)^2 K^2}{m\,K \cdot s \cdot 6000\,W} = 1.2544\times10^4\ K\,s^{-1}$$

10.26 Steel plate, 19 mm thick, is welded with 4 kW at 80 pct. efficiency with a welding speed of 2.5 mm s^{-1}. Assume that the edge of the molten pool corresponds to the isotherm of 1700 K. For this steel it is known that some martensite is found in the vicinity of the weld when the cooling rate exceeds 4.6 K s^{-1} at 866 K. a) If the plate is not preheated, would you expect to produce a weld with martensite? b) What is the peak temperature at 0.3 mm from the edge of the weld metal? Problem 10.24 gives properties of the steel.

a. Whether this is a "thick" or "thin" plate is not specified, so we calculate the cooling rate for each case (at r=0 and behind the weld).

For thin plate:

$$\frac{\partial T}{\partial t} = -2\pi k \rho c_p \left(\frac{V\delta}{Q}\right)^2 (T-T_0)^3 = -(2\pi)(35)(7690)(754)\left(\frac{2.5\times10^{-3}(0.019)}{3200}\right)^2 (866-300)^3$$

$$\frac{\partial T}{\partial t} = -50.9 \text{ K s}^{-1}$$

For thick plate:

$$\frac{\partial T}{\partial t} = -\frac{2\pi k V}{Q}(T-T_0)^2 = -\frac{(2\pi)(35)(2.5\times10^{-3})}{3200}(866-300)^2 = -55.0 \text{ K s}^{-1}$$

Notice that the thin plate solution gives a cooling rate that varies with δ, but the thick plate solution is independent of δ. Hence, the thick plate solution represents the maximum cooling rate in any real situation. So we use the lesser of the two and approximate the plate as a "thin plate". In either case, the critical cooling rate is exceeded so martensite will form during cooling.

b. From Problem 10.21: $\frac{1}{T_p - T_o} = \frac{2\pi k \alpha e}{QV}\left(\frac{Vr'}{2\alpha}\right)^2 + \frac{1}{T_M - T_o}$; $\alpha = \frac{35}{(7690)(754)} = 6.04\times10^{-6}\, m^2$

$$\frac{1}{T_p - T_o} = \frac{(2\pi)(35)(6.04\times10^{-6})}{(3200)(2.5\times10^{-3})}\left[\frac{(2.5\times10^{-3})(3\times10^{-4})}{(2)(6.04\times10^{-6})}\right]^2 + \frac{1}{1700-300} = 7.1493\times10^{-4}$$

$T_p = 1698.8$ K (only 3 μm from liquidus isotherm).

10.27 Nickel is "splat cooled" between two platens of copper which are rapidly accelerated toward each other. The process is sketched below.

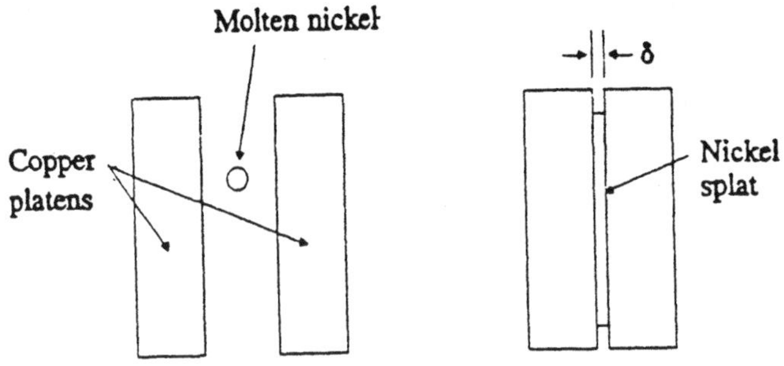

Before splatting After splatting

The splat thickness (δ) is 0.1 mm. Neglect superheat in the nickel droplet. a) Calculate the minimum freezing time of the splat by assuming no interfacial resistance to heat flow. b) Calculate the freezing time assuming that h = 2000 W m^{-2} K^{-1}, and the temperature of the copper platens is 293 K. Properties are listed at the beginning of this section.

a. $(T_M - T_o)\dfrac{C_p'}{H_f} = (1728-293)\dfrac{670}{2.91\times10^5} = 3.30$

$\left(\dfrac{k'\rho'C_p'}{k\rho C_p}\right)^{1/2} = \left[\dfrac{(35)(7850)(670)}{(390)(9000)(380)}\right]^{1/2} = 0.372$

Fig. (10.8) $\dfrac{T_s - T_o}{T_M - T_o} = 0.32$; $T_s = (1728-293)(0.32)+293$

$T_s = 752$ K

Then $(T_M - T_s)\dfrac{C_p'}{H_f\sqrt{\pi}} = (1728-752)\dfrac{670}{(2.91\times10^5)\sqrt{\pi}} = 1.27$

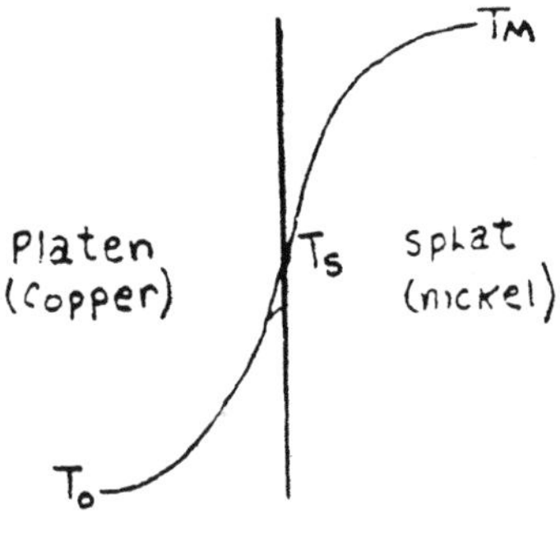

Fig. 10.5 $\beta = 0.82$: $M = 2\beta\sqrt{\alpha' t}$; $\alpha' = \dfrac{35}{(7850)(670)} = 6.66\times10^{-6}$ m^2 s^{-1}

$t = \left(\dfrac{M}{2\beta}\right)^2\dfrac{1}{\alpha'} = \left[\dfrac{0.05\times10^{-3}}{(2)(0.82)}\right]^2\dfrac{1}{6.66\times10^{-6}} = 1.40\times10^{-4}$ s

b. $M = \dfrac{h(T_M-T_o)}{\rho' H_f a}t - \dfrac{h}{2k'}M^2$: $t = \dfrac{\rho' H_f a}{h(T_M - T_o)}\left[M + \dfrac{h}{2k'}M^2\right]$

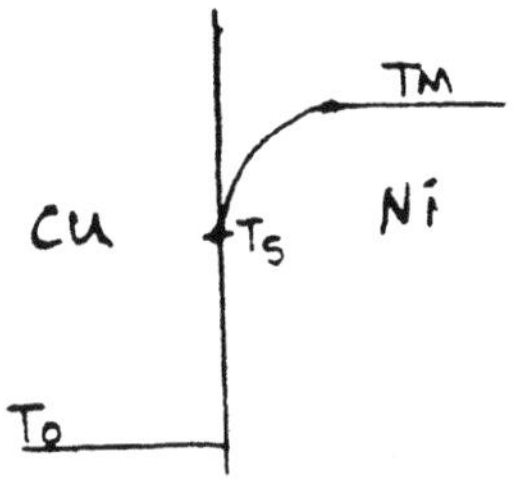

$a \equiv \dfrac{1}{2} + \sqrt{\dfrac{1}{4} + \dfrac{(670)(1728-293)}{3(2.91\times10^5)}} = 1.66$

$t = \dfrac{(7850)(2.91\times10^5)(1.66)}{(2000)(1728-293)}\left[0.05\times10^{-3} + \dfrac{2000}{(2)(35)}(0.05\times10^{-3})^2\right] = 6.6\times10^{-2}$ s

Maybe the thermal profile is better represented as:

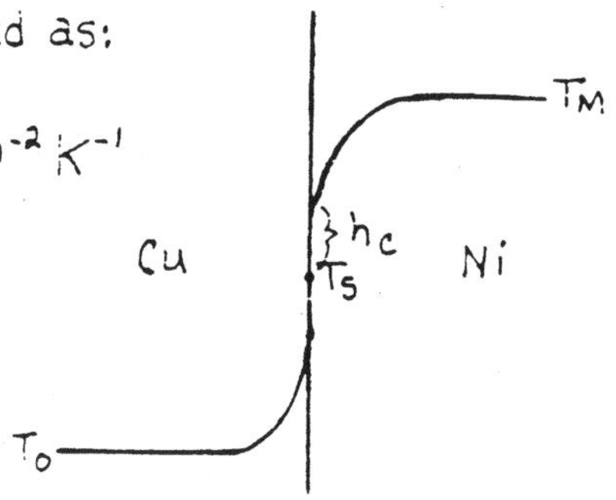

$$h_c = \left[1 + \sqrt{\frac{k'\rho' c_p'}{k\rho c_p}}\right] h = (1+0.372)(2000) = 2744 \ W m^{-2} K^{-1}$$

$T_s = 752 \ K$ (from part a)

$$a = \frac{1}{2} + \sqrt{\frac{1}{4} + \frac{(670)(1728-752)}{(3)(2.91\times10^5)}} = 1.50$$

$$t = \frac{\rho' H_f a}{h_c(T_M - T_s)}\left(M + \frac{h}{2k'}M^2\right) = \frac{(7350)(2.91\times10^5)(1.50)}{(2744)(1728-752)}\left[5\times10^{-5} + \frac{2000}{(2)(35)}(5\times10^{-5})^2\right]$$

$t = 6.4\times10^{-2} \ s$

$h_c > h$ but T_s (752 K) $> T_s$ (293 K) and the two practically balance each other.

10.28 In melt spinning, a jet of molten metal is directed to the surface of a rotating chill. Solidification takes place very rapidly, and a very thin ribbon of metal with thickness δ is produced. The process is shown on the next page.

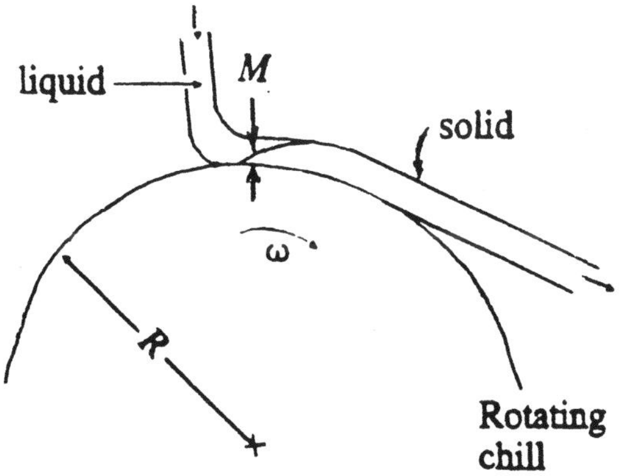

Calculate the rate of solidification (dM/dt in mm s^{-1}) when M is one-half of δ. Assume one-dimensional heat flow in the solidified metal and in the rotating chill, with $h = 1.7 \times 10^4$ W m^{-2} K^{-1}. The metal ribbon is aluminum with a thickness of 50 μm, and the rotating chill is copper. Properties are listed at the beginning of this section.

Metal data - Al, no superheat

$T_M = 933\,K$

$H_f = 3.91 \times 10^5\ J\ kg^{-1}$

$c_p' = 1050\ J\ kg^{-1}\ K^{-1}$

$\rho' = 2400\ kg\ m^{-3}$

$k' = 260\ W\ m^{-1}\ K^{-1}$

$S = 50 \times 10^{-6}\ m$

Rotating chill data

Cu - 293 K (bulk temperature)

$c_p = 380\ J\ kg^{-1}\ K^{-1}$

$\rho = 9000\ kg\ m^{-3}$

$k = 390\ W\ m^{-1}\ K^{-1}$

$h = 1.7 \times 10^4\ W\ m^{-2}\ K^{-1}$

$$(T_M - T_0)\frac{c_p'}{H_f} = (933-293)\frac{1050}{3.91\times10^5} = 1.72 \;:\; \left(\frac{k'\rho'c_p'}{k\rho c_p}\right)^{1/2} = \left[\frac{(260)(2400)(1050)}{(390)(9000)(380)}\right]^{\frac{1}{2}} = 0.70$$

$$\text{Fig. (10.8)}\quad \frac{T_s - T_0}{T_M - T_0} \simeq 0.55\,;\quad T_s = 0.55(933-293) + 293 = 645\,K$$

$$h_c = \left(1 + \sqrt{\frac{k'\rho'c_p'}{k\rho c_p}}\right)h = (1+0.70)\,1.7\times10^4 = 2.89\times10^4\ W\ m^{-2}\ K^{-1}$$

$$\text{Eq. (10.33)}\quad M = \frac{h_c(T_M - T_s)}{\rho' H_f a}t - \frac{h_c}{2k'}M^2 \;:\; a = \frac{1}{2} + \left[\frac{1}{4} + \frac{(1050)(933-645)}{(3)(3.91\times10^5)}\right]^{1/2} = 1.21$$

$$\frac{dM}{dt} = \frac{h_c(T_M - T_s)}{\rho' H_f a} - \frac{h_c M}{k'}\frac{dM}{dt} \quad \text{or} \quad \frac{dM}{dt} = \frac{[h_c(T_M - T_s)/\rho' H_f a]}{[1 + h_c M/k']}$$

$$\frac{dM}{dt} = \frac{2.89\times10^4 \dfrac{(933-645)}{(2400)(3.91\times10^5)(1.21)}}{1 + \dfrac{(2.89\times10^4)(25\times10^{-6})}{(260)}} = 7.3\times10^{-3}\ m\ s^{-1}$$

10.29 Silicon dies are attached to the lead frames in microelectronic devices by soldering (see sketch below).

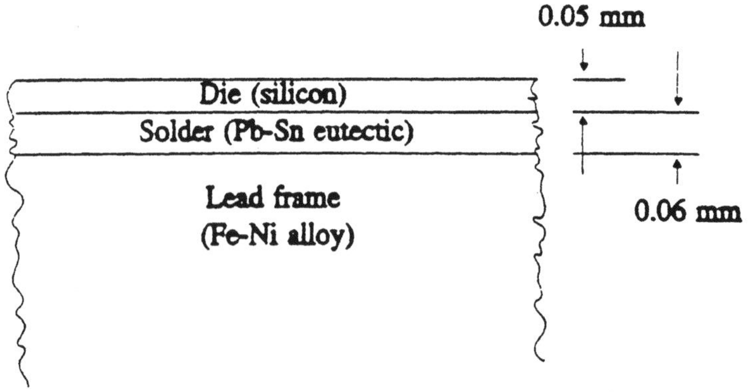

Assume that initially the die and the lead frame are at 423 K and that the solder is molten with no superheat (T_M = 456 K). The ambient above the die is at 293 K. a) What is the temperature at the solder/lead frame interface during solidification of the solder? b) What is the temperature at the solder/die interface early during the solidification of the solder? c) Estimate the time required for the top surface of the die to achieve 428 K. d) Has the solder completely solidified in the time calculated for part c)?

Data:

	T_M, K	H_f, J kg^{-1}	k, W m^{-1} K^{-1}	ρ, kg m^{-3}	C_p, J kg^{-1} K^{-1}
Fe-Ni alloy	—	—	12.6	8700	670
Pb-Sn eutectic	456	54×10^3	33.6	9200	209
Si		(See Example 10.8)			

a. $(T_M - T_0)\left(\dfrac{C_p'}{H_f}\right) = (456-423)\left(\dfrac{209}{54\times10^3}\right) = 0.128$

$$\left(\frac{k'\rho'C_p'}{k\rho C_p}\right)^{1/2} = \left[\frac{(33.6)(9200)(209)}{(12.6)(8700)(670)}\right]^{1/2} = 0.94; \text{ Fig. 10.8 } \frac{T_S - T_0}{T_M - T_0} \simeq 0.96$$

$T_S = (0.96)(T_M - T_0) + T_0 = (0.96)(456-423) + 423 = 455$ K

b. Thermal properties for Si from Exam. 10.8: $k = 31$ W m^{-1}K^{-1}; $\rho = 2300$ kg m^{-3};

$\alpha = 1.32\times10^{-5}$ m^2 s^{-1}; $C_p = \dfrac{k}{\rho\alpha} = 1021$ J kg^{-1} K^{-1}

$$\left(\frac{k'\rho'C_p'}{k\rho C_p}\right)^{1/2} = \left[\frac{(33.6)(9200)(209)}{(31)(2300)(1021)}\right]^{1/2} = 0.94; \text{ Fig. 10.8 } \frac{T_S - T_0}{T_M - T_0} \simeq 0.96$$

$T_S = (0.96)(T_M - T_0) + T_0 = (0.96)(456-423) + 423 = 455$ K

c. Neglect the heat loss from the top surface of the die and apply Fig. 9.8a with $Bi = 1000$.

$$\frac{T - T_f}{T_i - T_f} = \frac{423-455}{423-455} = 0.844$$

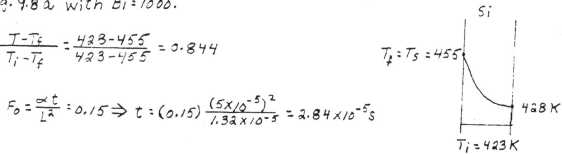

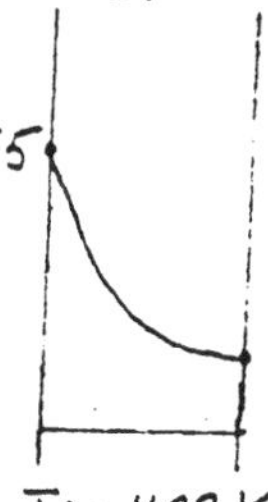

$$F_0 = \frac{\alpha t}{L^2} = 0.15 \Rightarrow t = (0.15)\frac{(5\times10^{-5})^2}{1.32\times10^{-5}} = 2.84\times10^{-5} \text{ s}$$

d. $(T_M - T_S)\dfrac{C_p'}{H_f\sqrt{\pi}} = (456-455)\dfrac{(209)}{(54\times10^3)(\sqrt{\pi})} = 0.0022$

Fig. 10.5 (difficult to read). We resort to the solution of

$\beta e^{\beta^2} \operatorname{erf}\beta = 0.0022$. $\beta = 0.044$.

$M = 2\beta(\alpha' t)^{1/2} = (2)(0.044)\left[(1.75\times10^{-5})(2.84\times10^{-5})\right]^{1/2}$

$M = 1.96\times10^{-6}$ m (from one side)

Total thickness solidified $= 2M = 3.92\times10^{-6}$ m. The solder is 6×10^{-5} m, so that in the time specified, it is not completely solid.

11.1 A radiation heat transfer coefficient is often defined as $q = h_r(T_1 - T_2)$.

a) For a gray solid at T_1, completely surrounded by an environment at T_2, show that h_r is given by

$$h_r = \sigma\varepsilon_1\left(T_1^2 + T_2^2\right)\left(T_1 + T_2\right).$$

b) Calculate the rate of heat transfer (by radiation and convection) from a vertical surface, 1.5 m high, and the percentage by radiation for the following cases:

Surface (ε = 0.8) at:	Air at:	q, W m^{-2}	% rad.
500 K	300 K	4.176×10^3	59.4
800 K	300 K	2.406×10^4	75.8
1100 K	300 K	7.718×10^4	86.0

a. $h_r = \dfrac{q}{(T_1 - T_2)} = \dfrac{\sigma \varepsilon_1 (T_1^4 - T_2^4)}{(T_1 - T_2)} = \sigma \varepsilon_1 (T_1^2 + T_2^2)(T_1 + T_2)$

b. Surface at 500 K

$$h_r = \frac{5.699 \times 10^{-8}\ \text{W}}{\text{m}^2\ \text{K}^4} \Big| 0.8 \Big| \left[(500^2 + 300^2)\text{K}^2 \Big| (500 + 300)\text{K}\right] = 12.4\ \text{W m}^{-2}\ \text{K}^{-1}$$

For h_c use simplified equation from problem 8.14

$$h_c = 1.45(\Delta T)^{1/3} = 1.45(500 - 300)^{1/3} = 8.48\ \text{W m}^{-2}\text{K}^{-1}$$

$$q = (h_r + h_c)(T_1 - T_2) = (12.4 + 8.48)(500 - 300) = 4176\ \text{W m}^{-2}$$

$$\%\ \text{rad.} = \frac{h_r}{h_r + h_c} = \left(\frac{12.4}{12.4 + 8.48}\right)(100) = 59.4\ \%$$

Surface at 800 K

$h_r = 36.6$ W m^{-2} K^{-1}; $h_c = 11.5$ W m^{-2} K^{-1}; $q = 2.406 \times 10^4$ W m^{-2}; % rad. = 75.8%

Surface at 1100 K

$h_r = 83.0$ W m^{-2}K^{-1}; $h_c = 13.47$ W m^{-2}K^{-1}; $q = 7.718 \times 10^4$ W m^{-2}; % rad. = 86.0%

The results have been added to the above table.

11.2 A metal sphere at 1255 K is suddenly placed in a vacuum space whose walls are at 355 K. The sphere is 50 mm in diameter, and we may assume that its surface is perfectly black. a) Calculate the time it takes to cool the sphere to 420 K. Assume that a uniform temperature exists in the sphere at each instant. *Data for sphere:* $\rho = 7000$ kg m^{-3}; $C_p = 1000$ J kg^{-1} K^{-1}. b) Discuss the validity of the assumption in part a) with quantitative reasoning using a conductivity of 50 W m^{-1} K^{-1}.

a. Rate of loss of energy = rate of heat transfer to surroundings.

$$-V\rho c_p \frac{dT_s}{dt} = A\sigma\left(T_s^4 - T_w^4\right) \; ; \; -\int_{T_s=T_i}^{T_s} \frac{dT_s}{\left(T_s^4 - T_w^4\right)} = \frac{A}{V}\frac{\sigma}{\rho c_p}\int_{t=0}^{t} dt$$

$$t = -\frac{V}{A}\frac{\rho c_p}{\sigma}\int_{T_i}^{T} \frac{dT_s}{\left(T_s^4 - T_w^4\right)}$$

Evaluation of the integral

$$T_s^4 - T_w^4 = T_w^4\left(\frac{T_s^4}{T_w^4} - 1\right); \text{ Let } x = \frac{T_s}{T_w}; \; \therefore \int \frac{dT_s}{T_s^4 - T_w^4} = \frac{1}{T_w^4}\int \frac{dT_s}{(x^4 - 1)}$$

$$\frac{1}{x^4 - 1} = \left(-\frac{1}{2}\right)\left(\frac{1}{x^2 + 1}\right) - \left(\frac{1}{4}\right)\left(\frac{1}{x + 1}\right) + \left(\frac{1}{4}\right)\left(\frac{1}{x - 1}\right)$$

$$\int \frac{dT_s}{T_s^4 - T_w^4} = \frac{1}{T_w^4}\left[-\frac{1}{2}\int \frac{dT_s}{\left(\frac{T_s^2}{T_w^2} + 1\right)} - \frac{1}{4}\int \frac{dT_s}{\left(\frac{T_s}{T_w} + 1\right)} + \frac{1}{4}\int \frac{dT_s}{\left(\frac{T_s}{T_w} - 1\right)}\right]$$

$$\therefore \int_{T_i}^{T_s} \frac{dT_s}{T_s^4 - T_w^4} = \frac{1}{T_w^3}\left[-\frac{1}{2}\tan^{-1}\frac{T_s}{T_w}\Bigg|_{T_i}^{T_s} - \frac{1}{4}\ln\left(\frac{T_s}{T_w} + 1\right)\Bigg|_{T_i}^{T_s} + \frac{1}{4}\ln\left(\frac{T_s}{T_w} - 1\right)\Bigg|_{T_i}^{T_s}\right] = -3.309 \times 10^{-11}\,\text{K}^{-3}$$

where $T_i = 1255$ K; $T_s = 420$ K; $T_w = 355$ K; $\frac{V}{A} = \frac{0.025}{3}$ m

$$t = -\frac{0.025\,\text{m}}{3}\left|\frac{7\times10^3\,\text{kg}}{\text{m}^3}\right|\frac{1\times10^3\,\text{J}}{\text{kg K}}\left|\frac{\text{m}^2\,\text{K}^4}{5.669\times10^{-8}\,\text{W}}\right|\frac{-3.309\times10^{-11}}{\text{K}^3} = 33.9\,\text{s}$$

b. Maximum Bi when sphere begins to cool

$$h_r = (5.669\times10^{-8})(1)\left(1255^2 + 355^2\right)(1255 + 355) = 155.3\ \text{W m}^{-2}\,\text{K}^{-1}$$

$$Bi = \frac{h_r R}{k} = \frac{155.3\,\text{W}}{\text{m}^2\,\text{K}}\left|\,0.025\,\text{m}\,\right|\frac{\text{m K}}{50\,\text{W}} = 0.078 \; \therefore \text{Newtonian Cooling}$$

11.3 Steel sheet is preheated in vacuum for subsequent vapor deposition. The steel sheet is placed between two sets of cylindrical heating elements (25 mm in diameter). The heating efficiency is kept high by utilizing radiation reflectors of polished brass. The steel sheet and radiation reflectors may be considered to be infinite parallel plates. a) Calculate the rate of heat transfer from the heating elements (maintained at 1810 K) to the steel when the steel is at 300, 390, 530, 670 and 810 K. The reflectors may be assumed to be at 310 K. b) Plot the results of part a) and determine by graphical integration the average value of the heat transfer rate as the steel is heated from 300 to 810 K. c) From part b) determine the time it takes to heat a steel sheet 6 mm thick from 300 to 810 K.

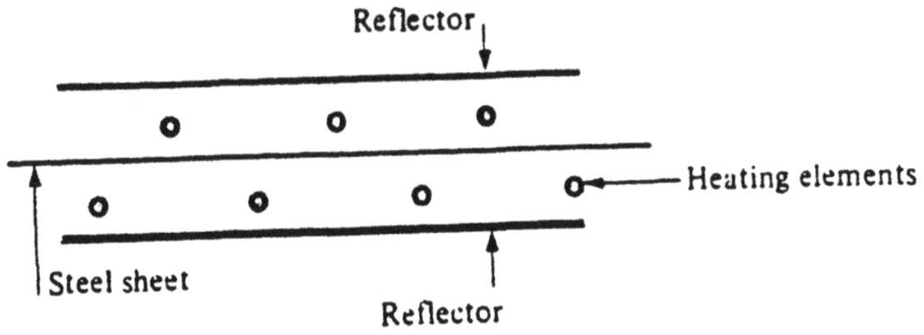

Data: ε (heating elements) = 0.9
ε (steel sheet) = 0.3
ε (reflectors) = 0.03
ρ of steel = 7690 kg m^{-3}
C_p of steel = 628 J kg^{-1} K^{-1}
Area of steel to area of elements = 10/1.

a. Subscripts: 1 for heating elements, 2 for steel, 3 for reflectors

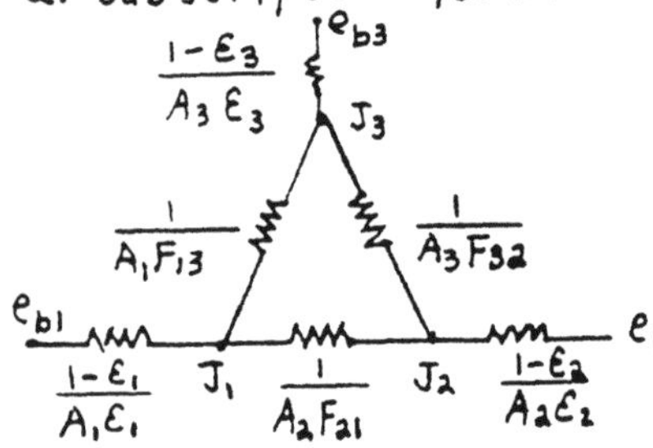

Since ε_3 is only 0.03 then the "current" from J_3 to e_{b3} is negligible and the network is simplified; that is, the reflectors are treated as "no-net-flux-surface". Therefore the circuit reduces to Fig. 11.18c and Eq. (11.45) applies.

$$F_{13} \simeq \frac{1}{2}\,;\quad F_{12} \simeq \frac{1}{2}\,;\quad A_2F_{21} = A_1F_{12}\,;\quad \therefore F_{21} = \frac{A_1}{A_2}F_{12} = \frac{1}{10}\,\frac{1}{2} = 0.05;$$

$$F_{21} + F_{23} = 1 \implies F_{23} = 1 - 0.05 = 0.95$$

$$A_1\bar{F}_{12} = 0.5\,A_1 + \cfrac{1}{\cfrac{1}{0.5A_1} + \cfrac{1}{(0.95)(10)A_1}} = A_1\left(0.5 + \cfrac{1}{2 + \cfrac{1}{(0.95)(10)}}\right) = 0.975\,A_1$$

$$\frac{1}{A_1\mathcal{F}_{12}} = \frac{1}{A_1}\,\frac{1-0.9}{0.9} + \frac{1}{0.975A_1} + \frac{1\,(1-0.3)}{(10A_1)(0.3)} = \frac{1}{0.730A_1}$$

$$Q_{1,net} = A_1 \mathcal{F}_{12} \sigma (T_1^4 - T_2^4) = A_1 (0.730)(5.699\times10^{-8})(1810^4 - 300^4)$$

$$\frac{Q_{1,net}}{A_1} = 4.46\times10^5 \text{ W m}^{-2}$$

T_1, K	T_2, K	q, W m^{-2}
1810	300	4.462×10^5
1810	390	4.456×10^5
1810	530	4.432×10^5
1810	670	4.381×10^5
1810	810	4.286×10^5

$$\bar{q} = 4.406\times10^5 \text{ W m}^{-2}$$

b.

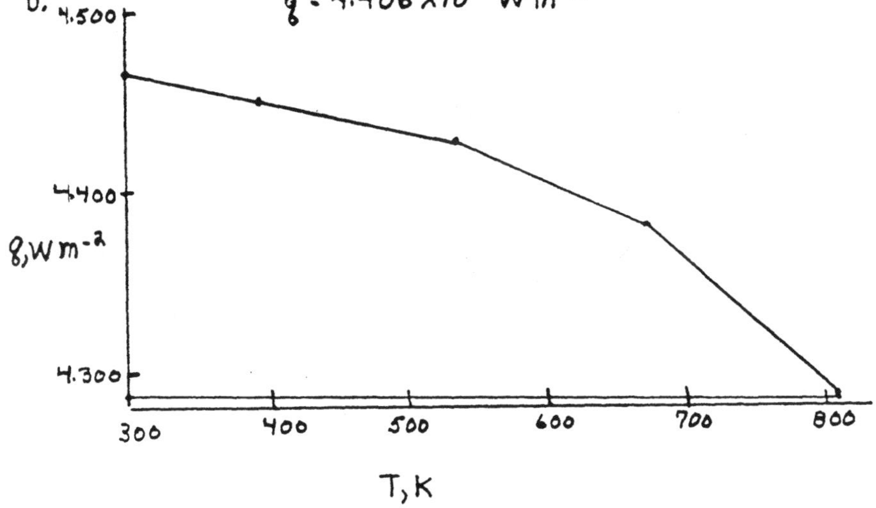

c. Since q does not vary significantly, it is acceptable to use $\bar{q}$ as follows:

$$V\rho c_p \Delta T = 2A\bar{q}t$$

$$t = \frac{V\rho c_p \Delta T}{2A\bar{q}} = \left(\frac{1}{2}\right)\left(\frac{0.006}{2}\right)\frac{(7690)(628)(810-310)}{(4.406\times10^5)} = 8.39 \text{ s}$$

11.4 A large furnace cavity has an inside surface temperature of 1090 K. The walls of the furnace are 0.3 m thick. A hole 150 mm × 150 mm is open through the furnace wall to the room at 295 K. a) Calculate the power loss by radiation through the open hole. b) Calculate the power loss if a sheet of nickel covers the hole on the outside surface. *Data*: emissivity of furnace refractories = 0.9; emissivity of nickel sheet = 0.4.

a. Assume the furnace refractories are perfect insulators. Treat "R" (refractory) as "no-net flux surface".

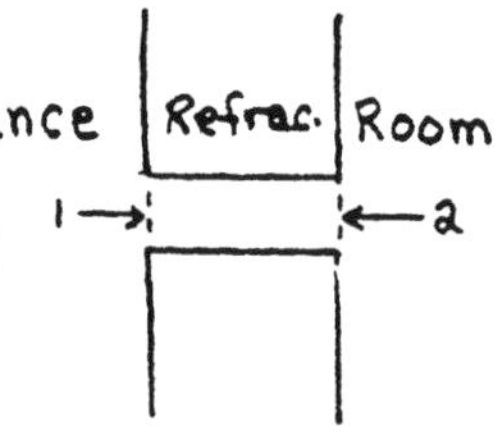

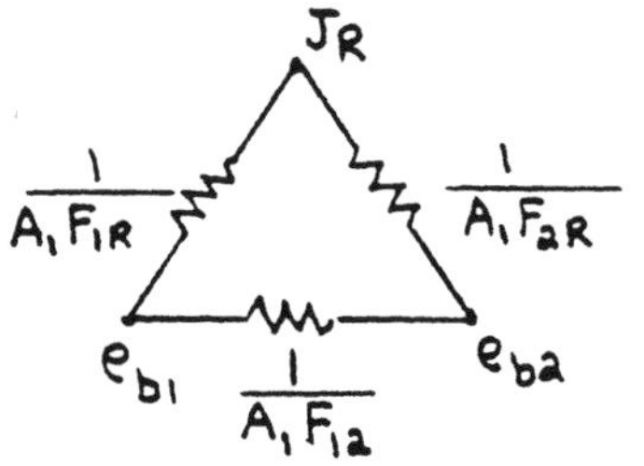

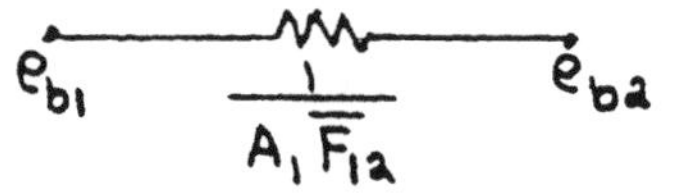

$\bar{F}_{12}$ (total exchange factor) can be found from Fig. 11.21 $\frac{D}{X} = \frac{0.15}{0.3} = 0.5$

$\bar{F}_{12} = 0.40$

$$Q = A_1\bar{F}_{12}\sigma(T_1^4 - T_2^4) = (0.15)^2(0.40)(5.669)\left[\left(\frac{1090}{100}\right)^4 - \left(\frac{295}{100}\right)^4\right] = 7.16\times10^2\ \text{W}$$

b.

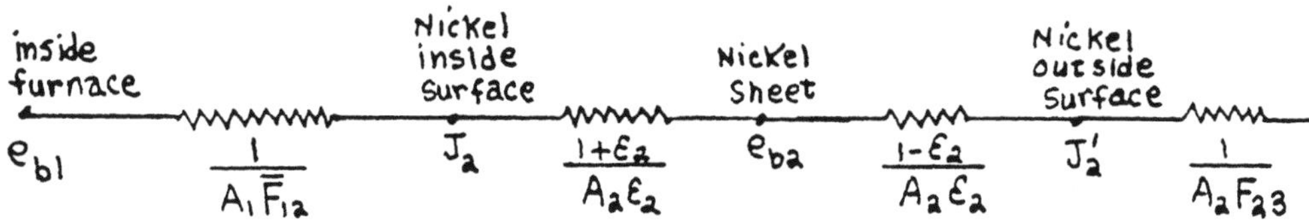

$$\frac{1}{A_1\mathscr{F}_{12}} = \frac{1}{A_1\bar{F}_{12}} + \frac{1-\varepsilon_2}{A_2\varepsilon_2} + \frac{1-\varepsilon_2}{A_2\varepsilon_2} + \frac{1}{A_2F_{23}}$$

$Q = A_1\mathscr{F}_{12}\sigma(T_1^4 - T_3^4)$: Evaluate $A_1\mathscr{F}_{12}$; $A_2 = A_1$, $\varepsilon_2 = 0.4$, $\bar{F}_{12} = 0.40$, $F_{23} = 1$

$$\frac{1}{A_1\mathscr{F}_{12}} = \frac{1}{A_1}\left[\frac{1}{0.40} + \frac{2(1-0.4)}{0.4} + 1\right] = \frac{6.50}{A_1} \ : A_1\mathscr{F}_{12} = 0.154A_1$$

$$Q = (0.154)(0.15)^2(5.699)\left[\left(\frac{1090}{100}\right)^4 - \left(\frac{295}{100}\right)^4\right] = 2.77\times10^2\ \text{W}$$

11.5 Cast iron is continually tapped from the bottom of a cupola into an open refractory channel. The metal enters at 1810 K and runs down the channel at a rate of 0.50 kg s^{-1}. The dimensions of the channel are shown below. Neglect the heat loss by conduction through the refractory and estimate the metal discharge temperature. *Data for molten cast iron*: C_p = 830 J kg^{-1} K^{-1}; ρ = 6890 kg m^{-3}; ε = 0.30.

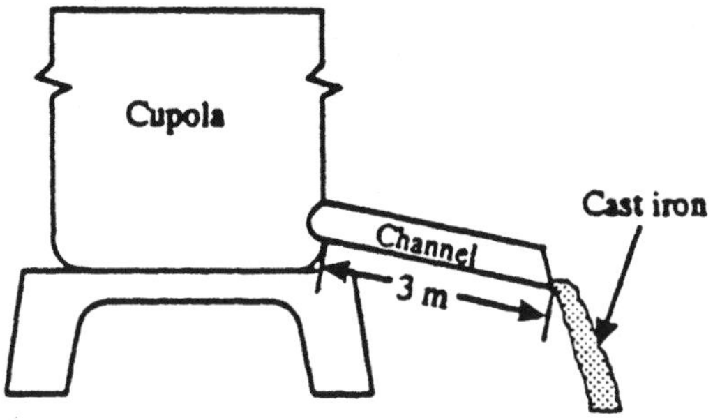

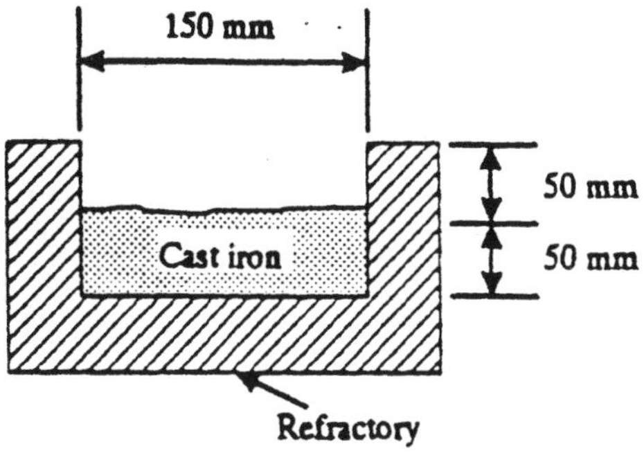

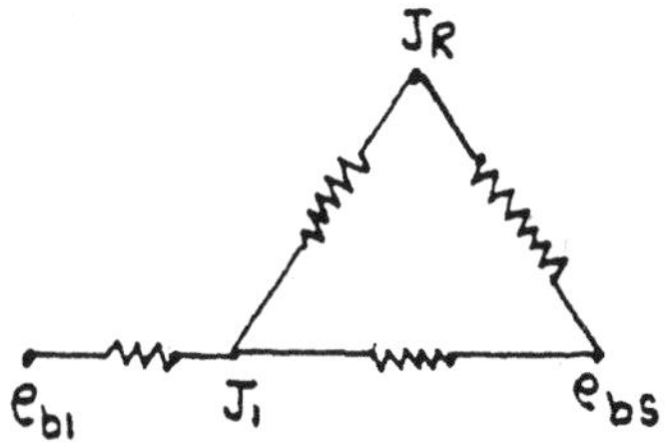

Eq. (11.44) $\dfrac{1}{A_1 \mathcal{F}_{1S}} = \dfrac{1}{A_1} \dfrac{1-\varepsilon_1}{\varepsilon_1} + \dfrac{1}{A_1 \bar{F}_{1S}}$ since $\varepsilon_S = 1$

$$A_1 \bar{F}_{1S} = A_1 F_{1S} + \frac{1}{\dfrac{1}{A_1 F_{1R}} + \dfrac{1}{A_S F_{SR}}}$$

Fig. 11.13: $F_{1S} = 0.70$; $F_{1R} = 1 - F_{1S} = 0.30$; $F_{SR} = F_{1R} = 0.30$

$$A_1 \bar{F}_{1S} = A_1\left(0.70 + \frac{1}{\dfrac{1}{0.30} + \dfrac{1}{0.30}}\right) = 0.85 A_1; \quad \frac{1}{A_1 \mathcal{F}_{12}} = \frac{1}{A_1}\left[\frac{1-0.3}{0.3} + \frac{1}{0.85}\right]$$

$A_1 \mathcal{F}_{1S} = 0.285 A_1$; $A_1 \mathcal{F}_{1S} = A_S \mathcal{F}_{S1} \Rightarrow \mathcal{F}_{S1} = \mathcal{F}_{1S}$

Apply Fig. 11.28 (b) $t = \dfrac{L}{\dfrac{\dot{W}}{\rho}\dfrac{1}{A}} = \dfrac{L\rho A}{\dot{W}} = \dfrac{(3)(6890)(0.150)(0.05)}{(0.50)} = 310$ s

$$\frac{\sigma \mathcal{F}_{S1} T_S^3 t}{\rho C_p l} = \frac{(5.669\times10^{-8})(0.285)(300)^3(310)}{(6890)(830)(0.05)} = 4.73\times10^{-4}; \quad \frac{T_o}{T_S} = \frac{1810\text{K}}{300\text{K}} = 6.03$$

$\dfrac{T}{T_o} = 0.92$: $T = (0.92)(1810) = 1665$ K

11.6 A method of melting metal so as to avoid crucible contamination is levitation melting. A metal sample is placed in an electromagnetic field from a coil wound as a cone. The field not only supplies power to melt the metal but also to levitate it. But with this setup, the strength of field necessary to keep the sample levitated is sometimes such that the metal is overheated. A means of preventing this is to flow cooling gas past the sample.

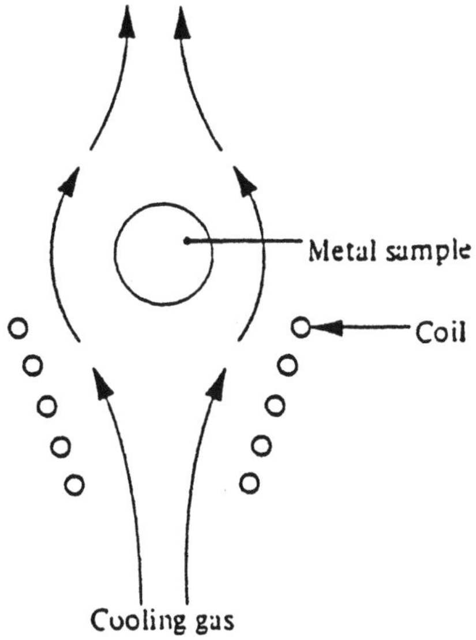

Develop an expression that relates the steady-state temperature of the metal to the heating power supplied by the field (Q_p), the gas temperature T_0, the gas velocity V, and any other parameters which you think are essential. Metal temperatures of interest are 1360-1920 K, and the convection heat transfer coefficient can be expressed as

$$h_c = KV^{0.7},$$

in which K is a known constant.

At steady state

Power to sample = Rate of heat loss by radiation + Rate of heat loss by convec.

$Q_p = A\left[\sigma\varepsilon(T_M^4 - T_S^4) + h_c(T_M - T_0)\right]$ where T_M = metal temp.; T_S = temp. of surroundings.; T_0 = gas temp.; A = surface area.

Since $T_M^4 \gg T_S^4$, we have:

$\frac{Q_p}{A} = \sigma\varepsilon T_M^4 + KV^{0.7}(T_M - T_0)$ which can be solved for T_M.

11.7 A steel sheet, 12.5 mm thick and having the shape of a square 1.5 m × 1.5 m, comes out of a heat-treating furnace at 1090 K. During heat treatment its surface was oxidized so that its emissivity is 0.8.

a) Calculate its initial cooling rate (K s^{-1}) if it is suspended freely by a wire in a room at 300 K. Neglect all heat transfer with convection, i.e., deal only with radiation heat transfer.

b) Calculate its initial cooling rate if it is supported vertically on a horizontal surface (also at 300 K) which has an emissivity of 0.2. Again deal only with radiation heat transfer.

Data (all units in SI):

	Supporting surface	Steel sheet
Thermal conductivity	50	20
Density	600	8010
Heat capacity	630	600
Emissivity	0.2	0.8

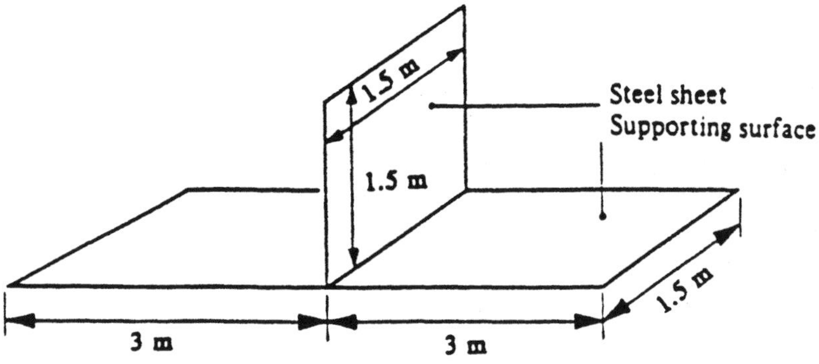

Setup for (b).

Subscripts: 1 for steel, 2 for the room, 3 for supporting surface

a. $q = \sigma \varepsilon_1 (T_1^4 - T_2^4)$

$$A_1 q = -\rho C_p V_1 \frac{dT_1}{dt}$$

$$\therefore \frac{dT}{dt} = -\frac{A_1}{V_1}\frac{1}{\rho C_p}\varepsilon_1 \sigma (T_1^4 - T_2^4) = -\frac{(160)(0.8)(5.699)}{(8010)(600)}\left[\left(\frac{1090}{100}\right)^4 - \left(\frac{300}{100}\right)^4\right] = -2.13 \text{ K s}^{-1}$$

b.

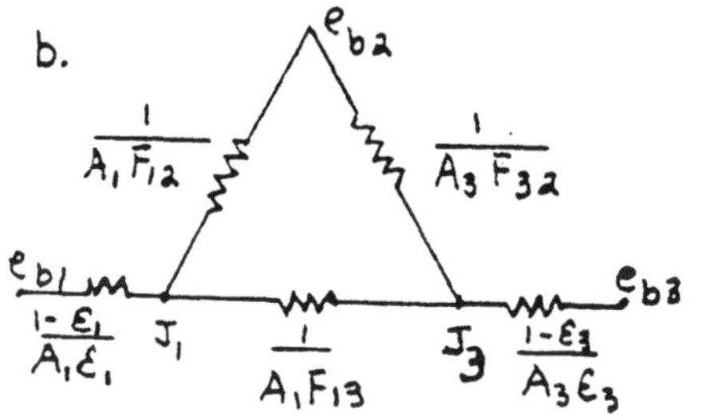

Solve for all the resistances

$$\frac{1-\varepsilon_1}{A_1\varepsilon_1} = \frac{(1-0.8)}{(2.25)(0.8)} = 0.1111 \text{ m}^{-2}$$

$$\frac{1}{A_1F_{13}} = \frac{1}{(2.25)(0.23)} = 1.9324 \text{ m}^{-2}$$

$$\frac{1-\varepsilon_3}{A_3\varepsilon_3} = \frac{(1-0.2)}{(4.5)(0.2)} = 0.8889 \text{ m}^{-2}$$

$$\frac{1}{A_1F_{12}} = \frac{1}{(2.25)(0.77)} = 0.5772 \text{ m}^{-2}$$

$$\frac{1}{A_3F_{32}} = \frac{1}{(4.5)(0.885)} = 0.2511 \text{ m}^{-2}$$

At each node, the sum of the current (flux) must be zero at steady state.

$$\text{Node 1: } \frac{e_{b1}-J_1}{0.1111} + \frac{J_3-J_1}{1.9324} + \frac{e_{b2}-J_1}{0.5772} = 0$$

$$\text{Node 3: } \frac{e_{b3}-J_3}{0.8889} + \frac{J_1-J_3}{1.9324} + \frac{e_{b2}-J_3}{0.2511} = 0$$

```
10 'Problem 11.7
20 'subscripts:  1 for steel sheet; 2 for room ; 3 for supporting surface
30  A1 = 2.25 : A3 = 4.5                       'areas in m^2
40  F13 = .23 : F12 = 1 -F13                   'view factors
50  F31 = A1*F13/A3 : F32 = 1 - F31            'view factors
60  E1 = .8 : E3 = .2                          'emissivities
70  T1 = 1090 : T2 = 300 : T3 = 300            'temperatures in K
80  K1 = 20 : RHO1 = 8010 : CP1 = 600          'thermal properties in SI units
90  X = .0125                                  'thickness of steel sheet
100 SIGMA = 5.699E-08                          'Stefan-Boltzmann's constant
110 'solve for the resistances
120  R1 = (1 - E1)/( A1*E1 ) : R2 = 1/( A1*F12 ) : R3 = 1/( A1*F13 )
130  R4 = 1/( A3*F32 ) : R5 = (1-E3)/( A3*E3 )
140 'solve for black surface emissive powers (really have units of fluxes)
150  EB1 = SIGMA*T1^4 :  EB2 = SIGMA*T2^4 : EB3 = SIGMA*T3^4
160 'solve for constants so that unknown radiosities can be determined
170  C = ( EB2/R2 + EB1/R1 )/( 1/R1 + 1/R2 + 1/R3 )
180  D = R3*( 1/R1 + 1/R2 + 1/R3 )
190  E = ( EB3/R5 + EB2/R4 )/( 1/R3 + 1/R4 + 1/R5 )
200  F = R3*( 1/R3 + 1/R4 + 1/R5 )
210  J1 = F*( C*D + E )/( D*F - 1 )       'radiosity
220  J3 = E + J1/F                        'radiosity
230  PRINT J1,J3
240 'heat loss from steel sheet
250  HEAT1 = (EB1 - J1)/R1                'in W  reminder for one side only
260  HEAT11 = (J1 - EB2)/R2 + (J1 - J3)/R3
270  PRINT HEAT1, HEAT11                             'HEAT1 should equal HEAT11
280  RATE = -2*HEAT1/(RHO1*CP1*A1*X)      'rate in K/s, HEAT1 is for one side
290  LPRINT : LPRINT : LPRINT "  The cooling rate is";RATE;"K/s"
300  END


    The cooling rate is-2.094041 K/s
```

11.8 Brass sheet, 1.6 mm thick, passes through a continuous tempering surface. The first portion of the furnace is filled with gases at 920 K, moving against the strip direction at a speed of 0.3 m s^{-1}. The strip itself moves at a speed of 0.05 m s^{-1}. The walls are of SiO_2 brick, each 0.3 m from the brass sheet as it moves vertically, and the gases are 40% H_2O, 10% CO_2, and 50% N_2. How many meters of strip would have to be in the first portion of the furnace so that the strip could exit into the second section at a temperature of 700 K?

Data:

$C_{P_{brass}} = 428$ J kg^{-1} K^{-1} $\quad C_{P_{gas}} = 1110$ J kg^{-1} K^{-1} $\quad \eta_{gas} = 4 \times 10^{-5}$ N s m^{-2}

$\rho_{brass} = 9200$ kg m^{-3} $\quad \rho_{gas} = 0.48$ kg m^{-3}

$k_{brass} = 170$ W m^{-1} K^{-1} $\quad k_{gas} = 0.086$ W m^{-1} K^{-1}

The sheet loses energy by radiation and convection.

Convection: Relative velocity between the strip and gas is $V_\infty = 0.35$ m s^{-1}.

Taking $L = 3$ m as an estimate, then $Re_L = \dfrac{L V_\infty \rho}{\eta}$

$$Re_L = \frac{(3)(0.35)(0.48)}{4\times 10^{-5}} = 12.6\times 10^{3}$$

For flow parallel to the flat plate, use Eq. (7.27):

$$Nu_L = 0.664\, Pr^{0.343}\, Re_L^{0.5}; \quad Pr = \frac{\eta C_P}{k} = \frac{(4\times10^{-5})(1110)}{0.086} = 0.516$$

$$Nu_L = 0.664(0.516)^{0.343}(12.6\times 10^3)^{0.5} = 59.4$$

$$h_c = \frac{k\, Nu_L}{L} = \frac{(0.086)(59.4)}{3} = 1.70 \text{ W m}^{-2}\text{K}^{-1}$$

Radiation: $F_{12} = F_{21} = 1$ $\quad \epsilon_1 (800\text{K}) = 0.6$ (Fig. 11.8)

$$h_r = \epsilon_1 \left[\frac{\sigma (T_2^4 - T_1^4)}{T_2 - T_1}\right]$$

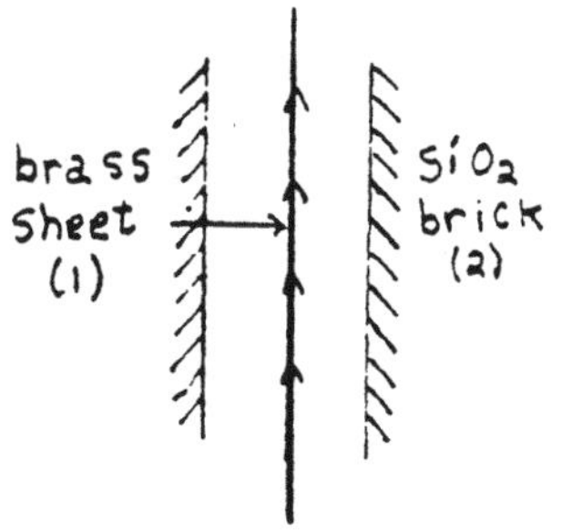

When strip enters furnace, $T_1 = 300$ K, and when it leaves, $T_1 = 700$K. Assume $T_2 = T_{gas} = 920$K.

$$\text{Enters: } h_r = (0.6)(5.699\times10^{-8})\left[\frac{920^4 - 300^4}{920-300}\right] = 39.06 \text{ W m}^{-2}\text{K}^{-1}$$

$$\text{Leaves: } h_r = (0.6)(5.699\times10^{-8})\left[\frac{920^4 - 700^4}{920-700}\right] = 74.03 \text{ W m}^{-2}\text{K}^{-1}$$

Obviously radiation dominates, so we can ignore h_c and use Fig. 11.28 (a).

$\frac{T}{T_0} = \frac{700}{300} = 2.33$; $\frac{T_0}{T_2} = \frac{300}{900} = 0.326$; $\mathcal{F}_{21} = \epsilon_1 = 0.6$; $\ell = 0.8$mm (heated from both sides)

$$\frac{\sigma \epsilon_1 T_2^3 t}{\rho_1 c_{p_1} \ell} = 0.5 \Rightarrow t = \frac{(0.5)(9200)(428)(0.8\times10^{-3})}{(5.699\times10^{-8})(0.6)(920^3)} = 59.2 \text{ s} \Rightarrow L = (0.05)(59.2) = 2.96 \text{ m}$$

11.9 Estimate the length of the furnace, shown below, that is used for forming sheet glass (3 mm thick). The glass enters the furnace at 1500 K, and it should leave at 1100 K. The liquid tin and the furnace are kept at 1000 K. Assume that the heat transfer coefficient at the tin-glass interface is infinite. The emissivity of the glass is shown in Fig. 11.27. Other properties of the glass are $k = 4.2$ W m^{-1} K^{-1}; $C_p = 420$ J kg^{-1} K^{-1}; $\rho = 3320$ kg m^{-3}. The emissivity of the molten tin is 0.1.

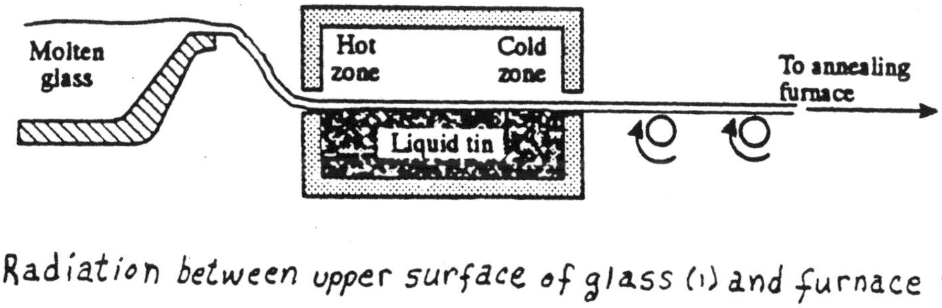

Radiation between upper surface of glass (1) and furnace refractories (2).

Assume $\mathcal{F}_{12} = \epsilon_1 (1300\text{K}) = 0.53$

Then $h_r = \epsilon_1 \left[\frac{\sigma (T_1^4 - T_2^4)}{T_1 - T_2}\right]$ where $\epsilon_1 = 0.53$, T_1 varies from 1500K to 1100K and $T_2 = 1000$ K. $\therefore h_r(1500\text{K}) = 245.4$; $h_r(1100\text{K}) = 140.2$

If we ignore the thermal conduction in the direction of the motion of the glass, then the temperature in the glass must satisfy:

$$\frac{\partial \theta}{\partial t} = \alpha \frac{\partial^2 \theta}{\partial x^2}$$ where: Initial condition; $\theta(x,0) = \theta_i$ (entering furnace)

Boundary conditions; $\theta(0,t) = 0$ (glass/tin interface)

$$\frac{\partial \theta}{\partial x}(L,t) = -\frac{h_r}{k}\theta(L,t)$$ [upper glass surface]

where $\theta = T - T_2$ (T_2 = furnace and tin temperatures).

Solution by separation of variables is:

$$\theta = 2\theta_i \sum_{n=1}^{\infty} \exp(-\lambda_n^2 \alpha t) \frac{(h^2/k^2 + \lambda_n^2)\sin \lambda_n x}{(h^2/k^2 + \lambda_n^2)L + (h/k)}\left[\frac{1}{\lambda_n} - \frac{\cos \lambda_n L}{\lambda_n}\right]$$

where λ_r are the positive roots of $\lambda_n \cot \lambda_n L + \frac{h}{k} = 0$

To simplify the problem, we assume that during most of the cooling period there is a linear temperature distribution in the glass.

$$-\rho_1 C_{P_1} L \frac{d\bar{T}}{dt} = k_1 \frac{T_L - T_2}{L} + h_r (T_L - T_2)$$

but $\bar{T} = \frac{T_L + T_2}{2}$; $T_L = 2\bar{T} - T_2$

$$-\frac{d\bar{T}}{dt} = \frac{2\alpha_1}{L^2}(\bar{T} - T_2) + \frac{2h_r}{\rho_1 C_{P_1} L}(\bar{T} - T_2) = \beta(\bar{T} - T_2)$$

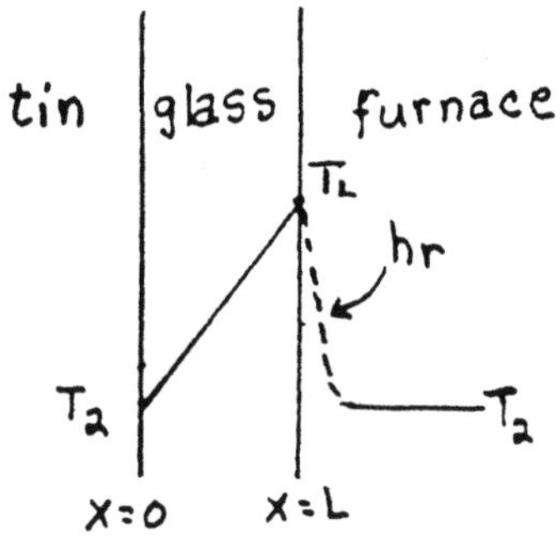

where $\beta = \frac{2\alpha_1}{L^2} + \frac{2h_r}{\rho_1 C_{P_1} L}$

with $\bar{T} = \bar{T}_i$ at $t = 0$, this gives $t = -\frac{1}{\beta} \ln \frac{\bar{T} - T_2}{\bar{T}_i - T_2}$

$\alpha_1 = \frac{4.2}{(420)(3320)} = 3.01 \times 10^{-6}\ m^2 s^{-1}$; $L = 0.003\ m$; $h_r(T_L = 1500\ K) = 245\ W m^{-2} K$;

$h_r(T_L = 1100\ K) = 140.2$; Use $h_r = \frac{245 + 140.2}{2} = 193\ W m^{-2} K^{-1}$

$$\beta = \frac{(2)(3.01 \times 10^{-6})}{(0.003)^2} + \frac{(2)(193)}{(3320)(420)(0.003)} = 0.761\ s^{-1}$$

$\bar{T}_i = \frac{1500 + 1000}{2} = 1250\ K$; $\bar{T} = 1100\ K$

$$t = -\frac{1}{0.761} \ln \frac{1100 - 1000}{1250 - 1000} = 1.20\ s$$

Let L_f = length of furnace, m; R = production rate, m s^{-1}

R, m s^{-1}	$L_f = Rt$, m
0.1	0.12
1	1.2
10	12

11.10 A slab of steel is placed into a reheat furnace so that it can be heated to 1500 K and subsequently hot-rolled to plate. The gas temperature in the furnace has been carefully measured and found to be 1520 K. Energy to the furnace is by burning gas, and its flame behaves as though it is a gray surface ($\varepsilon = 0.9$) at 2000 K. The convective heat transfer coefficient is 85 W m^{-2} K^{-1} and the emissivity of the steel is 0.8. Estimate the temperature of the slab after the system comes to thermal equilibrium. For simplicity assume that the flame and slab surface are parallel planes that are 5 m × 5 m, and the gas is transparent. Consider only heat exchange among the slab, flame and gas.

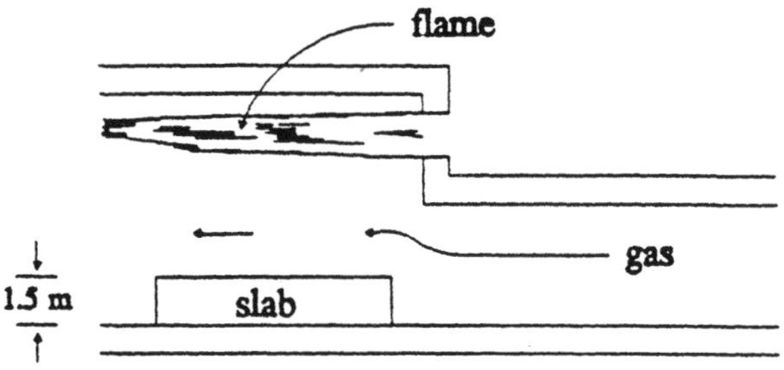

f ≙ flame; g ≙ gas; s ≙ slab; r ≙ refractory

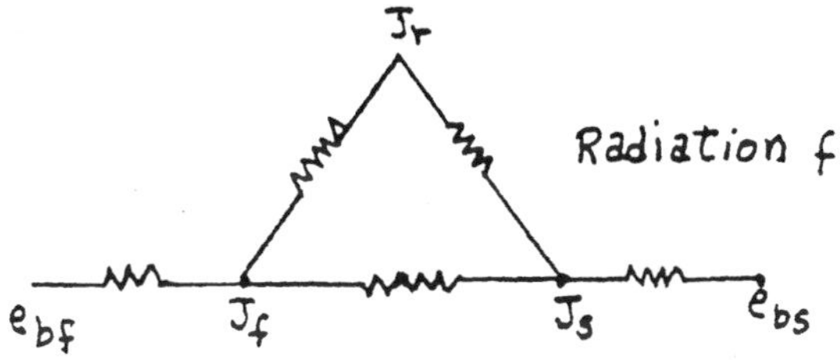

Radiation from flame to slab:

This circuit reduces to:

$$\frac{1}{A_f \mathfrak{F}_{fs}} = \frac{1}{A_f}\frac{1-\varepsilon_f}{\varepsilon_f} + \frac{1}{A_f \bar{F}_{fs}} + \frac{1}{A_s}\frac{1-\varepsilon_s}{\varepsilon_s} \; ; \; A_s = A_f$$

$$\frac{1}{A_f \mathfrak{F}_{fs}} = \frac{1}{A_f}\left[\frac{1-\varepsilon_f}{\varepsilon_f} + \frac{1}{\bar{F}_{fs}} + \frac{1-\varepsilon_s}{\varepsilon_s}\right]$$

$$A_f \bar{F}_{fs} = A_f F_{fs} + \frac{1}{\dfrac{1}{A_f F_{fr}} + \dfrac{1}{A_s F_{sr}}} \; ; \; \bar{F}_{fs} = F_{fs} + \frac{1}{\dfrac{1}{F_{fr}} + \dfrac{1}{F_{sr}}}$$

Fig. 11.13 ($a = b = 5$ m and $c = 2.5$ m from the sketch).

$F_{fs} = F_{sf} = 0.38$; $F_{sr} = 1 - F_{sf} = 0.62$; $F_{fr} = 1 - F_{fs} = 0.62$

$$\bar{F}_{fs} = 0.38 + \frac{1}{\frac{1}{0.62} + \frac{1}{0.62}} = 0.690$$

$$\frac{1}{A_f \mathcal{F}_{fs}} = \frac{1}{A_f}\left[\frac{1-0.9}{0.9} + \frac{1}{0.690} + \frac{1-0.8}{0.8}\right] = \frac{1.81}{A_f}\ ; \therefore\ \mathcal{F}_{fs} = 0.552$$

At steady state (with $A_f = A_s$)

$$\mathcal{F}_{fs}\,\sigma\,(T_f^4 - T_s^4) = h_c\,(T_s - T_g)$$

where $T_f = 2000$ K ; $T_g = 1520$ K ; $h_c = 85$ W m^{-2} K^{-1}

Solve by trial and error. $T_s \simeq 1962$ K

11.11 One side of a flat ceramic shell mold (30 mm thick) is maintained at 1500 K while the opposite side radiates into a large room at 300 K. The emissivity of the radiating surface is 0.5, and the thermal conductivity of the ceramic mold material is 0.7 W m^{-1} K^{-1}. Calculate the steady-state flux and the surface temperature of the mold.

Consider radiation from right side of shell mold (1) to the room (2).

$$Q = A_1 \mathcal{F}_{12}\,\sigma\,(T_1^4 - T_2^4) = A_1 \varepsilon_1 \sigma\,(T_1^4 - T_2^4)$$

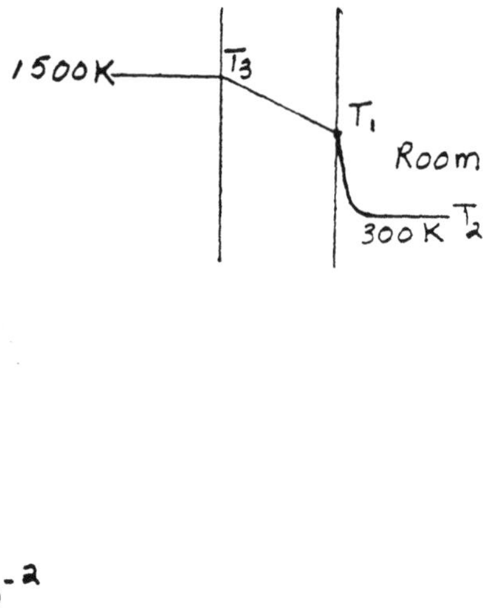

Consider conduction through the plate.

$$Q = Q_{cond.} = A_1 \frac{k}{L}(T_3 - T_1)$$

At steady state: $\mathcal{F}_{12}\,\sigma\,(T_1^4 - T_2^4) = \frac{k}{L}(T_3 - T_1)$

$$(0.5)(5.699)\left[\left(\frac{T_1}{100}\right)^4 - \left(\frac{300}{100}\right)^4\right] = \frac{0.7}{0.03}(1500 - T_1)$$

By trial and error $T_1 \simeq 855$ K

$$\therefore\ q = \frac{k}{L}(T_3 - T_1) = \frac{0.7}{0.03}(1500 - 855) = 1.505 \times 10^4 \text{ W m}^{-2}$$

11.12 A plate is heated uniformly to 810 K, and then it is suspended vertically next to and parallel to a cooler plate in a large room. Both plates are 1.3 m × 1.3 m × 12.5 mm thick, and they are separated by 0.3 m. The cooler plate and the room are at 300 K. What is the initial cooling rate (K s^{-1}) of the hotter plate? *Data:* $\varepsilon = 0.7$ and $\rho C_p = 8 \times 10^6$ J m^{-3} K^{-1} for both plates.

Subscripts: 1 for hotter plate, 2 for room, 3 for cooler plate

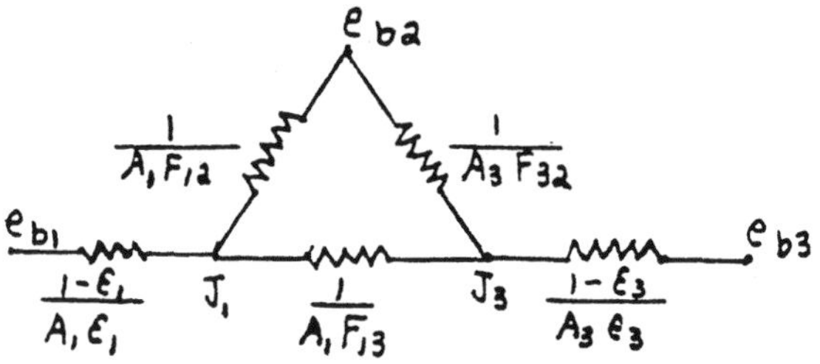

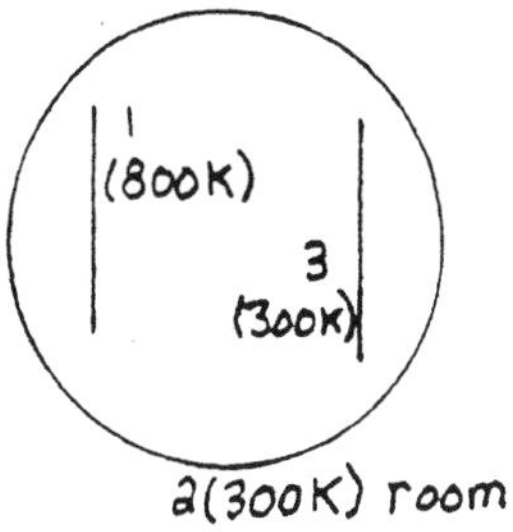

The following program calculates the initial cooling rate.

```
10 'Problem 11.12
20 'subscripts:  1 for hotter plate ; 2 for room ; 3 for cooler plate
30 A1 = 2.69 : A3 = 2.69                   'areas in m^2
40 F13 = .62 : F12 = 1 -F13                'view factors
50 F31 = A1*F13/A3 : F32 = 1 - F31         'view factors
60 F31 = A1*F13/A3 : F32 = 1 - F31         'view factors
70 E1 = .7 : E3 = .7                       'emissivities
80 T1 = 810 : T2 = 300 : T3 = 300          'temperatures in K
90 RHOCP1 = 8000000!                       'product of density and Cp in SI units
100 X = .0125                              'thickness of steel sheet
110 SIGMA = 5.699E-08                      'Stefan-Boltzmann's constant
120 'solve for the resistances
130 R1 = (1 - E1)/( A1*E1 ) : R2 = 1/( A1*F12 ) : R3 = 1/( A1*F13 )
140 R4 = 1/( A3*F32 ) : R5 = (1-E3)/( A3*E3 )
150 'solve for black surface emissive powers (really have units of fluxes)
160 EB1 = SIGMA*T1^4 :  EB2 = SIGMA*T2^4 : EB3 = SIGMA*T3^4
170 'solve for constants so that unknown radiosities can be determined
180 C = ( EB2/R2 + EB1/R1 )/( 1/R1 + 1/R2 + 1/R3 )
190 D = R3*( 1/R1 + 1/R2 + 1/R3 )
200 E = ( EB3/R5 + EB2/R4 )/( 1/R3 + 1/R4 + 1/R5 )
210 F = R3*( 1/R3 + 1/R4 + 1/R5 )
220 J1 = C + ( C + E*F )/( D*F - 1 )    'radiosity
230 J3 = E + J1/F                        'radiosity
240 PRINT J1,J3
250 'heat loss from surface of hotter sheet facing the cooler sheet, W
260 HEAT1 = (EB1 - J1)/R1
270 HEAT11 = (J1 - EB2)/R2 + (J1 - J3)/R3
280 PRINT HEAT1, HEAT11                          'HEAT1 should equal HEAT11
290 'heat loss from hotter sheet facing the room
300 HEAT2 = A1*SIGMA*E1*( T1^4 - T2^4 )  ' in W
310 HEAT = HEAT1 + HEAT2                  'total heat loss from hotter plate, W
320 RATE = -HEAT/(RHOCP1*A1*X)            'rate in K/s
330 LPRINT : LPRINT : LPRINT "   The cooling rate is ";RATE;"K/s"
340 END
```

The cooling rate is -.322901 K/s

11.13 A light bulb (100 watts) can be approximated to be a sphere. It is held at the top level of a large amount of snow; after steady state is achieved, a hole with the form of a hemisphere of radius R is made in the snow. Consider only radiation heat transfer from the bulb to the snow and to the surroundings. Within the snow, however, there is conduction heat transfer. The snow and the surroundings are at 0°F. The surface of the snow forming the hole is at 32°F. *Thermal properties of snow*: $k = 0.2$ Btu h^{-1} ft^{-1} °F^{-1}; $\rho C_p = 10$ Btu ft^{-3} °F^{-1}; $\varepsilon = 0.7$. a) Calculate the radius of the hole. b) Calculate the surface temperature of the light bulb. Assume that this surface is black and that the radius of the spherical bulb is 0.2 ft.

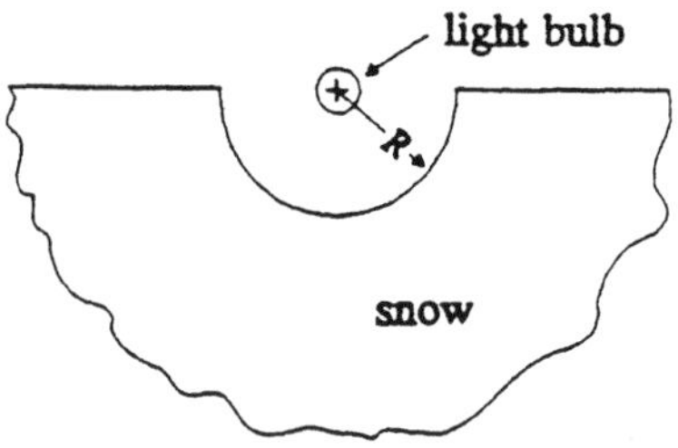

a. Steady-state for snow $\frac{d}{dr}\left(r^2 \frac{dT}{dr}\right) = 0$

Boundary Conditions: at $r = R$, $T = T_R = 32°F$

$r = \infty$, $T = T_\infty = 0°F$

$r^2 \frac{dT}{dr} = C_1 \Rightarrow \frac{dT}{dr} = \frac{C_1}{r^2}$; $T = -\frac{C_1}{r} + C_2$

at $r = R$: $T_R = -\frac{C_1}{R} + C_2$

at $r = \infty$: $T_\infty = C_2$ $\therefore -C_1 = (T_R - T_\infty) R$

Then $\frac{T - T_\infty}{T_R - T_\infty} = \frac{R}{r}$

At surface of hole, $r = R$

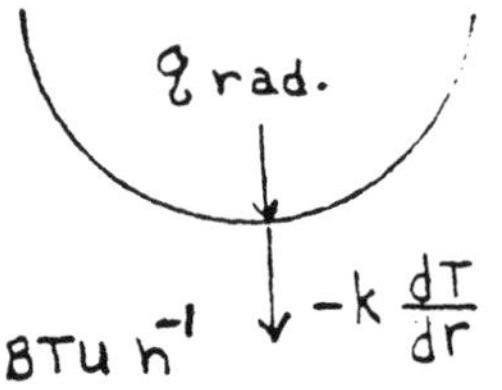

$q_{rad.} = -k\left(\frac{dT}{dr}\right)_{r=R}$; $2\pi R^2 q_{rad.} = \frac{1}{2} Q_{bulb}$

$\therefore \frac{Q_{bulb}}{4\pi R^2} = k(T_R - T_\infty)\frac{R}{R^2}$ $\therefore R = \frac{Q_{bulb}}{4\pi k (T_R - T_\infty)}$; 1 W = 3.41 BTU h^{-1}

$$R = \frac{341 \text{ BTU}}{\text{h}} \left| \frac{}{4\pi} \right| \frac{\text{h ft °F}}{0.2 \text{ BTU}} \left| \frac{}{(32-0)°F} \right. = 4.24 \text{ ft}$$

b. Subscripts: Surroundings-3, snow surface-2, light bulb-1

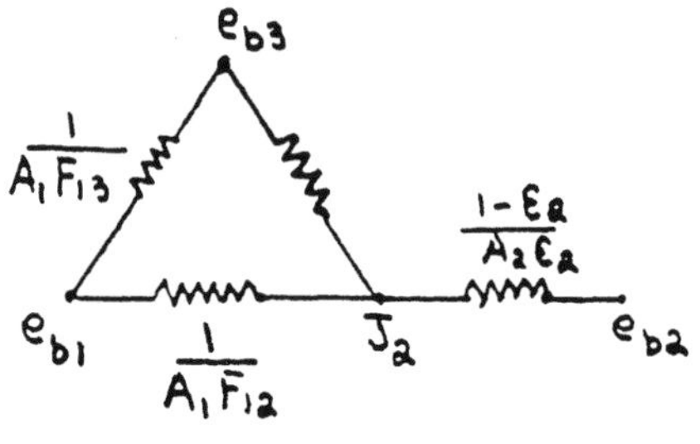

Because snow is poor thermal conduction we expect $J_2 \approx e_{b2}$. Also the surface temp. of the bulb can be expected to be $\gg T_3$ and $\gg T_2$. Therefore, e_{b1} will be divided 1/2 to surroundings and 1/2 to snow.

11.13 cont.

$$Q = \frac{1}{2} Q_{bulb} = A_1 F_{13} \sigma (T_1^4 - T_3^4)$$

$$T_1^4 - T_3^4 = \frac{\frac{1}{2} Q_{bulb}}{A_1 F_{13} \sigma} = \frac{\frac{341}{2}}{(4\pi)(0.2)^2(0.5)(0.1713\times10^{-8})} = 3.9603\times10^{11}\ {}^\circ R^4$$

with $T_3 = 460^\circ R$; $T_1 = 815^\circ R = 355^\circ F$

11.14 Thin sheet of alloy is prepared by "roller casting." In this process, a melt solidifies between two rolls that are water-cooled to produce sheet of thickness δ. *Assume*: (1) that temperature in the sheet is independent of x; (2) that conduction in the y-direction can*not* be ignored; (3) heat loss from the surface is *only* by radiation; (4) the sheet temperature at $y = 0$ is T_0 and the room temperature is T_R, and (5) sheet velocity is U. a) Give a differential equation for the temperature, T, in the sheet that applies for $x \geq 0$. Be sure that your equation is consistent with the assumptions. b) Write appropriate boundary conditions for T. c) If this had been presented as a numerical problem, how would you solve it?

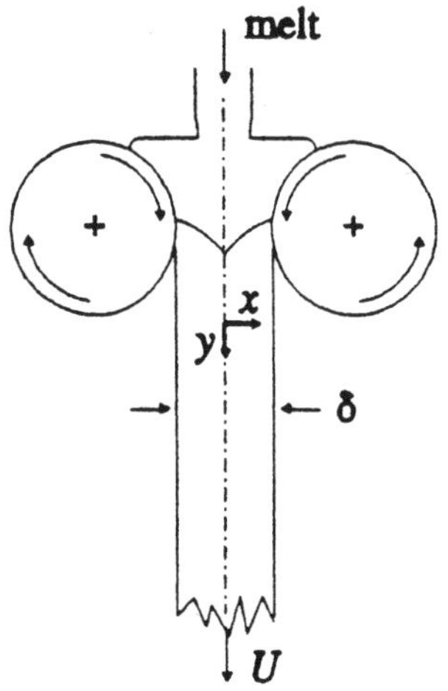

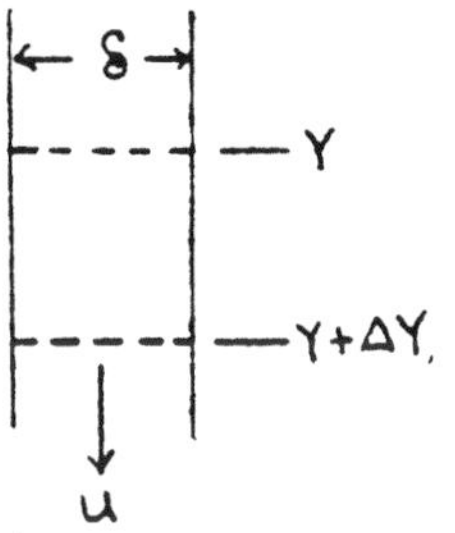

a. Assume steady-state, In = Out

In: $\delta q_Y|_Y + (uc_pT|_Y)\delta\rho$

Out: $\delta q_Y|_{Y+\Delta Y} + (uc_pT|_{Y+\Delta Y})\delta\rho + 2\Delta Y q_{rad}$.

$$\delta q_Y|_Y + (uc_pT|_Y)\delta\rho = \delta q_Y|_{Y+\Delta Y} + (uc_pT|_{Y+\Delta Y})\delta\rho + 2\Delta Y q_{rad}$$

$$\lim_{\Delta Y\to 0} \frac{q_Y|_{Y+\Delta Y} - q_Y|_Y}{\Delta Y} + uc_p\rho \frac{T|_{Y+\Delta Y} - T|_Y}{\Delta Y} + \frac{2q_{rad}}{\delta} = 0$$

$$\frac{dq_Y}{dY} + uc_p\rho\frac{dT}{dY} + \frac{2q_{rad}}{\delta} = 0$$

But $\dfrac{dq_Y}{dY} = -k\dfrac{d^2T}{dY^2}$ and $q_{rad} = \sigma\epsilon(T^4 - T_R^4)$

$$\therefore -k\delta\frac{d^2T}{dY^2} + uc_p\rho\delta\frac{dT}{dY} + 2\sigma\epsilon(T^4 - T_R^4) = 0$$

$$\text{or } \frac{d^2T}{dY^2} - \frac{u}{\alpha}\frac{dT}{dY} - \frac{2\sigma\epsilon}{k\delta}(T^4 - T_R^4) = 0$$

b. Boundary Conditions: $T(0) = T_o$; $T(\infty) = T_R$

c. Use finite difference approximations as presented in Chap. 16.

11.15 A sphere is suspended by a wire and is lowered at a constant velocity into a very long cylindrical tube as indicated in the sketch. Assume that, over the length L, the view factor is constant, the tube wall has a constant temperature T_w, and the rate of descent (V) is constant. If the sphere has an emissivity ε and enters the length L with a temperature T_0, derive an equation which gives its temperature as it leaves the length L. $T_w >> T_0$ and the sphere radius is R; consider only radiation heat transfer.

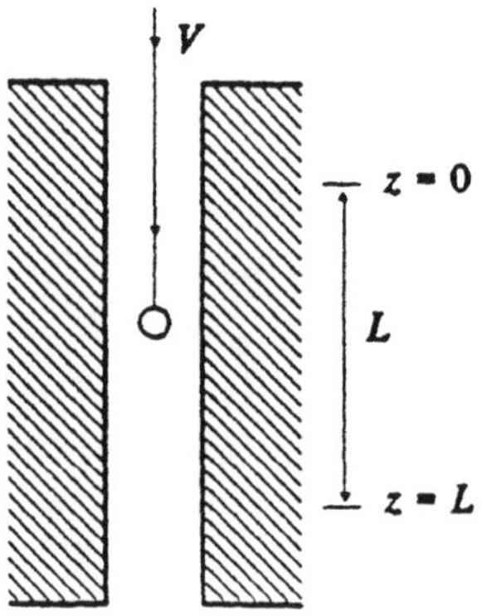

Subscripts: W for wall, s for sphere

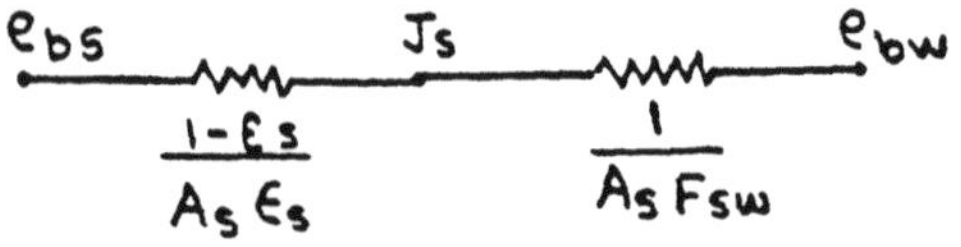

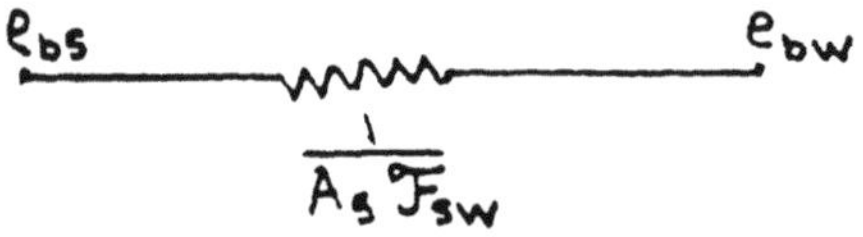

$$Q_{ws} = A_s \mathcal{F}_{sw} \sigma (T_w^4 - T_s^4)$$

$$\mathcal{F}_{sw}^{-1} = F_{sw}^{-1} + \left(\frac{\varepsilon_s}{1-\varepsilon_s}\right)^{-1} ; \quad \mathcal{F}_{sw}^{-1} = 1 + \frac{1-\varepsilon_s}{\varepsilon_s} = \varepsilon_s^{-1} \text{ or } \mathcal{F}_{sw} = \varepsilon_s$$

Then $Q_{ws} = A_s \varepsilon_s \sigma (T_w^4 - T_s^4)$

Since $T_w >> T_o$, we may assume $T_w^4 >> T_s^4$ over length L and

with $Q_{ws} = Q$ and $\varepsilon_s = \varepsilon$, we have $Q = A_s \varepsilon \sigma T_w^4$

This energy heats the sphere by an amount:

$$Q = V_s \rho c_p \frac{dT}{dt} \text{ and with } \frac{dT}{dt} = V \frac{dT}{dx}$$

$$Q = V_s \rho c_p V \frac{dT}{dx}$$

Then $A_s \varepsilon \sigma T_w^4 = V_s \rho c_p V \frac{dT}{dx}$; $\frac{A_s}{V_s} = \frac{3}{R}$

$$\int_{T_o}^{T_L} dT = \frac{3 \varepsilon \sigma T_w^4}{R \rho c_p V} \int_0^L dx$$

$$T_L - T_o = \frac{3 \varepsilon \sigma T_w^4 L}{R \rho c_p V}$$

11.16 A ladle of hot metal is used to feed an atomizer to produce metal powder. By means of a controlled stopper rod, the metal slowly leaves the ladle at a constant mass flow rate, W. Assume that the ladle refractory can be treated as a "no-net-flux surface" and that the only heat loss from the ladle is by radiant heat transfer.

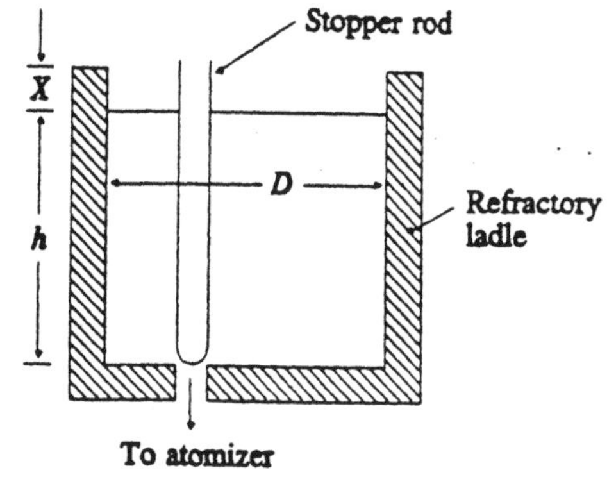

a) When the ladle is full (i.e., $X = 0$), write an equation that gives the radiant heat loss (Q, energy/time) from the top surface of the melt.

b) When the ladle is 3/4 empty and $D/X = 4$, write an equation for Q. Use the following variables: T_1 = temperature of metal; T_2 = temperature of surroundings; h = depth of melt; D = inside diameter of ladle; $A = \pi D^2/4$ = area of top surface of melt; ε_1 = emissivity of melt surface; and X = distance from top of ladle to top of melt (see diagram).

In your answers give numerical values for view factors and total exchange factors.

a. $Q = \frac{\pi D^2}{4} \varepsilon_1 \sigma (T_M^4 - T_0^4)$ where T_M = metal temp.; T_0 = Atmos. temp.

b. 1-metal; 2-plane across opening; 3-refractory

Equation (11.45) can be used: $Q_{1,net} = A \mathcal{F}_{12} \sigma (T_1^4 - T_2^4)$

$$\text{where } \frac{1}{A_1 \mathcal{F}_{12}} = \frac{1}{A_1}\frac{1-\varepsilon_1}{\varepsilon_1} + \frac{1}{A_1 \bar{F}_{12}} + \frac{1}{A_2}\frac{1-\varepsilon_2}{\varepsilon_2}$$

$$A_1 = A_2 = A = \frac{\pi D^2}{4}; \quad \varepsilon_2 = 1$$

Also, Fig. 11.21 can be used to get $\bar{F}_{12}$; $\frac{D}{X} = 4$, then $\bar{F}_{12} = 0.79$

$$\text{Hence, } \frac{1}{\mathcal{F}_{12}} = \frac{1-\varepsilon_1}{\varepsilon_1} + \frac{1}{0.79}; \quad \mathcal{F}_{12} = \left[\frac{1-\varepsilon_1}{\varepsilon_1} + \frac{1}{0.79}\right]^{-1}$$

$$\therefore Q_{1,net} = \frac{\pi D^2}{4}\left(\frac{1-\varepsilon_1}{\varepsilon_1} + \frac{1}{0.79}\right)^{-1} \sigma (T_1^4 - T_2^4)$$

11.17 A long rod ($\varepsilon = 0.8$) is heated to 1255 K and is placed near a well insulated reflector ($\varepsilon = 0.01$) as shown. The purpose of the reflector is to retard the cooling of the rod. The diameter of the rod is 75 mm, and the diameter of the reflector is 500 mm. The room surrounding the reflector and the rod are at 300 K. a) Calculate the radiant power loss per unit length of the rod when its temperature is 1255 K. b) Repeat if there is no reflector.

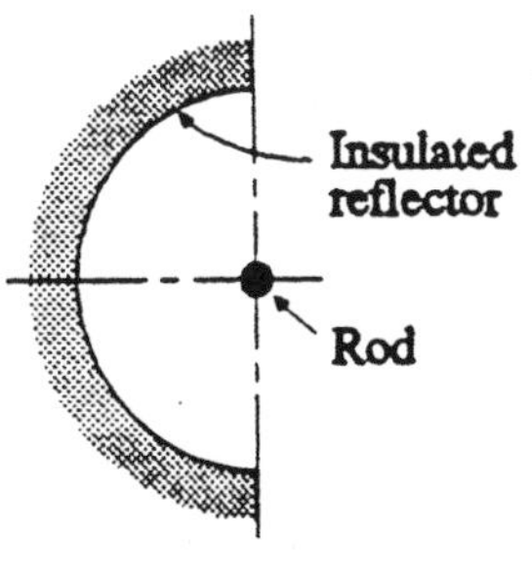

a. Subscripts: 1 for long rod (gray), 2 for room (black), R for reflector

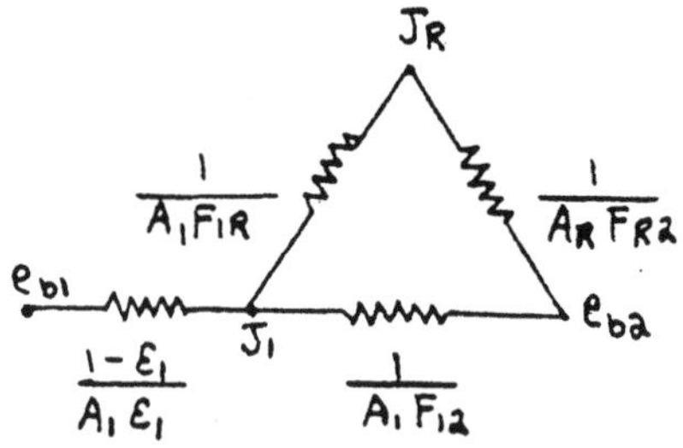

View Factors

$F_{12} = F_{1R} = 0.5$ (by inspection); $F_{R1} = \frac{A_1}{A_R} F_{1R}$; $F_{R2} = 1 - F_{R1} = 1 - \frac{A_1}{A_R} F_{1R}$

$A_1 = \pi d_1 l$; $A_R = \frac{\pi}{2} d_R l$; $\therefore \frac{A_1}{A_R} = \frac{2 d_1}{d_R} = \frac{(2)(75)}{(500)} = 0.3$

$F_{R2} = 1 - (0.3)(0.5) = 0.85$

Total Exchange Factor (Eq. 11.43)

$$A_1 \bar{F}_{12} = A_1 F_{12} + \left(\frac{1}{A_1 F_{1R}} + \frac{1}{A_R F_{R2}}\right)^{-1} = A_1\left[F_{12} + \left(\frac{1}{F_{1R}} + \frac{A_1}{A_R}\frac{1}{F_{R2}}\right)^{-1}\right]$$

$$A_1 \bar{F}_{12} = A_1\left[0.5 + \left(\frac{1}{0.5} + \frac{0.3}{0.85}\right)^{-1}\right] = 0.925 A_1$$

$A_1 \mathcal{F}_{12}$

$$\frac{1}{A_1 \mathcal{F}_{12}} = \frac{1-\varepsilon_1}{A_1 \varepsilon_1} + \frac{1}{A_1 \bar{F}_{12}} \text{ or } \frac{1}{\mathcal{F}_{12}} = \frac{1-\varepsilon_1}{\varepsilon_1} + \frac{1}{\bar{F}_{12}} = \frac{1-0.8}{0.8} + \frac{1}{0.925}$$

$\mathcal{F}_{12} = 0.751$

Heat Loss per meter of length

$$Q = A_1 \mathcal{F}_{12}(e_{b1} - e_{b2}) = A_1 \mathcal{F}_{12}\sigma(T_1^4 - T_2^4)$$

$$Q = (\pi)(0.075)(1)(0.751)(5.699)\left[\left(\frac{1255}{100}\right)^4 - \left(\frac{300}{100}\right)^4\right] = 2.49 \times 10^4 \text{ W}$$

b. $Q = A_1 \varepsilon_1 \sigma (T_1^4 - T_2^4)$

$$Q = (\pi)(0.075)(1)(0.8)(5.699)\left[\left(\frac{1255}{100}\right)^4 - \left(\frac{300}{100}\right)^4\right]$$

$Q = 2.66 \times 10^4$ W per meter of length

In this configuration, cooling is not retarded very much.

11.18 Two parallel plates, 0.3 × 0.6 m, are spaced 0.3 m apart in a large heat treating furnace and heated uniformly to 810 K. They are then removed from the furnace and cooled in a room maintained at 300 K. The emissivities of the plates are 0.4 and 0.8, respectively. Other thermal properties are equal. Calculate the ratio of the initial cooling rates (K s^{-1}) of the plates. Consider only radiation heat transfer.

```
10 'Problem 11.18
20 'subscripts:  1 for plate with 0.4 emissivity ; 2 for room ; 3 for plate
                 with 0.8 emissivity
30 A1 = .18  : A3 =  .18                   'areas in m^2
40 F13 = .35 : F12 = 1 -F13                'view factors
50 F31 = A1*F13/A3 : F32 = 1 - F31         'view factors
60 F31 = A1*F13/A3 : F32 = 1 - F31         'view factors
70 E1 = .4 : E3 = .8                       'emissivities
80 T1 = 810 : T2 = 300 : T3 = 810          'temperatures in K
90 RHOCP1 = 8000000!: RHOCP3 = 8000000!
   'product of density and Cp in SI units; any value will do because they
    cancel when the ratio of cooling rates is calculated.
100 X1 = .0125 : X3 = .0125                'thicknesses of plates are equal
110 SIGMA = 5.699E-08                      'Stefan-Boltzmann's constant
120 'solve for the resistances
130  R1 = (1 - E1)/( A1*E1 ) : R2 = 1/( A1*F12 ) : R3 = 1/( A1*F13 )
140  R4 = 1/( A3*F32 ) : R5 = (1-E3)/( A3*E3 )
150 'solve for black surface emissive powers (really have units of fluxes)
160  EB1 = SIGMA*T1^4 :  EB2 = SIGMA*T2^4 : EB3 = SIGMA*T3^4
170 'solve for constants so that unknown radiosities can be determined
180  C = ( EB2/R2 + EB1/R1 )/( 1/R1 + 1/R2 + 1/R3 )
190  D = R3*( 1/R1 + 1/R2 + 1/R3 )
200  E = ( EB3/R5 + EB2/R4 )/( 1/R3 + 1/R4 + 1/R5 )
210  F = R3*( 1/R3 + 1/R4 + 1/R5 )
220  J1 = C + ( C + E*F )/( D*F - 1 )   'radiosity
230  J3 = E + J1/F                      'radiosity
240  PRINT J1,J3
250  heat loss from surface of plate 1 facing plate 3, W
260  HEAT1 = (EB1 - J1)/R1
270  HEAT11 = (J1 - EB2)/R2 + (J1 - J3)/R3
280  PRINT HEAT1, HEAT11                           'HEAT1 should equal HEAT11
290 'heat loss from surface of plate 1 facing the room
300  HEAT1R = A1*SIGMA*E1*( T1^4 - T2^4 )  ' in W
310  HEAT(1) = HEAT1 + HEAT1R                  'total heat loss from plate 1, W
320  RATE1 = -HEAT(1)/(RHOCP1*A1*X1)             'rate in K/s
330  LPRINT : LPRINT : LPRINT "  The cooling rate of plate 1 is ";RATE1;" K/s"
350 'heat loss from surface of plate 3 facing plate 1, W
360  HEAT3 = (EB3 - J3)/R5
370  HEAT33 = (J3 - EB2)/R4 + (J3 - J1)/R3
380  PRINT HEAT3, HEAT33                           'HEAT3 should equal HEAT33
390 'heat loss from surface of plate 3 facing the room
400  HEAT3R = A3*SIGMA*E3*( T3^4 - T2^4 )  ' in W
410  HEAT(3) = HEAT3 + HEAT3R                  'total heat loss from plate 1, W
420  RATE3 = -HEAT(3)/(RHOCP3*A3*X3)             'rate in K/s
430  LPRINT :LPRINT "  The cooling rate of plate 3 is ";RATE3;" K/s"
440  LPRINT :LPRINT "  The ratio of rates, plate 3 to plate 1, is";RATE3/RATE1
450 END
```

```
The cooling rate of plate 1 is -.1642466  K/s

The cooling rate of plate 3 is -.3462781  K/s

The ratio of rates, plate 3 to plate 1, is 2.108282
```

11.19 A long rod, with a diameter of 3 mm, is placed in a vacuum parallel to a large and flat heated surface. The other side of the rod is exposed to vacuum chamber walls at 300 K. Assume that the heated surface and the surface of the rod are both gray with equal emissivities of 0.7. The rod has a volumetric heat capacity of $\rho C_p = 6.7 \times 10^6$ J m^{-3} K^{-1}. Calculate the heating rate of the rod at 335 K, 500 K, 700 K, and 800 K.

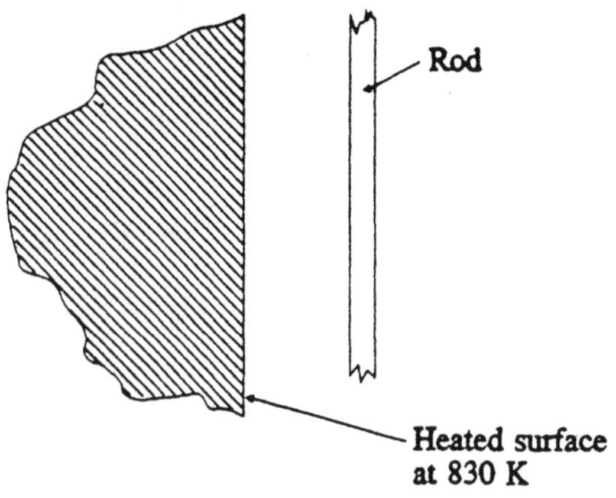


```
10 'Problem 11.19
20 'subscripts:  1 for rod ; 2 for vacuum chamber ; 3 for heated surface
30  D1 = .003 : L = 10 : PI = 3.1416    'diameter and length of the rod in m
40  A1 = PI*D1*L : A3 = 100             'The large heated surface is 100 m^2.
50  F13 = .5  : F12 = 1 -F13            'view factors
60  F31 = A1*F13/A3 : F32 = 1 - F31     'view factors
70  F31 = A1*F13/A3 : F32 = 1 - F31     'view factors
80  E1 = .7 : E3 = .7                   'emissivities
90  T2 = 300 : T3 = 830                 'temperatures in K
100 RHOCP1 = 6700000!                   'product of density and Cp in SI units
110 SIGMA = 5.699E-08                   'Stefan-Boltzmann's constant
120 'solve for the resistances
130  R1 = (1 - E1)/( A1*E1 ) : R2 = 1/( A1*F12 ) : R3 = 1/( A1*F13 )
140  R4 = 1/( A3*F32 ) : R5 = (1-E3)/( A3*E3 )
145 '
150 'solve for black surface emissive powers (really have units of fluxes)
160  FOR I = 1 TO 4
170    READ T1
180    EB1 = SIGMA*T1^4 :  EB2 = SIGMA*T2^4 : EB3 = SIGMA*T3^4
190   'solve for constants so that unknown radiosities can be determined
200    C = ( EB2/R2 + EB1/R1 )/( 1/R1 + 1/R2 + 1/R3 )
210    D = R3*( 1/R1 + 1/R2 + 1/R3 )
220    E = ( EB3/R5 + EB2/R4 )/( 1/R3 + 1/R4 + 1/R5 )
230    F = R3*( 1/R3 + 1/R4 + 1/R5 )
240    J1 = C + ( C + E*F )/( D*F - 1 )    'radiosity
250    J3 = E + J1/F                       'radiosity
260    PRINT J1,J3                         'check that these are positive
270   'heat gain to rod
280    HEAT1 = (EB1 - J1)/R1
290    HEAT11 = (J1 - EB2)/R2 + (J1 - J3)/R3
300    PRINT HEAT1, HEAT11  : PRINT        'HEAT1 should equal HEAT11
310    VOL1 = PI*D1*D1*L/4                 'volume of the rod, m^3
320    RATE  = -HEAT1/(RHOCP1*VOL1)        'rate in K/s
325    LPRINT : LPRINT "   When the temperature of the rod is";T1;" K,"
330    LPRINT "   its heating rate is ";RATE;" K/s."
340  NEXT I
350 END
360 '
370 DATA 335, 500, 700, 800
```

When the temperature of the rod is 335 K,
its heating rate is 1.260526 K/s.

When the temperature of the rod is 500 K,
its heating rate is .8643503 K/s.

When the temperature of the rod is 700 K.
its heating rate is -.5455295 K/s.

When the temperature of the rod is 800 K,
its heating rate is -1.891107 K/s.

11.20 For a cylindrical ingot of steel, 1 m diameter by 2 m long, placed in a refractory-lined furnace at 1500 K, calculate the radiation heat transfer coefficient to the steel surfaces. Repeat when the surface of the steel has reached 500 K and 1000 K.

The steel is surrounded by the furnace walls.

Hence, $q = \sigma \varepsilon_s (T_r^4 - T_s^4)$ r ≙ refractory ; s ≙ steel

or; $q = h_r (T_r - T_s)$

Hence, $h_r = \dfrac{\sigma \varepsilon_s (T_r^4 - T_s^4)}{(T_r - T_s)}$

use $T_r = 1500$ K and estimate of ε_s ($\varepsilon_s = 0.8$) from Figure 11.9

T_s, K	h_r, W m^{-2} K^{-1}
300	192
500	228
1000	370

12.1 Show that the units in Eq. (12.1) are as indicated. Do the same for Eqs. (12.2) and (12.3).

Eq. (12-1) $\quad W_{Ax} = -D_A\left(\frac{\partial \rho_A}{\partial x}\right)$ with ρ_A in kg m^{-3}, x in m and D_A in $m^2\ s^{-1}$.

Units of W_{Ax} are $\frac{m^2}{s}\Big|\frac{kg}{m^3}\Big|\frac{1}{m}$ = kg (of A) $m^{-2}\ s^{-1}$ (mass flux).

Eq. (12-2) $\quad J_{Ax} = -D_A\left(\frac{\partial C_A}{\partial x}\right)$ with C_A in mol (of A) m^{-3}, x in m and D_A in $m^2 s^{-1}$.

Units of J_{Ax} are $\frac{m^2}{s}\Big|\frac{mol}{m^3}\Big|\frac{1}{m}$ = mol (of A) $m^{-2}\ s^{-1}$ (molar flux).

Eq. (12.3) $\quad W_{Ax} = -\rho D_A\left(\frac{\partial \rho_A^*}{\partial x}\right)$ with ρ in kg m^{-3}, D_A in $m^2\ s^{-1}$, ρ_A^* no units

and x in m

Units of J_{Ax} are kg $m^{-2} s^{-1}$ (mass flux).

12.2 Discuss the reasons why self-diffusion data must apply to homogeneous materials only.

Self-diffusion refers to atomic motion in the absence of a concentration gradient. Thus, self-diffusion applies to homogeneous materials.

12.3 Read one of the references of Footnote 1 and derive Eq. (12.5).

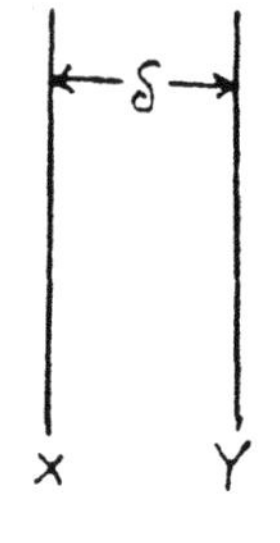

Consider exchange of atoms between two atomic planes X and Y with atomic spacing δ.

Rate at which atoms jump from plane X to plane Y is $J_{X\to Y}$ expressed as number of atoms s^{-1}.

$J_{X\to Y} = \frac{1}{6}\nu c^* \delta A$ where 6 is the number of jump directions, ν is the jump frequency (s^{-1}), c^* is the concentration of radioactive atoms (m^{-3}), δ is the interplanar spacing (m), and A is the area normal to the flow (m^2).

$$J_{Y\to X} = \frac{1}{6}\nu\left(c^* + s\frac{dc^*}{dx}\right)sA$$

Thus, net rate is $J = J_{X\to Y} - J_{Y\to X} = -\frac{s^2 A}{6}\nu\frac{\partial c^*}{\partial x}$

Fick 1st law gives $J = -D^* A\frac{\partial c^*}{\partial x}$

$\therefore D^* = \frac{1}{6}s^2\nu$

12.4 Look up the article by Compaan and Haven (Table 12.1) and summarize the method used to derive the correlation coefficients in Table 12.1.

If all atoms were radioactive, then Eq. (12.5) would apply directly to tracer diffusion coefficients and $D_i^* = D_i^T$. As discussed on pp. 423 and 424, however, we cannot directly measure D_i^* in the laboratory, so dilute concentrations of a radioactive tracer are used to set up measurable gradients.

Keeping in mind that the intent of such an experiment is to get data that can be used to calculate D_i^*, then we must realize that D_i^* relates directly to the exchange of positions between vacancies and their neighboring atoms. Now with tracers introduced, we have a system in which vacancies and tracers both diffuse, so there must be a means to track both to accurately calculate D_i^*. We can measure D_i^T but not D (vacancies), so we rely on the correlation coefficients (*viz.*, Table 12.1). The correlation coefficients are calculated (or were calculated), using random walk theory restricted to directions in the crystal.

12.5 Work out the units in Eq. (12.20).

$$\dot{n}_{Ax} = \frac{-n_A B_A}{N_0}\frac{\partial \bar{G}_A}{\partial x}$$

Units: $\dot{n}_{Ax}$ = atoms $m^{-2}\,s^{-1}$

n_A = atoms m^{-3}

$B_A = m^2\,s^{-1}\,J^{-1}$ (or m^2 atoms $s^{-1}\,J^{-1}$)

$N_0 = mol^{-1}$ (or atoms mol^{-1})

$\bar{G}_A = J\,mol^{-1}$

$x = m$

12.6 Derive Eq. (12.28), after reading Zener's article.

Jumps = (Attempts)(Probability of success)

Hence, $\nu = \nu_0 Z \exp\left(-\frac{\Delta G^\ddagger}{RT}\right)$

$\nu_0 Z$ = attempts to jumps into neighboring site, s^{-1}, where Z is the number of nearest neighbors and ν_0 is the vibrational frequency, s^{-1}. $\exp\left(-\frac{\Delta G^\ddagger}{RT}\right)$ = probability that an attempt is successful, where $\Delta G^\ddagger$ is the free energy barrier, J mol^{-1}.

12.7 Using the method of Fuller, Schettler, and Giddings, estimate the diffusion coefficient for a CO_2-O_2 gas mixture at 1-atm pressure and 700 K and compare the result to the data in Fig. 12.21.

$$D_{AB} = \frac{(1\times10^{-3})\,T^{1.75}}{P(\nu_B^{1/3} + \nu_A^{1/3})^2}\left(\frac{1}{M_A} + \frac{1}{M_B}\right)^{1/2}$$

$$D_{AB} = \frac{(1\times10^{-3})(700)^{1.75}}{(1)(16.6^{1/3} + 26.9^{1/3})^2}\left(\frac{1}{32} + \frac{1}{44}\right)^{1/2} = 0.719\ cm^2\ s^{-1}$$ which compares very well to the value in Fig. 12.21.

12.8 Given a tortuosity of 2.0, a void fraction of 0.25, and $r = 5 \times 10^{-3}$ cm, calculate the effective diffusion coefficient for CO-CO_2 in a reduced iron oxide pellet at 800 K.

$$D_{AB} = \frac{10^{-3}\,T^{1.75}}{P(\nu_A^{1/3} + \nu_B^{1/3})^2}\left(\frac{1}{M_A} + \frac{1}{M_B}\right)^{1/2}$$ where $\nu_A(CO) = 18.9$, $\nu_B(CO_2) = 26.9$

$$D_{AB} = \frac{10^{-3}(800)^{1.75}}{(1)(18.9^{1/3} + 26.9^{1/3})^2}\left(\frac{1}{28} + \frac{1}{44}\right)^{1/2} = 0.909\ cm^2\ s^{-1}$$

$$D_K = 9700\, r\left(\frac{T}{M}\right)^{1/2} = (9700)(5\times10^{-3})\left(\frac{800}{28}\right)^{1/2} = 259\ cm^2\ s^{-1}\ (CO)$$

$D_K(CO) = 259\ cm^2\ s^{-1}$; $D_K(CO_2) = 206\ cm^2\ s^{-1}$

Since $D_{AB} \ll D_K$, ordinary diffusion prevails.

$$D_{eff} = \frac{D_{AB}\,\omega}{\tau} = \frac{(0.908)(0.25)}{2} = 0.144\ cm^2\ s^{-1}$$

12.9 PbS has a NaCl-type structure. Would you expect the self-diffusion coefficient of Pb or S to be higher? How would you expect the addition of Ag_2S to affect the diffusion coefficient of Pb, knowing that the defects in PbS are predominantly Frenkel defects on the Pb sublattice, and that the undoped PbS is an *n*-type semiconductor? How would Bi_2S_3 additions affect it? [Ref.: G. Simovich and J. B. Wagner, Jr., *J. Chem. Phys.* **38**, 1368 (1963).]

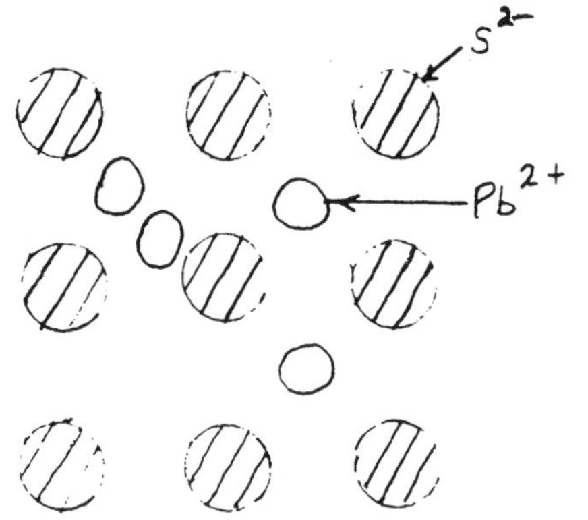

The Pb^{2+} ions diffuse faster than do the S^{2-} ions. The ionic radius of Pb^{2+} is 0.84Å, whereas that of S^{2-} is 1.84Å. The Frenkel defects comprise interstitials of Pb^{2+} and corresponding vacant sites. Diffusion of the Pb^{2+} ions readily occurs by the Frenkel defects, whereas there are no corresponding defects for the diffusion of the S^{2-} ions.

When Ag_2S is added to the structure, we can expect that some of the Ag^{1+} ions will fill the Pb-site vacancies. Hence, the diffusion coefficient of Pb^{2+} will decrease.

When Bi_2S_3 is added, two Bi^{3+} ions can only substitute for two Pb^{2+} ions, while producing a Pb vacancy. Hence, the diffusion coefficient of Pb^{2+} will increase.

12.10 Describe the conditions under which the following terms are applicable: (1) self-diffusion, (2) tracer diffusion, (3) chemical diffusion, (4) interstitial diffusion, (5) substitutional diffusion, (6) interdiffusion coefficient, and (7) intrinsic diffusion.

(1) Pure metals and homogeneous alloy.

(2) Used to determine self-diffusion coefficients with radioactive atoms (i.e., tracers). Eq. (12.6) relates tracer diffusion to self diffusion.

(3) Diffusion due to a concentration gradient.

(4) Diffusion of interstitial elements via instertices.

(5) Diffusion of substitutional elements via vacancies of lattice sites.

(6) Satisfies the diffusion equation when bulk motion due to the diffusion is ignored. Commonly reported simply as diffusion coefficient and used for engineering calculations.

(7) Basic for a given element in an alloy in the sense that substitutional elements diffuse at their respective rates, when bulk motion is taken into account.

12.11 Calculate the self-diffusion coefficient in liquid lead using the Sutherland hydrodynamical model, and the Eyring activated state model. Data for viscosity of liquid metals is given in Fig. 1.9. Calculate D^* at 873 K, 1073 K, and 1273 K. a) Do the calculated diffusion coefficients vary linearly with temperature according to the fluctuation model of Reynik? according to the Arrhenius relation (Eq. (12.27))? b) What is the error in the calculated values? Experimental results for liquid lead are given in Fig. 12.18.

Sutherland Model $D^* = \frac{K_B T}{4\pi R \eta}$, $K_B = 1.380 \times 10^{-23}\, JK^{-1}$, $R = 1.75 \times 10^{-10}\, m$ (Pb)

T, K	η, cP	η, $N s m^{-2}$	D^*, $m^2 s^{-1}$
873	1.40	1.4×10^{-3}	3.91×10^{-9}
1073	1.20	1.20×10^{-3}	5.61×10^{-9}
1273	1.0	1×10^{-3}	7.99×10^{-9}

Eyring Model $D^* = \frac{K_B T}{2 R \eta}$

T, K	D^*, $m^2 s^{-1}$
873	1.66×10^{-8}
1073	3.25×10^{-8}
1273	5.02×10^{-8}

a. D^* linear with T - Only the Sutherland values are plotted because the Eyring values are too great. Both curves are approximately linear.

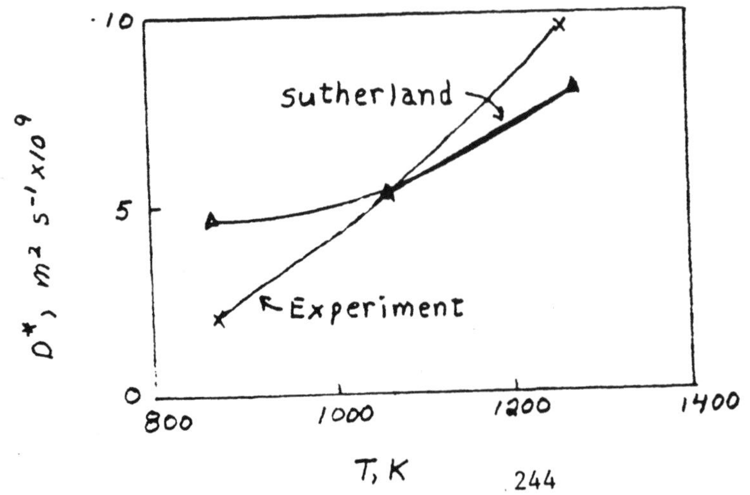

Ahhrenius Behavior $D^* = D_0 \exp\left(-\frac{Q}{RT}\right)$

plot of $\ln D^*$ vs. $\frac{1}{T}$ should be linear.

T, K	T^{-1}, K^{-1}	ln D (Suther.)	ln D (expt.)
873	11.4×10^{-4}	−19.36	−20.03
1073	9.32×10^{-4}	−19.00	−19.02
1273	7.86×10^{-4}	−18.65	−18.43

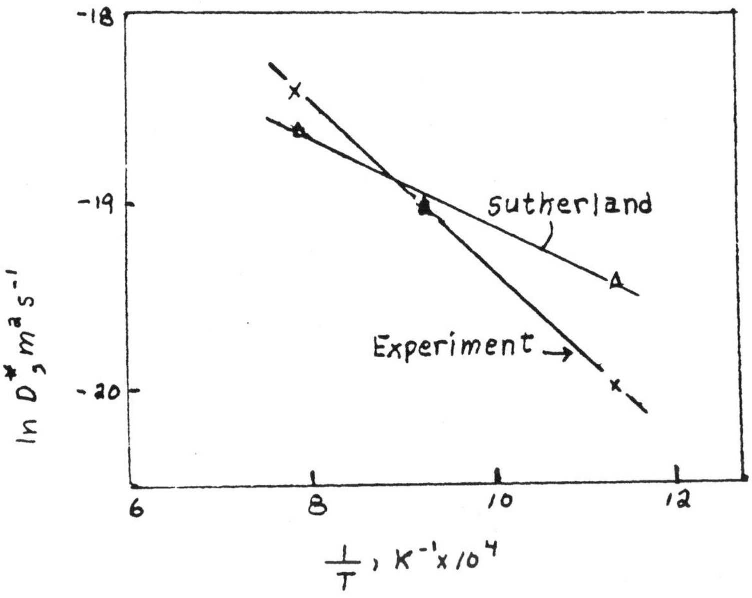

Both can be represented by Ahhrenius plots. Therefore it is diffucult to state whether linear or Ahhrenius is obeyed.

b. Errors with the Eyring model are very large. The errors with the Sutherland model are as follows:

T, K	Error, %
873	+95
1073	2.2
1273	−6.2

12.12 The self-diffusion coefficients of gold and nickel in Au-Ni alloys are shown in Fig. 12.4. Show that the alloys of Au and Ni are not ideal solutions.

$\tilde{D} = X_A D_B^* + X_B D_A^*$. One of the requirements for this relationship is that the alloy be an ideal solution. According to Fig. 12.4 however, $\tilde{D} < D_{Ni}^* < D_{Au}^*$ for $0 \le X_{Ni} \le 0.87$ so this equation can not be obeyed. Hence, Au-Ni alloys do not exhibit ideal solution behavior.

12.13 The diffusion coefficient of acetone vapor in air is 1.1×10^{-5} m^2 s^{-1} at 273 K and 1 atm. What is the diffusion coefficient at 500 K and 0.5 atm?

Eq. (12.54) $D_{AB} = A\dfrac{T^{1.75}}{P}$ where A is constant for particular gas.

Then $\dfrac{D_1}{D_2} = \dfrac{T_1^{1.75} P_2}{P_1 T_2^{1.75}} = \dfrac{(273)^{1.75}(0.5)}{(1)(500)^{1.75}} = 0.173$

$\therefore D_2 = \dfrac{1.1 \times 10^{-5}}{0.173} = 6.36 \times 10^{-5}$ m^2 s^{-1}

12.14 The diffusion coefficient of He in Pyrex* is 2.6×10^{-16} m^2 s^{-1} at 373 K and 1.6×10^{-14} m^2 s^{-1} at 873 K. What is it at 1000 K? *Pyrex® is a type of glass commonly used in laboratories.

This involves diffusion in a solid, so we assume that $D = D_0 \exp\left(-\dfrac{Q}{RT}\right)$

Given D_1 and D_2 at T_1 and T_2, we have:

$$\frac{D_1}{D_2} = \exp\left[-\frac{Q}{R}\left(\frac{1}{T_1} - \frac{1}{T_2}\right)\right] \text{ or } \frac{Q}{R} = \frac{\ln\left(\frac{D_1}{D_2}\right)}{\frac{1}{T_2} - \frac{1}{T_1}} = \frac{\ln\left(\frac{2.6 \times 10^{-16}}{1.6 \times 10^{-14}}\right)}{\frac{1}{873} - \frac{1}{373}} = 2.683 \times 10^3 \text{ K}$$

$$\text{and } D_0 = \frac{D_1}{\exp\left(-\frac{Q}{R}\frac{1}{T}\right)} = \frac{2.6 \times 10^{-16}}{\exp\left(-\frac{2.683 \times 10^3}{373}\right)} = 3.458 \times 10^{-13} \text{ m}^2 \text{ s}^{-1}$$

$$\therefore D(1000 \text{ K}) = 3.458 \times 10^{-13} \exp\left(-\frac{2683}{1000}\right) = 2.36 \times 10^{-14} \text{ m}^2 \text{ s}^{-1}$$

12.15 The diffusion coefficient of oxygen in liquid copper is not linear with temperature. The following data are available:

at 1150°C, $D = 7 \times 10^{-9}$ m^2 s^{-1}

at 1250°C, $D = 10^{-8}$ m^2 s^{-1}.

Predict the diffusion coefficient at 1350°C.

Not linear, therefore use $D = D_0 \exp\left(-\frac{Q}{RT}\right)$

$$\therefore -\frac{Q}{R} = \frac{\ln D_1 - \ln D_2}{\frac{1}{T_1} - \frac{1}{T_2}} = \frac{\ln\left(\frac{7\times10^{-9}}{10^{-8}}\right)}{\frac{1}{1423} - \frac{1}{1523}}$$

$$\frac{Q}{R} = 7.730\times10^{-3}\ \text{K}$$

Then at $T_3 = 1350°\text{C} = 1623\ \text{K}$, we have

$$\ln\left(\frac{D_1}{D_3}\right) = -7.730\times10^{3}\left[\frac{1}{1423} - \frac{1}{1623}\right] = -0.6694$$

$$\text{or } \frac{D_3}{D_1} = 1.953 \Rightarrow D_3 = (1.953)(7\times10^{-9}) = 1.37\times10^{-8}\ \text{m}^2\ \text{s}^{-1}$$

12.16 What is the diffusivity of Mg vapor in helium gas at 1200 K? The gas has a composition of 90 He and 10 Mg (by volume).

For Mg: $\frac{\varepsilon}{K_B} = 1.15\ T_b = (1.15)(1376) = 1582\ \text{K}$ Eq. (12.52)

$\frac{\varepsilon}{K_B} = 1.92\ T_M = (1.92)(924) = 1774\ \text{K}$ Eq. (12.53)

$\frac{\varepsilon}{K_B}(\text{avg.}) = 1678\ \text{K}$

$\sigma \simeq \frac{1}{2}(3.00 + 3.18) = 3.09\ \text{Å}$

For He: $\frac{\varepsilon}{K_B} = 10.22\ \text{K}$ and $\sigma = 2.551\ \text{Å}$

For the mixture: $\left(\frac{\varepsilon}{K_B}\right)_{AB} = \sqrt{(1678)(10.2)} = 131\ \text{K}$; $\sigma_{AB} \simeq \frac{1}{2}(3.08 + 2.55) = 2.82\ \text{Å}$

$$T^*_{AB} = \left(\frac{K_B}{\varepsilon}\right)_{AB} T = \frac{1200}{131} = 9.16.$$

Eq. (12.51)
$$\Omega_{D,AB} = \frac{1.06036}{(9.16)^{0.15610}} + \frac{0.19300}{\exp(0.47635)(9.16)} + \frac{1.03587}{\exp(1.52996)(9.16)} + \frac{1.76474}{\exp(3.89411)(9.16)} = 0.755$$

$$\text{Eq. (12.50)} \; D_{AB} = \frac{(1.8583\times10^{-3})(1200)^{1.5}}{(1)(2.83)^2(0.755)}\left(\frac{1}{24.3}+\frac{1}{4}\right)^{1/2} = 6.94\ \text{cm}^2\ \text{s}^{-1}$$

$$D_{AB} = 6.94\times10^{-4}\ \text{m}^2\ \text{s}^{-1}$$

12.17 At 1500°C, the diffusivity of Cr in molten nickel is 5×10^{-9} m² s⁻¹. At 1600°C, it is 7×10^{-9} m² s⁻¹. Estimate the diffusivity at 1700°C.

Assume an Arrhenius relationship $D = D_o\, e^{-Q/RT}$

$$\frac{Q}{R} = \frac{\ln\left(\frac{D_1}{D_2}\right)}{\frac{1}{T_2}-\frac{1}{T_1}} = \frac{\ln\left(\frac{5\times10^{-9}}{7\times10^{-9}}\right)}{\frac{1}{1873}-\frac{1}{1773}} = 1.122\times10^4$$

$$\text{and } D_o = \frac{D_1}{\exp\left(-\frac{Q}{R}\frac{1}{T}\right)} = \frac{5\times10^{-9}}{\exp\left(\frac{-1.122\times10^4}{1773}\right)} = 2.301\times10^{-6}$$

$$\therefore D(1700^\circ C) = 2.301\times10^{-6}\exp\left(-\frac{1.122\times10^4}{1973}\right) = 9.497\times10^{-9}\ \text{m}^2\ \text{s}^{-1}$$

Assume a linear law $D = a + bT$

$$b = \frac{7\times10^{-9}-5\times10^{-9}}{1600-1500} = 2\times10^{-11}\ \text{m}^2\ \text{s}^{-1}\ \text{K}^{-1}$$

$$a = 7\times10^{-9} - 2\times10^{-11}(1873) = -3.046\times10^{-8}$$

$$\therefore D(1700^\circ C) = -3.046\times10^{-8} + (2\times10^{-11})(1973) = 9.00\times10^{-9}\ \text{m}^2\ \text{s}^{-1}$$

12.18 Calculate the diffusion coefficient D_{AB} for zinc vapor diffusing through helium at 773 K.

$$\sigma_{Zn} \simeq 2.55\,\text{Å};\ \left(\frac{\varepsilon}{K_B}\right)_{Zn} = (1.15)(906+273) = 1356\ \text{K} \text{ or } \left(\frac{\varepsilon}{K_B}\right)_{Zn} = (1.92)(419.5+273)\text{K} = 1330\text{K}$$

$$\text{Use} \left(\frac{\varepsilon}{K_B}\right)_{Zn} \simeq 1343\text{K}$$

$$\sigma_{He} = 2.551\ \text{Å};\ \left(\frac{\varepsilon}{K_B}\right)_{He} = 10.22\ \text{K}$$

$$\text{Then } \sigma_{AB} = \frac{1}{2}(\sigma_{Zn}+\sigma_{He}) = 2.55\,\text{Å};\ \left(\frac{\varepsilon}{K_B}\right)_{AB} = \left[\left(\frac{\varepsilon}{K_B}\right)_{Zn}\left(\frac{\varepsilon}{K_B}\right)_{He}\right]^{1/2} = 117\ \text{K}$$

$$T^*_{AB} = \left(\frac{K_B}{\varepsilon}\right)_{AB} T = \frac{773}{117} = 6.61$$

$$\Omega_{D,AB} = \frac{1.06036}{(6.61)^{0.15610}} + \frac{0.19300}{\exp(0.47635)(6.61)} + \frac{1.03587}{\exp(1.52996)(6.61)} + \frac{1.76474}{\exp(3.89411)(6.61)}$$

$$= 0.798$$

Eq. (12.50) $D_{AB} = \dfrac{(1.8583\times10^{-3})(773)^{1.5}}{(1)(2.55)^2(0.798)} \left(\dfrac{1}{4.003} + \dfrac{1}{65.4}\right)^{1/2} = 3.96 \ cm^2 s^{-1}$

$$D_{AB} = 3.96 \times 10^{-4} \ m^2 s^{-1}$$

12.19 At 1 atm. pressure and 1900 K, the diffusion coefficient for a gaseous suboxide (Al_2O) in argon is 5×10^{-4} m^2 s^{-1}. Predict the diffusion coefficient for Al_2O in argon at 1900 K and 10^{-6} atm.

We can apply Eq. (12.54) which is of the form: $D_{AB} = \dfrac{AT^{1.75}}{P}$ where A = constant

for given combinations of A and B.

Therefore $\dfrac{(D_{AB})_2}{(D_{AB})_1} = \left(\dfrac{T_2}{T_1}\right)^{1.75}\left(\dfrac{P_1}{P_2}\right)$

$(D_{AB})_2 = \left(\dfrac{1900}{1900}\right)^{1.75}\left(\dfrac{1}{10^{-6}}\right)(5\times10^{-4} \ m^2 s^{-1})$

$\therefore D_{AB} (10^{-6} \ atm) = 500 \ m^2 s^{-1}$

12.20 At 273 K and 1 atm. the diffusion coefficient for cadmium vapor in nitrogen is 1.5×10^{-5} m^2 s^{-1}. Calculate the diffusion coefficient for cadmium vapor in helium at 1500 K and 0.5 atm.

Eq. (12.54) $D_{AB} = \dfrac{(1\times10^{-3})T^{1.75}}{P(v_B^{1/3}+v_A^{1/3})^2}\left(\dfrac{1}{M_A}+\dfrac{1}{M_B}\right)^{1/2}$; Let $B = N_2$ and $A = Cd$

$(v_{N_2}^{1/3}+v_{Cd}^{1/3})^2 = \dfrac{(1\times10^{-3})T^{1.75}}{P \ D_{AB}}\left(\dfrac{1}{M_{Cd}}+\dfrac{1}{M_{N_2}}\right)^{1/2}$

$(17.9^{1/3}+v_{Cd}^{1/3})^2 = \dfrac{(1\times10^{-3})(273)^{1.75}}{(1)(0.15)}\left(\dfrac{1}{112.4}+\dfrac{1}{28}\right)^{1/2} = 25.82$

$\therefore v_{Cd} = 14.99$ (not given in Table 12.10)

Now for Cd-He at 1500K

$$D_{AB} = \frac{(1\times10^{-3})(1500)^{1.75}}{(0.5)(2.88^{1/3}+15.0^{1/3})^2}\left(\frac{1}{112.4}+\frac{1}{4}\right)^{1/2} = 24.3\ \text{cm}^2\ \text{s}^{-1}$$

$D_{AB} = 2.43\times10^{-3}\ \text{m}^2\ \text{s}^{-1}$ very high because of low P, high T and one component is He.

12.21 To quantify the nucleation theory of the liquid to solid transformation, it is necessary to have a value of the self-diffusion coefficient for the liquid at temperatures significantly less than the freezing point. Given that D^* of liquid metal A is 5×10^{-9} m² s⁻¹ at 1200 K and 10^{-8} m² s⁻¹ at 1500 K, estimate D^* at 900 K. Use two distinctly different methods and compare the two estimates.

Method 1 – Assume activated type process

$$D = D_0 \exp\left(-\frac{Q}{R}\frac{1}{T}\right);\quad \frac{D_1}{D_2} = \exp\left[-\frac{Q}{R}\left(\frac{1}{T_1}-\frac{1}{T_2}\right)\right];\quad \ln\left(\frac{D_1}{D_2}\right) = -\frac{Q}{R}\left(\frac{1}{T_1}-\frac{1}{T_2}\right)$$

$$\therefore \frac{Q}{R} = \frac{\ln\left(\frac{D_2}{D_1}\right)}{\left(\frac{1}{T_1}-\frac{1}{T_2}\right)} = \frac{\ln(10^{-8}/5\times10^{-9})}{\frac{1}{1200}-\frac{1}{1500}} = 4159$$

$$D_0 = \frac{10^{-8}}{\exp\left(-\frac{4159}{1500}\right)} = 1.600\times10^{-7}\ \text{m}^2\ \text{s}^{-1}$$

Then $D^*(900\text{K}) = (1.600\times10^{-7})\exp\left(-\frac{4159}{900}\right) = 1.58\times10^{-9}\ \text{m}^2\ \text{s}^{-1}$

Method 2 – Assume linear

$$D^* = a + bT;\quad D_2^* - D_1^* = b(T_2 - T_1);\quad b = \frac{D_2^* - D_1^*}{T_2 - T_1} = \frac{10^{-8} - 5\times10^{-9}}{1500-1200} = 1.667\times10^{-11}$$

$$D^*_{900} = D_2^* + b(900 - T_2) = 10^{-8} + 1.667\times10^{-11}(900-1500)$$

$$D^*_{900} = 0\ \text{m}^2\text{s}^{-1}$$

Obviously the linear method gives an absurd result, so at least in this case the "activated type" equation is better.

12.22 At 1500°C the diffusion coefficient of Mo in an Fe melt is 3×10^{-9} m^2 s^{-1} and at 1600°C it is 3.5×10^{-9} m^2 s^{-1}. Predict the diffusion coefficient of Mo in an Fe melt which has been supercooled to 1200°C.

Assume $D = D_0 \exp\left(-\frac{Q}{RT}\right)$

Given D_1 and D_2 at T_1 and T_2, we have:

$$\frac{D_1}{D_2} = \exp\left[-\frac{Q}{R}\left(\frac{1}{T_1} - \frac{1}{T_2}\right)\right] \text{ or } \frac{Q}{R} = \frac{\ln\left(\frac{D_1}{D_2}\right)}{\frac{1}{T_2} - \frac{1}{T_1}} = \frac{\ln\left(\frac{3\times10^{-9}}{3.5\times10^{-9}}\right)}{\frac{1}{1873} - \frac{1}{1773}} = 5.12\times10^{3}\text{ K}$$

and $$D_0 = \frac{D_1}{\exp\left(-\frac{Q}{R}\cdot\frac{1}{T}\right)} = \frac{3\times10^{-9}}{\exp\left(-\frac{5.12\times10^{3}\text{ K}}{1873}\right)} = 4.616\times10^{-8}\text{ m}^2\text{ s}^{-1}$$

$$\therefore\ D = 4.616\times10^{-8}\exp\left(-\frac{5.12\times10^{3}}{1473}\right) = 1.428\times10^{-9}\text{ m}^2\text{ s}^{-1}$$

12.23 A binary gas of *A* and *B* molecules has a diffusion coefficient of 10 cm^2 s^{-1} at 0°C and 1 atm. Predict the diffusion coefficient in the same gas at 500°C and 10 atm.

Referring to Eq. (12.54): $D_{AB} = \frac{AT^{1.75}}{P}$

$$(D_{AB})_2 = (D_{AB})_1\left(\frac{T_2}{T_1}\right)^{1.75}\left(\frac{P_1}{P_2}\right) = (10)\left(\frac{773}{273}\right)^{1.75}\left(\frac{1}{10}\right) = 6.18\text{ cm}^2\text{ s}^{-1} \text{ or } 6.18\times10^{-4}\text{ m}^2\text{ s}^{-1}$$

12.24 A chamber containing $O_2(g)$ and $N_2(g)$ is divided by a porous solid, 1 cm thick. On one side of the porous solid, there is continuous flow keeping the composition at 80 mole percent O_2. On the other side, there is also continuous flow which maintains the composition of O_2 at 40 mole percent. The pressures and temperatures in both parts of the chamber are uniform and constant at 1 atm. and 1000 K, respectively. The porous solid has pores with an average radius of 1 μm, a porosity of 0.2 and a tortuosity of 5. Calculate the flux of oxygen which diffuses through the porous solid.

$$D_{AB} = \frac{10^{-3}\, T^{1.75}}{P(V_B^{1/3} + V_A^{1/3})^2}\left(\frac{1}{M_A} + \frac{1}{M_B}\right)^{1/2}$$ where $P = 1$ atm.; $T = 1000$ K; $M_A = 32$; $M_B = 28$;

$V_A = 16.6$; $V_B = 17.9$

$$D_{AB} = \frac{10^{-3}(1000)^{1.75}}{(1)(17.9^{1/3} + 16.6^{1/3})^2}\left(\frac{1}{32} + \frac{1}{28}\right)^{1/2} = 1.72 \text{ cm}^2 \text{ s}^{-1}$$

Check for Knudsen flow, Eq. (12.56)

$$D_K = (9700)(1 \times 10^{-4})\left(\frac{1000}{28}\right)^{1/2} = 5.80 \text{ cm}^2 \text{ s}^{-1}$$

$$\frac{D_{AB}}{D_K} = \frac{1.72}{5.80} = 0.297 \text{ so we use } D_{AB}$$

Then $D_{eff.} = \dfrac{D_{AB}\,\omega}{\delta} = \dfrac{(1.72)(0.2)}{5} = 0.0689$

$j = D_{eff.} \dfrac{C_2 - C_1}{\delta}$, where $C_2 = \dfrac{80 \text{ moles } O_2}{100 \text{ moles}} \left| \dfrac{1 \text{ mole}}{22{,}400 \text{ cm}^3 (STP)} \right| \dfrac{273 \text{ K}}{1000 \text{ K}}$

$= 9.75 \times 10^{-6}$ moles O_2 cm^{-3}

$C_1 = 4.88 \times 10^{-6}$ moles O_2 cm^{-3}

$\delta = 1$ cm

$j = 3.36 \times 10^{-7}$ moles s^{-1} cm^{-2}

13.1 One side of an iron sheet, 0.01 cm thick, is subjected to a carburizing atmosphere at 1200 K such that a surface concentration of 1.2% carbon is maintained. The opposite face is maintained at 0.1% carbon. At steady state, determine the flux (mol cm^{-2} s^{-1}) of carbon through the sheet: a) if the diffusion coefficient is assumed to be independent of concentration ($D = 2 \times 10^{-7}$ cm^2 s^{-1}); b) if the diffusion coefficient varies as shown to the right.

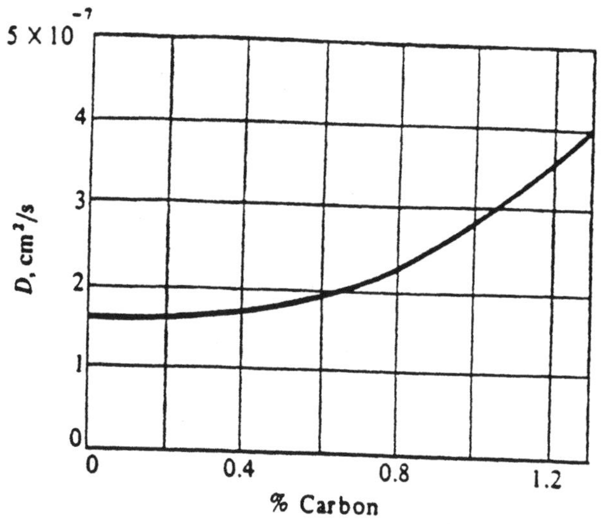

a. $j = -D\frac{dC}{dx} = D\frac{C_1 - C_2}{\delta}$

$C_1 = \frac{1.2\text{g}}{100\text{ g alloy}} \left| \frac{1\text{ gmol}}{12\text{g}} \right| \frac{7.58\text{ g alloy}}{\text{cm}^3} = 7.58 \times 10^{-3}$ gmol cm^{-3} ; $C_2 = 0.63 \times 10^{-3}$ gmol cm^{-3}

$j = (2 \times 10^{-7}) \frac{(7.58 - 0.63) \times 10^{-3}}{0.01} = 1.39 \times 10^{-7}$ gmol cm^{-2} s^{-1}

b. At steady state j is constant.

$j = \frac{1}{\delta}\int_{C_2}^{C_1} D\,dc$. The integral is evaluated numerically by the area under the curve between 0.1% C and 1.2% C

$\int_{C_2}^{C_1} D\,dc = 2.42 \times 10^{-7}$ % C cm^2 s^{-1} $= 1.53 \times 10^{-9}$ gmol cm^{-1} s^{-1}

$j = \frac{1.53 \times 10^{-9}}{0.01} = 1.53 \times 10^{-7}$ gmol cm^{-2} s^{-1}

13.2 A composite foil made of metal *A* bonded to metal *B*, each 0.01 cm thick, is subjected to 0.5 atm of pure hydrogen on metal *A*'s face; the other side, metal *B*'s face, is subjected to a perfect vacuum. At the temperature of interest and 1 atm of hydrogen, the solubility of hydrogen in metal *A* is 4×10^{-4} g per cm^3 of *A* and in *B* it is 1×10^{-4} g per cm^3 of *B*. It is also known that hydrogen diffuses four times faster in *A* than *B* and that *A* and *B* do not diffuse in each other. Draw the concentration profile of hydrogen across the composite foil at steady state.

No discontinuity in chemical potential of hydrogen across the foil. Hence, in terms of concentration, there is a discontinuity.

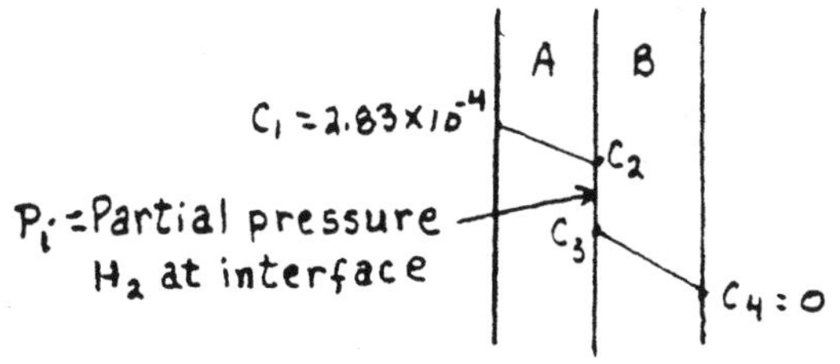

$C_1 = 4 \times 10^{-4}\sqrt{0.5} = 2.83 \times 10^{-4}$ g cm^{-3}

At steady state: $j_A = j_B$

$$D_A \frac{C_1 - C_2}{\delta_A} = D_B \frac{C_3 - C_4}{\delta_B}; \text{ Since } \delta_A = \delta_B \text{ and } D_A = 4D_B \text{ then}$$

$$(1)\quad 4(C_1 - C_2) = C_3 - C_4$$

Also $C_2 = 4\times10^{-4}\sqrt{P_i}$; $C_3 = 1\times10^{-4}\sqrt{P_i}$ then

$$(2)\quad C_2 = 4C_3$$

Combining (1) and (2)

$$C_3 = \frac{4}{17} C_1 = 6.66\times10^{-5}\,\text{g cm}^{-3}$$

$$C_2 = 2.66\times10^{-4}\,\text{g cm}^{-3}$$

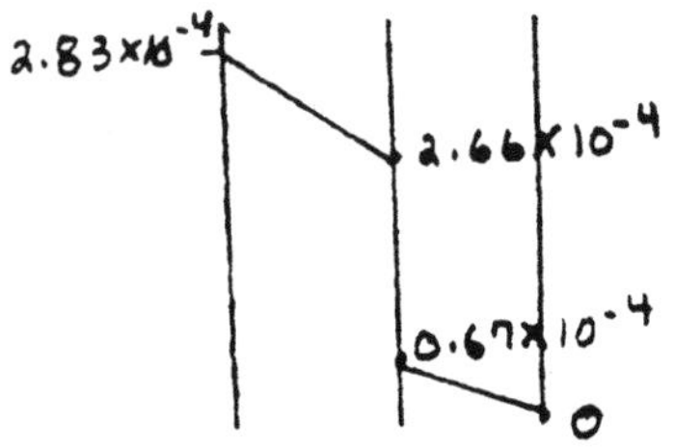

13.3 A thin sheet of iron at 800°C is subjected to different gaseous atmospheres on both of its surfaces such that the composition of one face is at 4 atom percent carbon and the other is at zero atom fraction carbon. At steady state, make a plot of the composition profile in the sample indicating *clearly* compositions and respective distances.

The thickness is 1 mm and density changes during the experiment may be neglected. At 800°C, it is known that the diffusion coefficient of carbon in iron is given by:

$$D = 10^{-6}\ \text{cm}^2\ \text{s}^{-1} \text{ in ferrite } (\alpha),$$

$$D = 10^{-8}\ \text{cm}^2\ \text{s}^{-1} \text{ in austenite } (\gamma).$$

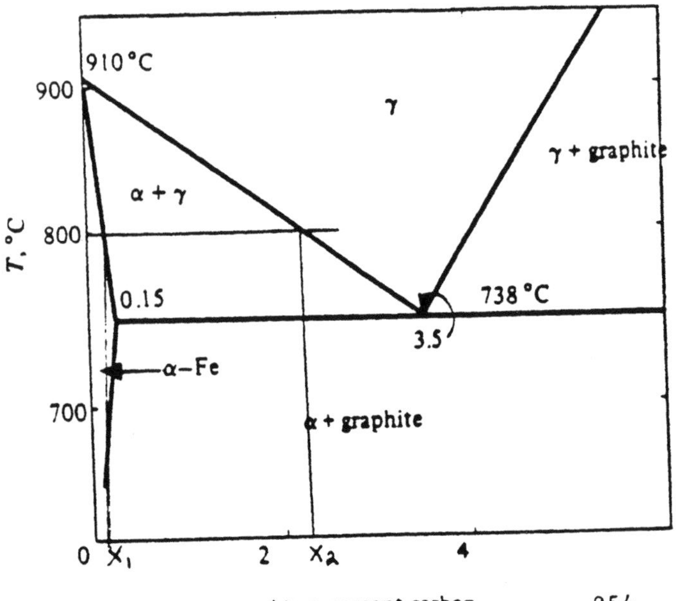

The phase diagram indicates that α and γ are in equilibrium at 800°C. Interpolating, we get:

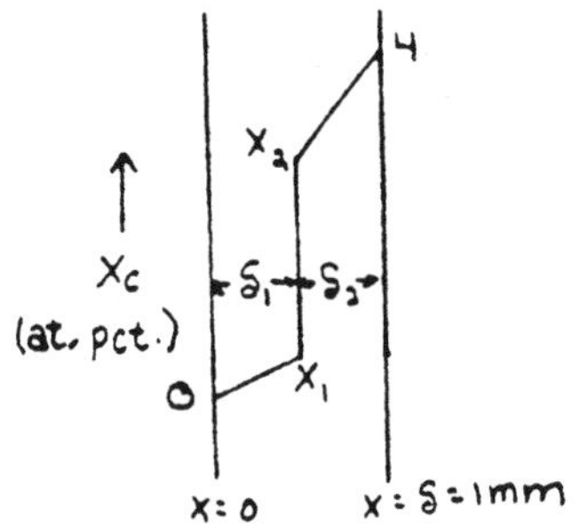

$$\frac{738-800}{738-9} = \frac{0.15-X_1}{0.15-0} \;;\; X_1 = 0.096\ \%\ C$$

$$\frac{738-800}{738-910} = \frac{3.5-X_2}{3.5-0} \;;\; X_2 = 2.238\ \%\ C$$

<u>At steady state</u>: $D_\gamma\left(\frac{dc}{dx}\right)_\gamma = D_\alpha\left(\frac{dc}{dx}\right)_\alpha \;;\; 10^{-8}\,\frac{(4-2.238)}{\delta_2} = 10^{-6}\,\frac{(0.096-0)}{\delta_1}$

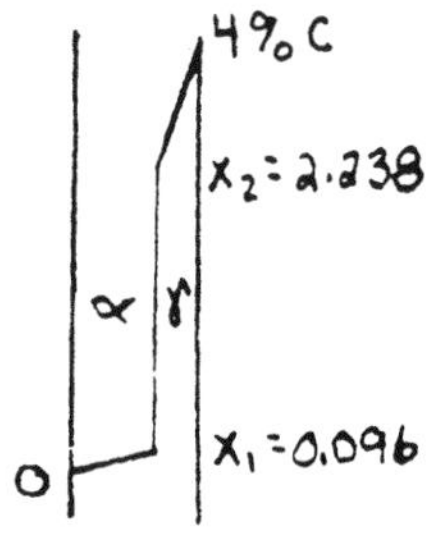

Where δ_1 and δ_2 are the thicknesses of α and γ in mm.

Also $\delta_1 + \delta_2 = 1\text{mm}$ $\therefore$ $\delta_1 = 0.845$ mm, $\delta_2 = 0.155$ mm

$\delta(\alpha) = 0.845$ mm; $\delta(\gamma) = 0.155$ mm

13.4 Often electronic packages are hermetically sealed with polymers, but after being put in service corrosion is sometimes observed. This happens because H_2O molecules can diffuse through polymers. Assume that the equilibrium between water vapor and water dissolved (or absorbed) by the polymer is simply represented by the reaction:

$$H_2O(g) \;=\; H_2O\ \text{(dissolved)},$$

with the equilibrium constant

$$K = \frac{C}{p_{H_2O}} = 10^{-4}.$$

where C is the concentration of H_2O in the polymer (moles cm^{-3}) and p_{H_2O} is the pressure of $H_2O(g)$ in atm.

a) Assume equilibrium at the surfaces and calculate the flux of H_2O through the polymer (in moles cm^{-2} s^{-1}); assume steady state.

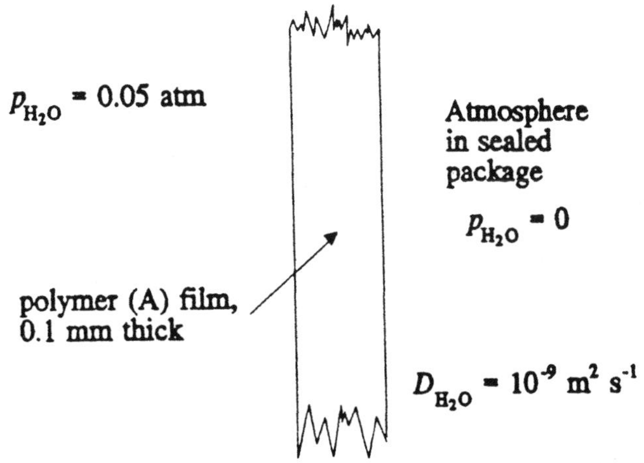

b) Now two films, polymer A and B, are used. Each is 0.1 mm thick.

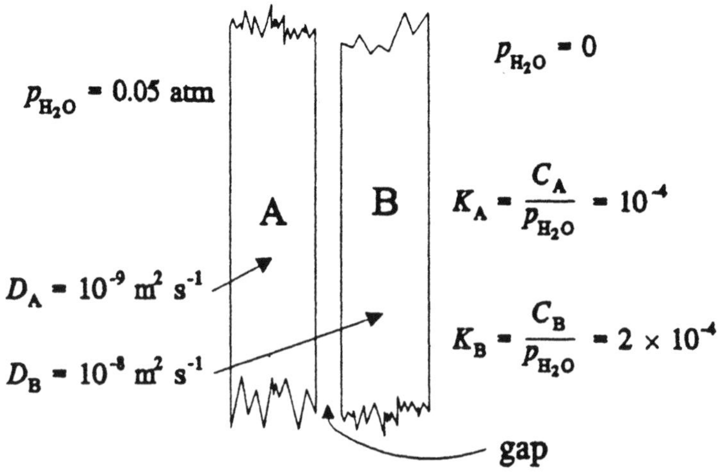

Assume steady state and equilibrium at all interfaces. What is the pressure of $H_2O(g)$ in the gap?

a. $j_{H_2O} = j_A = D\frac{C_1 - C_2}{\delta}$

$C_1 = K\,P_{H_2O} = (10^{-4})(0.05) = 5\times10^{-6}$ mol cm^{-3}

$j_A = \frac{10^{-9}\,m^2}{s}\left|\frac{(5\times10^{-6} - 0)\text{ moles}}{cm^3}\right|\frac{}{10^{-4}\,m} = 5\times10^{-9}$ moles H_2O $cm^{-2}\,s^{-1}$

b. At steady state: $j_A = j_B$

$\therefore \frac{D_A(C_{A1} - C_{A2})}{\delta} = \frac{D_B(C_{B1} - C_{B2})}{\delta}$

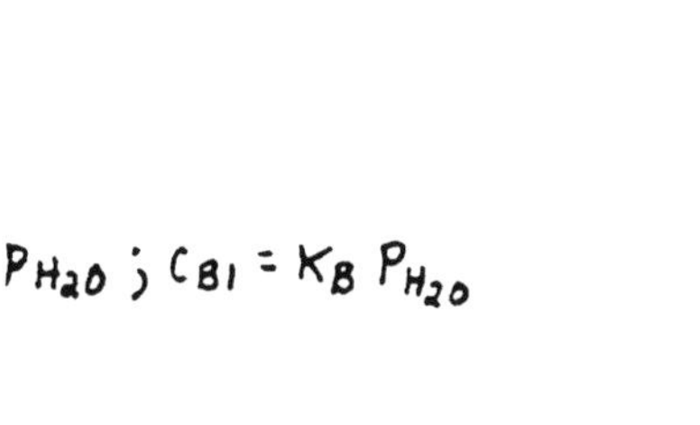

$C_{A1} = (10^{-4})(0.05) = 5\times10^{-6}$ mol cm^{-3}; $C_{B2} = 0$

$(10^{-9})(5\times10^{-6} - C_{A2}) = (10^{-8})(C_{B1} - 0)$

$5\times10^{-6} - C_{A2} = 10\,C_{B1}$

In the gap, thermodynamics requires: $C_{A2} = K_A P_{H_2O}$; $C_{B1} = K_B P_{H_2O}$

$\therefore \frac{C_{A2}}{C_{B1}} = \frac{1}{2}$

$5\times10^{-6} - 0.5\,C_{B1} = 10\,C_{B1}$; $C_{B1} = 4.76\times10^{-7}$ mol cm^{-3}

$C_{A2} = 2.38\times10^{-7}$ mol cm^{-3}

$P_{H_2O} = \frac{C_{B1}}{K_B} = \frac{4.76\times10^{-7}}{2\times10^{-4}} = 2.38\times10^{-3}$ atm

or $P_{H_2O} = \frac{C_{A2}}{K_A} = \frac{2.38\times10^{-7}}{10^{-4}} = 2.38\times10^{-3}$ atm

13.5 Hydrogen gas is maintained at 3 bar and 1 bar on opposite sides of a plastic membrane which is 0.3 mm thick. The temperature is 25°C, and the diffusion coefficient of hydrogen in the plastic is $8.7 \times 10^{-10}\ m^2\ s^{-1}$. The solubility of hydrogen in the membrane is 1.5×10^{-3} kmol m^{-3} bar^{-1}. What is the mass diffusion flux of hydrogen through the membrane? Give your result in kmol s^{-1} m^{-2}.

At 1 bar, c_2 = Solubility = 1.5×10^{-3} kmol m^{-3}

3 bar, $c_1 = (3)(1.5\times10^{-3}) = 4.5\times10^{-3}$ kmol m^{-3}

$$j = -D\frac{dc}{dx} = -\frac{8.7\times10^{-10}\ m^2}{s}\cdot\frac{(1.5-4.5)\times10^{-3}\ kmol}{m^3}\cdot\frac{1}{0.3\times10^{-3}\ m}$$

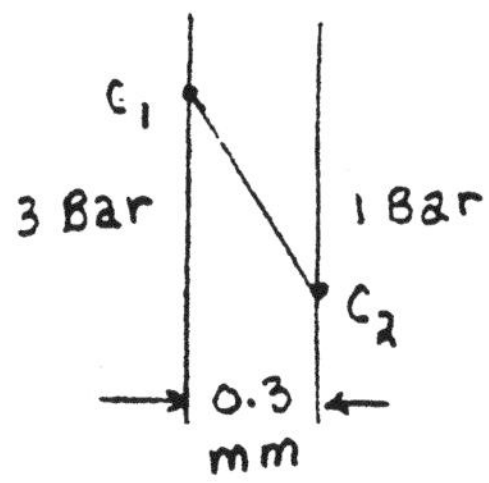

$j = 8.70\times10^{-9}$ kmol m^{-2} s^{-1}

13.6 The Grube solution is used to analyze diffusion data for a diffusion couple in which the solid is semi-infinite on both sides and when the chemical diffusion coefficient is uniform. Now consider a diffusion couple made from two thin solids so that the Grube solution for semi-infinite solids is not applicable. Such a couple is shown to the right with the initial condition for C_A shown. a) Assume that $\tilde{D}$ is uniform. Write a partial differential equation for $C_A(x,t)$ where x is the space coordinate and t is time. b) For $0 \le x \le L$, write appropriate boundary conditions and an appropriate initial condition for C_A. You may assume that component A is not volatile; i.e., no A is lost from the diffusion couple.

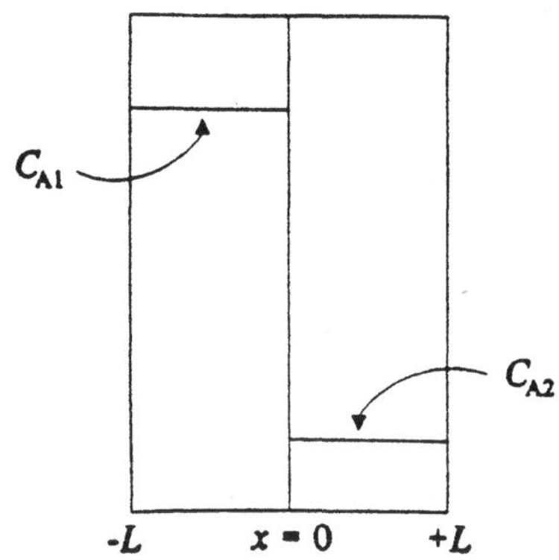

a. $\frac{\partial C_A}{\partial t} = \tilde{D}\frac{\partial^2 C_A}{\partial x^2}$

b. I.C.: $C_A(x,0) = C_{A2}$

B.C.: $C_A(0,t) = C_A'$ (constant)

$\frac{\partial C_A}{\partial x}(L,t) = 0$

The first B.C. arises from the fact that the concentration at $x=0$ is simply the average and remains so during diffusion.

$C_A' = \frac{1}{2}(C_{A1} + C_{A2})$

The second B.C. simply states that there is no loss of A from the surface.

Added note: If we shift the origin, then the solution must be Eq. (13.41), where $C_s = C_A'$; $C_i = C_{A2}$, and $D = \tilde{D}$.

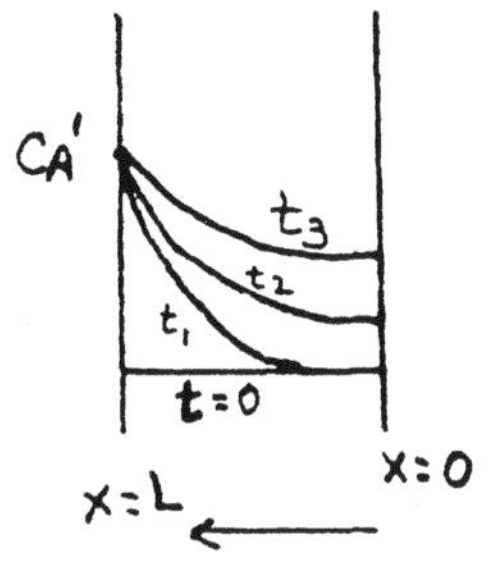

13.7 An unknown amount of radioactive gold is deposited as a thin layer on the ends of two rods of gold. The two rods are then joined to form a specimen having a planar source of radioactive gold (Au^*) atoms at the origin $x = 0$. After diffusion for 100 hours at 920°C, the distribution of Au^* is as shown below. Calculate the self-diffusion coefficient of gold in pure gold, based on the data at 0.3 mm and 0.6 mm as indicated in the plot of the relative concentration of gold.

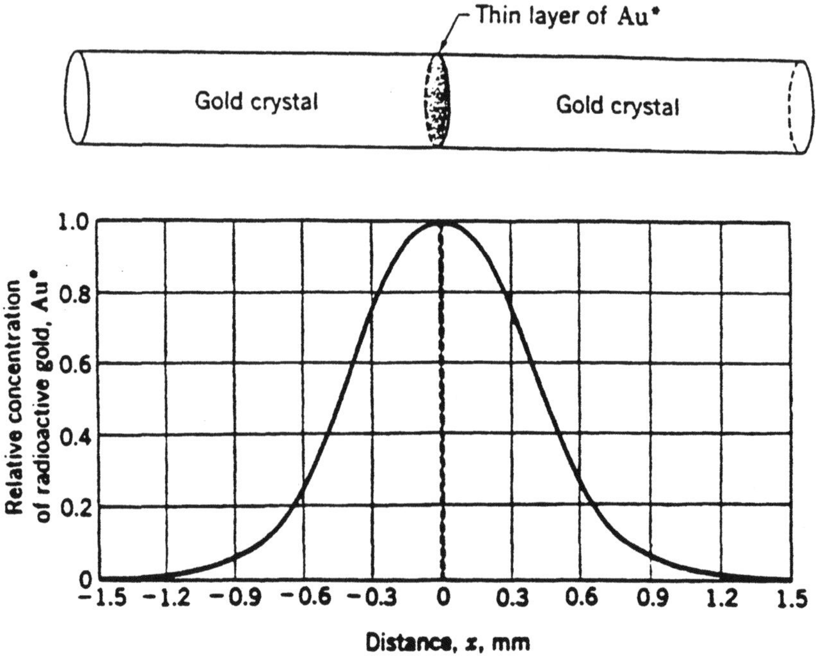

Point	x, mm	C_A^*
1	0.3	0.78
2	0.6	0.28

Apply Eq. (13.22) $\frac{C_{A1}^*}{C_{A2}^*} = \exp\left[-\left(\frac{x_1^2}{4D_A^* t}\right) + \left(\frac{x_2^2}{4D_A^* t}\right)\right] = \frac{0.78}{0.28} = 2.79$

$$-\frac{1}{D_A^*}\left[\frac{x_1^2}{4t} - \frac{x_2^2}{4t}\right] = 1.024$$

or

$$D_A^* = \left(\frac{1}{1.024}\right)\left(\frac{1}{4}\right)\left(\frac{1}{3.60\times10^5}\right)\left[(6\times10^{-4})^2 - (3\times10^{-4})^2\right] = 1.83\times10^{-13}\ m^2\ s^{-1}$$

13.8 Silicon is exposed to a gas that establishes a concentration of 10^{18} atoms (Al) cm^{-3} on the surface of the silicon. The process is carried out at 1473 K. a) After 30 min, at what depth below the surface of the Si will the concentration be 10^{16} atoms cm^{-3}? b) Calculate the amount of Al (in atoms cm^{-2}) that diffuses into the Si after 30 min of treatment at 1473 K.

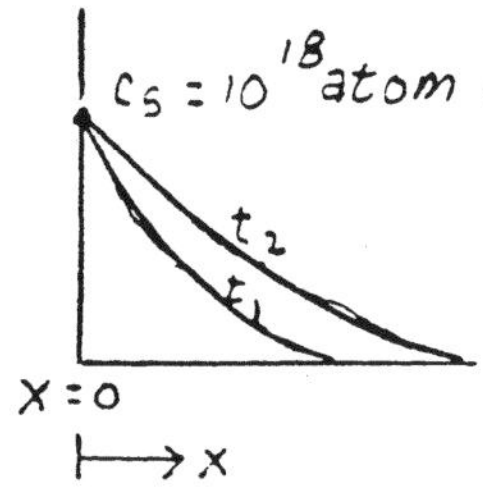

This is an application of the "error function solution" used for diffusion into a $\frac{\infty}{2}$ solid.

$$\frac{C - C_s}{C_i - C_s} = \text{erf}\,\frac{x}{2\sqrt{\tilde{D}t}}$$

a. $\tilde{D}(1473\,K) = 8\times10^{-4}\exp\left[\frac{-335{,}000}{(8.314)(1473)}\right] = 1.05\times10^{-15}\ m^2 s^{-1}$ (Table 13.4)

$$\text{erf}\,\frac{x}{2(\tilde{D}t)^{1/2}} = \frac{10^{16} - 10^{18}}{0 - 10^{18}} = 0.99 \;:\; \frac{x}{2(\tilde{D}t)^{1/2}} = 1.82$$

$$x = (1.82)(2)\left[(1.05\times10^{-15})(1800)\right]^{1/2} = 5.0\times10^{-6}\ m = 5.0\ \mu m$$

b. $j|_{x=0} = (C_s - C_i)\left(\frac{\tilde{D}}{\pi t}\right)^{1/2}$, atoms $cm^{-2}\ s^{-1}$

We want $\frac{J}{A} = \int_0^t j|_{x=0}\, dt$, atom cm^{-2}

The result of the integration is $\frac{J}{A} = \left(\frac{\tilde{D}t}{\pi}\right)^{1/2}(C_s - C_i) = \sqrt{\frac{(1.05\times10^{-15})(1800)}{\pi}}\,(10^{18} - 10^{16})$

$$\frac{J}{A} = 7.69\times10^{11}\ \text{atoms}\ m^{-2} = 7.69\times10^{15}\ \text{atoms}\ cm^{-2}$$

13.9 The Matano-Boltzmann analysis is used to calculate the interdiffusion coefficient, $\tilde{D}$, from diffusion couple data. It can also be used to determine the intrinsic diffusion coefficients in a binary by inserting inert markers at the original interface.

a) The distance moved by the markers is proportional to the square root of time. Show that υ_x in Eq. (12.14) is given by

$$\upsilon_x = \frac{S}{2t},$$

where S is the distance moved by the markers and t is the diffusion time.

b) Assuming that $\tilde{D}$ and S are determined in a diffusion-couple, what two equations are needed to simultaneously solve for the intrinsic diffusion coefficients in the binary. [Hint: Review Section 12.2.2.]

a. $S = B t^{1/2}$, where B = constant

$$V_x = \frac{dS}{dt} = \frac{B}{2}\frac{1}{t^{1/2}} = \frac{S}{2t^{1/2}}\frac{1}{t^{1/2}} = \frac{S}{2t}$$

b. $V_x = \frac{1}{c}(D_1 - D_2)\frac{\partial c_1}{\partial x}$

$\tilde{D} = X_1 D_2 + X_2 D_1$

In the course of an experiment, determine $\tilde{D}$ and $\frac{\partial c_1}{\partial x}$ at a particular point. Hence X_1, X_2 and c are also known. By measuring S, then V_x is also known. All this leaves are two unknowns, D_1 and D_2, which are calculated from the simultaneous equations.

13.10 A gold-nickel diffusion couple of limiting compositions $X_{Ni} = 0.0974$ and $X_{Ni} = 0.4978$ is heated at 925°C for 2.07×10^6 s. Layers 0.003 in. (0.0762 mm) thick and parallel to the original interface are machined off and analyzed. a) Using the data tabulated below, calculate the diffusion coefficient at 20, 30, and 40 at.% nickel. b) Suppose that markers are inserted at the original interface and move along during the diffusion process at a composition of 0.30 atom fraction nickel. From this, determine the intrinsic coefficients of gold and nickel at 0.30 atom fraction nickel.

Slice No.	at% Ni	Slice No.	at% Ni	Slice No.	at% Ni	Slice No.	at% Ni
11	49.78	21	35.10	29	21.38	38	13.26
12	49.59	22	33.17	30	20.51	39	12.55
14	47.45	23	31.40	31	19.12	41	11.41
16	44.49	24	29.74	32	17.92	43	10.48
18	40.58	26	25.87	33	16.86	45	9.99
19	38.01	27	24.11	35	15.49	47	9.74
20	37.01	28	22.49	37	13.90		

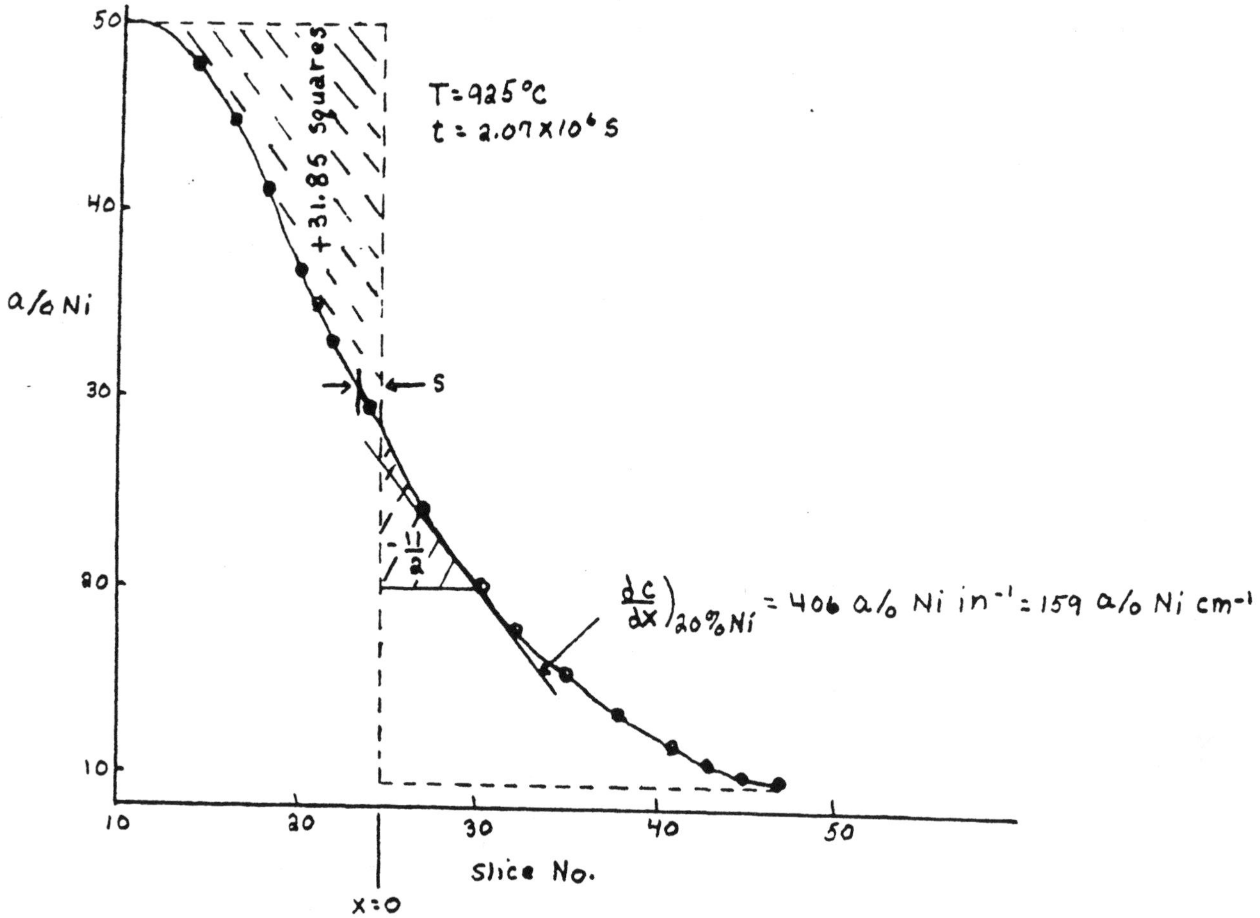

Matano Interface (Satisfies Eq. 13.33)

a. $\tilde{D} = -\frac{1}{2t} \frac{1}{\left(\frac{dc}{dx}\right)} \int_{c_1}^{c} x\,dc$

at 20 a/o Ni (slice 30) $\frac{dc}{dx} = -159$ a/o cm^{-1}

$\int_{c_1}^{c} x\,dc = \frac{26.35 \text{ squares}}{} \left| \frac{2 \text{ slices} \times 2 \text{ a/o Ni}}{1 \text{ square}} \right| \frac{0.003 \text{ in}}{1 \text{ slice}} \left| \frac{2.54 \text{ cm}}{1 \text{ in}} \right. = 0.803$ a/o Ni-cm

$\tilde{D}_{20\,a/o\,Ni} = + \frac{(1)(0.803)}{(2)(2.07 \times 10^{6})(159)} = 1.22 \times 10^{-9}\ \text{cm}^2\,\text{s}^{-1}$

at 30 a/o Ni $\frac{dc}{dx} = -157$ a/o cm^{-1}

$\int_{c_1}^{c} x\,dc = 31.55$ squares $= 0.955$ a/o Ni-cm

$\tilde{D}_{30\,a/o\,Ni} = + \frac{0.955}{(2)(2.07 \times 10^{6})(157)} = 1.47 \times 10^{-9}\ \text{cm}^2\,\text{s}^{-1}$

at 40 a/o Ni $\frac{dc}{dx} = -239$ a/o cm^{-1}

$\int_{c_1}^{c} x\,dc = 22.85$ squares $= 0.696$ a/o Ni-cm

$\tilde{D}_{40\ a/o\ Ni} = + \dfrac{0.696}{(2)(2.07\times10^{6})(239)} = 7.03\times10^{-10}\ cm^2\ s^{-1}$

b. S (marker motion) $\simeq 9.1\times10^{-3}$ cm

$V_x = \dfrac{S}{2t} = \dfrac{9.1\times10^{-3}\ cm}{(2)(2.07\times10^{6})\ s} = 2.21\times10^{-9}\ cm\ s^{-1}$

(A) $\tilde{D} = X_{Ni} D_{Au} + X_{Au} D_{Ni}$

$V_x = \dfrac{1}{C}(D_{Au} - D_{Ni})\dfrac{\partial C_{Au}}{\partial X}$

(B) $V_x = (D_{Au} - D_{Ni})\dfrac{\partial X_{Au}}{\partial t}$ where $\dfrac{\partial X_{Au}}{\partial X} = -\dfrac{\partial X_{Ni}}{\partial X} = \dfrac{1}{100}\dfrac{\partial C}{\partial X}$

Solving (A) and (B) simultaneously at $X_{Ni} = 0.30$

$1.47\times10^{-9} = 0.30\ D_{Au} + 0.70\ D_{Ni}$

$2.21\times10^{-9} = (D_{Au} - D_{Ni})\ 1.57$

$D_{Au} = 2.45\times10^{-9}\ cm^2\ s^{-1}$

$D_{Ni} = 1.05\times10^{-9}\ cm^2\ s^{-1}$

13.11 Metals A and B form alloys of fcc structure at 1200°C. They are allowed to interdiffuse as a diffusion couple for 10^5 s, and the concentration profile obtained is given in the accompanying figure. Determine the value of the interdiffusion coefficient, $\tilde{D}$, at a concentration C_B = 0.02 mol cm^{-3}.

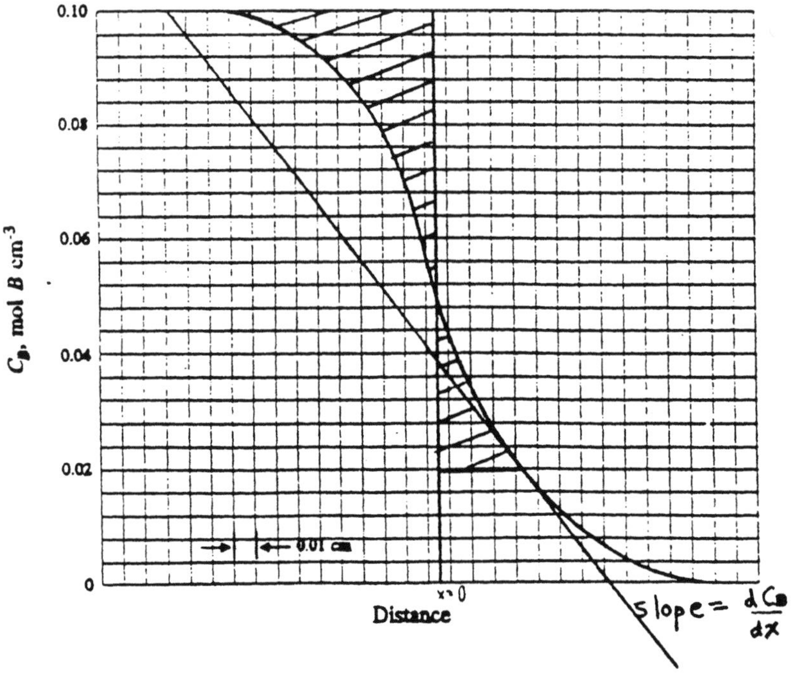

First, locate $x=0$ such that we satisfy $\int_{x=-\infty}^{\infty} x\, dC_B = 0$

Second, determine slope at $C_B = 0.02$ mol cm^{-3}. $\frac{dC_B}{dx} \simeq -0.483$ mol cm^{-4}

Third, determine the area from $C_B(x=-\infty)$ to $C_B = 0.02$ mol cm^{-3}.

$$\int_{C_{B1}}^{C_B} x\, dC_B \simeq -1.084 \times 10^{-3} \text{ mol cm}^{-2}$$

$$\tilde{D} = \frac{1}{2t}\frac{1}{\left(\frac{dC_B}{dx}\right)_{C_B}}\int_{C_{B1}}^{C_B} x\, dC_B = \frac{1}{(2)(10^5)}\left(\frac{1}{0.483}\right)1.084 \times 10^{-3} = 1.12 \times 10^{-8} \text{ cm}^2 \text{ s}^{-1}$$

13.12 Intrinsic silicon (i.e., pure Si) is processed in a gas which establishes a concentration of 10 ppm (1 ppm = 10^{-4} wt.%) of boron at the surface of the silicon. Distances from the left vertical face are given in the following table.

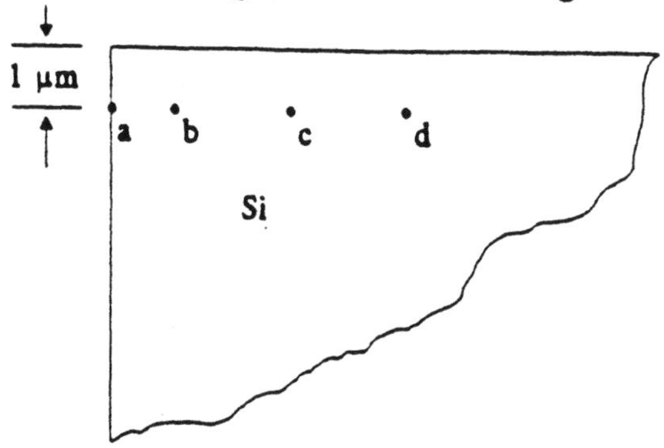

Point	Distance, μm
a	0
b	1
c	3
d	5

After 10 h exposure to the gas, what are the concentrations of boron at points a, b, c and d? At the process temperature, the diffusion coefficient for B in Si is $D = 10^{-12}$ cm^2 s^{-1} = 10^{-4} μm^2 s^{-1}.

Concentration depends on two dimensional diffusion.

$$\frac{C - C_s}{C_i - C_s} = \left[\text{erf}\left(\frac{x}{2\sqrt{Dt}}\right)\text{erf}\left(\frac{y}{2\sqrt{Dt}}\right)\right]$$

Point a: $y = 1$ μm, $x = 0$

$$\frac{C - C_s}{C_i - C_s} = (\text{erf } 0)\left(\text{erf } \frac{1}{2\sqrt{(10^{-4})(3.6 \times 10^4)}}\right) = 0$$

$\therefore C = C_s = 10$ ppm

Point b: $y = 1$ μm, $x = 1$ μm

$$\frac{C - C_s}{C_i - C_s} = \text{erf}\left(\frac{1}{2\sqrt{3.6}}\right)\text{erf}\left(\frac{1}{2\sqrt{3.6}}\right) = (0.290)(0.290) = 0.0841$$

$\therefore C = (0.0841)(0-10) + 10 = 9.2\ ppm$

Point c: $Y = 1\ \mu m,\ X = 3\ \mu m$

$$\frac{C - C_s}{C_i - C_s} = \text{erf}\left(\frac{3}{2\sqrt{3.6}}\right)\text{erf}\left(\frac{1}{2\sqrt{3.6}}\right) = (0.736)(0.290) = 0.214$$

$C = (0.214)(0-10) + 10 = 7.9\ ppm$

Point d: $Y = 1\ \mu m,\ X = 5\ \mu m$

$$\frac{C - C_s}{C_i - C_s} = \text{erf}\left(\frac{5}{2\sqrt{3.6}}\right)\cdot(0.290) = (0.937)(0.290) = 0.272$$

$C = (0.272)(0-10) + 10 = 7.3\ ppm$

13.13 A fine steel wire of 0.2 wt.% C is passed through a tube furnace at 1200°C which contains a carburizing gas. The composition of the carburizing gas is adjusted so that it fixes 0.8 wt.% C on the surface of the wire. By neglecting diffusion in the axial direction of the wire, calculate the average composition of the wire after it passes through the tube. At this temperature the steel is a single phase (austenite). *Data*: diameter of wire, 0.01 cm; length of furnace, *L*, 1.5 m; velocity of wire, *V*, 15 cm s^{-1}.

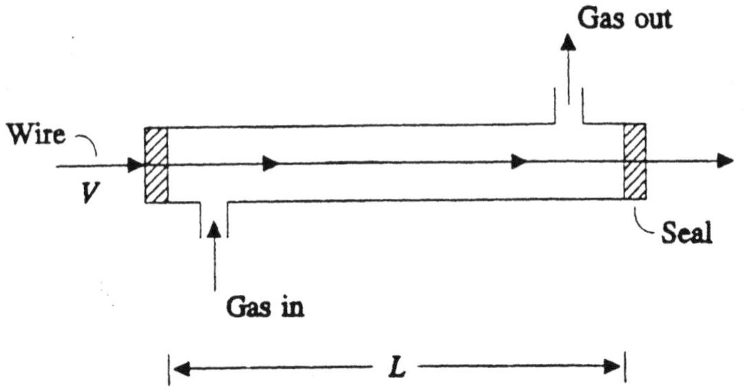

$C = f(r, z)$, but if we neglect diffusion in the z-direction we can treat diffusion such that $C = C(r,t)$ where $t = \frac{z}{V}$.

To pass through the tube: $t = \frac{L}{V} = \frac{15\ m\ s}{0.15\ m} = 100\ s$

Apply Fig. 13.6 for ∞-cylinder. Fig. 12.10 gives $D(1200°C) = 1.6 \times 10^{-6}\ cm^2\ s^{-1}$

$$\frac{Dt}{R^2} = \frac{(1.6\times10^{-6})(100)}{(5\times10^{-3})^2} = 6.4$$

Then $\frac{\bar{C} - C_s}{C_i - C_s} = 0$; $\bar{C} = C_s = 0.8$ wt. pct. C

The process could be carried out at a much higher velocity. Extrapolate the curve to $\frac{\bar{c} - c_s}{c_i - c_s} = 0.001$; $\bar{c} = 0.7994$ wt. pct. C

Then $\frac{Dt}{R^2} \simeq 1.14$; $t = (1.14)\frac{(5\times10^{-3})^2}{(1.6\times10^{-6})} = 17.8$ s

$V = \frac{15\text{ m}}{17.8\text{ s}} = 0.842 \text{ m s}^{-1}$

13.14 A thin layer of Au is plated on to the end of a Ni bar. The bar is annealed at 900°C for 10 h; at 900°C the interdiffusion coefficient of Au in Ni is 10^{-11} cm^2 s^{-1}. It is known that Au and Ni are completely soluble at 900°C. After the treatment, the concentration of Au at a distance of 0.05 cm from the end is 0.1 atom fraction of Au. At what distance from the end is the atom fraction of Au equal to 0.05?

Eq. (13.20) $C = \frac{B}{2(\pi Dt)^{1/2}} \exp\left(-\frac{x^2}{4Dt}\right)$; $\frac{C_1}{C_2} = \frac{\exp\left(-\frac{x_1^2}{4Dt}\right)}{\exp\left(-\frac{x_2^2}{4Dt}\right)}$

$\exp\left(-\frac{x_2^2}{4Dt}\right) = \frac{C_2}{C_1}\exp\left(-\frac{x_1^2}{4Dt}\right)$ or $\frac{x_2^2}{4Dt} = \ln\left(\frac{C_1}{C_2}\right) + \frac{x_1^2}{4Dt}$

$\therefore x_2^2 = (4Dt)\ln\frac{C_1}{C_2} + x_1^2$

$C_1 = 0.1$ at $x_1 = 0.05$ and $C_2 = 0.05$ at $x = x_2$

$x_2^2 = (4)(10^{-11})(3.6\times10^4)\ln\left(\frac{0.1}{0.05}\right) + (0.05)^2$

$x_2 = 5.001\times10^{-2}$ cm

13.15 A long cylindrical bar of steel which contains 3 ppm of hydrogen is dehydrogenated by a two-step vacuum process. The first step is treatment at 150°C for time period t_1, followed by the second step at 300°C for time period t_2. If $t_1 = 2t_2$, calculate: a) the total time ($t = t_1 + t_2$) to reduce the average composition to 1.5 ppm of hydrogen, and b) the center composition after the two-step treatment. *Data*: $D_H = 1.0 \exp(-4000/T)$ with D_H in cm^2 s^{-1} and T in K. The diameter of the bar is 2 cm.

a. $\frac{\bar{c} - c_s}{c_i - c_s} = \frac{1.5 - 0}{3.0 - 0} = 0.5$

Fig. (13.6) gives $\frac{Dt}{R^2} \simeq 0.072$

But $\frac{Dt}{R^2} = \frac{D_1t_1}{R^2} + \frac{D_2t_2}{R^2} = \frac{\left[D_1t_1 + \frac{D_2t_1}{2}\right]}{R^2} = \frac{\left[D_1 + \frac{D_2}{2}\right]t_1}{R^2}$

$D_1 = \exp\left(-\frac{4000}{423}\right) = 7.82\times10^{-5}$ cm^2 s^{-1}

$D_2 = \exp\left(-\frac{4000}{573}\right) = 9.30\times10^{-4}$ cm^2 s^{-1}

$$t_1 = \frac{(0.072)(1)^2}{\left(7.82\times10^{-5} + \frac{9.30\times10^{-4}}{2}\right)} = 132.5\ \text{s}; \quad t_2 = \frac{t_1}{2} = 66.3\ \text{s}$$

Total time $(t_1+t_2) = 199$ s

b. For the two-step process, $\frac{Dt}{R^2} = 0.072$. use Fig. 9.8a for a cylinder with the highest Bi No.

$$\frac{C-C_s}{C_i-C_s} = 0.97; \quad C = 0.97(3-0)+0 = 2.91\ \text{ppm (in center)}$$

13.16 The solubility of hydrogen in solid copper at 1000°C is 1.4 ppm (by mass) under a pressure of hydrogen of 1 atm. At 1000°C, $D_{\underline{H}} = 10^{-6}$ cm^2 s^{-1}. a) Determine the time for hydrogen to reach a concentration of 1.0 ppm at a depth of 0.1 mm in a large chunk of copper initially with null hydrogen if the copper is subjected to 2 atm pressure of H_2 at 1000°C. b) Copper foil, 0.2 mm thick, is equilibrated with hydrogen at a pressure of 4 atm at 1000°C. The same foil is then placed in a perfect vacuum at 1000°C and held for 60 s. Calculate the concentration of hydrogen at the center of the foil after the 60 s period.

a. At the surface, $C_s = kp^{1/2} = (1.4)(2)^{1/2} = 1.98$ ppm

$$\frac{C-C_s}{C_2-C_s} = \text{erf}\frac{x}{2(Dt)^{1/2}}; \quad \frac{1-1.98}{0-1.98} = \text{erf}\frac{x}{2\sqrt{Dt}} \Rightarrow \text{From Table 9.3}, \frac{x}{2\sqrt{Dt}} = 0.47$$

$$\text{Then } t = \frac{1}{(0.47)^2}\frac{(0.01)^2}{4\times10^{-6}} = 113\ \text{s}$$

b. After equilibriation at $P_{H_2} = 4$ atm.

$$C = C_i = kp^{1/2} = (1.4)(4)^{1/2} = 2.8\ \text{ppm}$$

$C_s = 0$ vacuum

$$\frac{Dt}{L^2} = \frac{(10^{-6})(60)}{(0.01)^2} = 0.6 \quad \text{Apply Fig. 9.8a}: \frac{C-C_s}{C_i-C_s} = \frac{C-0}{2.8-0} = 0.3$$

$\therefore C = 0.84$

13.17 The term "banding" is used to describe chemical heterogeneity in rolled alloys that shows up as closely spaced light and dark bands in the microstructure of steel. These bands represent areas of segregation of alloying elements that formed during freezing of the ingot. During rolling the segregated areas are elongated and compressed into narrow bands. Assume that the alloy concentration varies sinusoidally with distance after rolling according to the sketch below.

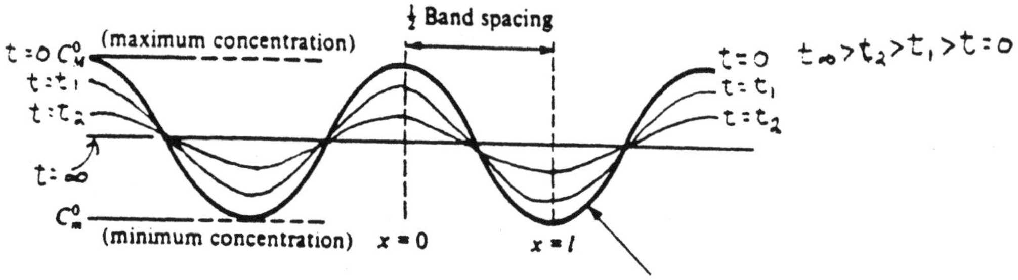

$$C = C_0 + (C_M^0 - C_0)\cos\frac{\pi x}{l}$$

If the steel is now heated to the austenite range and held at some constant temperature, then a) schematically sketch the concentration profile as time passes; b) write a differential equation for concentration (state assumptions) and c) write the boundary conditions (for time and space) that apply; d) solve for the concentration as a function of time and x; e) derive an equation for the residual segregation index.

a. See the concentration profiles sketched in the problem statement.

b. No flux into or out of volume element contained between $x=0$ and $x=L$.

One dimensional diffusion.

$\tilde{D}$ is uniform and constant.

$$\frac{\partial c}{\partial t} = \tilde{D}\frac{\partial^2 c}{\partial x^2}$$

c. B.C.: $\frac{\partial c}{\partial x}(0,t) = 0$; $\frac{\partial c}{\partial x}(\ell,t) = 0$

I.C.: $c(x,0) = C_0 + (C_M^0 - C_0)\cos\left(\frac{\pi x}{\ell}\right)$

d. Solution given by Eqs. (13.62) and (13.63)

$$c(x,t) = C_0 + \sum_{n=1}^{\infty} A_n \exp\left(-n^2\pi^2\frac{Dt}{\ell^2}\right)\cos\frac{n\pi x}{\ell}$$

where $A_n = \frac{2}{\ell}\int_0^{\ell} f(x)\cos\frac{n\pi x}{\ell}\,dx$

Let's evaluate the A_n with $f(x)$ as the initial distribution

$$A_n = \frac{2}{\ell}\left[C_0\int_0^{\ell}\cos\left(\frac{n\pi x}{\ell}\right)dx + (C_M^0 - C_0)\int_0^{\ell}\cos\left(\frac{n\pi x}{\ell}\right)\cos\left(\frac{\pi x}{\ell}\right)dx\right]$$

Examine the first integral:

$$\int_0^{\ell} \cos\left(\frac{n\pi x}{\ell}\right) dx = \frac{\ell}{n\pi} \sin \frac{n\pi x}{\ell} \Big|_0^{\ell} = \frac{\ell}{n\pi}\left(\sin n\pi - \sin 0\right) = 0 \text{ since } n \text{ is an integer.}$$

Examine the second integral:

$$\int_0^{\ell} \cos \frac{n\pi x}{\ell} \cos \frac{\pi x}{\ell}\, dx = 0 \text{ except for } n = 1\text{, where we have,}$$

$$\int_0^{\ell} \cos^2 \frac{\pi x}{\ell}\, dx = \frac{\ell}{\pi} \int_{\frac{\pi x}{\ell}=0}^{\pi} \cos^2\left(\frac{\pi x}{\ell}\right) d\left(\frac{\pi x}{\ell}\right) = \left(\frac{\ell}{\pi}\,\frac{\pi}{2}\right) = \frac{\ell}{2} \therefore \text{All } A_n = 0 \text{ except } n = 1$$

$$A_1 = \frac{2}{\ell}\left[\left(C_M^0 - C^0\right)\frac{\ell}{2}\right] = C_M^0 - C_0$$

$$\therefore C(x,t) = C_0 + \left(C_M^0 - C_0\right) \exp\left(-\pi^2 \frac{Dt}{\ell^2}\right) \cos\left(\frac{\pi x}{\ell}\right)$$

$$\text{or } \frac{C - C_0}{C_M^0 - C_0} = \exp\left(-\pi^2 \frac{Dt}{\ell^2}\right) \cos\left(\frac{\pi x}{\ell}\right)$$

$$\therefore C = C_M \text{ at } x = 0; \quad \frac{C_M - C_0}{C_M^0 - C_0} = \exp\left(-\frac{\pi^2 Dt}{\ell^2}\right)$$

$$C = C_m \text{ at } x = \ell; \quad \frac{C_m - C_0}{C_M^0 - C_0} = -\exp\left(-\frac{\pi^2 Dt}{\ell^2}\right)$$

$$\text{We want } \delta \equiv \frac{C_M - C_m}{C_M^0 - C_m^0} \text{ but } C_M - C_m = \left(C_M^0 - C_0\right)\exp\left(-\frac{\pi^2 Dt}{\ell^2}\right) + C_0$$

$$+\left(C_M^0 - C_0\right)\exp\left(-\frac{\pi^2 Dt}{\ell^2}\right) - C_0$$

$$= 2\left(C_M^0 - C_0\right)\exp\left(-\pi^2 \frac{Dt}{\ell^2}\right)$$

$$\text{or } C_M - C_m = \left(C_M^0 - C_m^0\right)\exp\left(-\pi^2 \frac{Dt}{\ell^2}\right)$$

$$\delta = \exp\left(-\pi^2 \frac{Dt}{\ell^2}\right)$$

13.18 Assume that the banding in a wrought cupronickel alloy (single phase) is described by the cosine function in Problem 13.17. The average composition of the alloy is 10% Ni-90% Cu, and the segregation ratio before homogenization is 1.4. Segregation ratio is defined as $S = C_M^0/C_m^0$. a) What are the maximum and minimum compositions of nickel? b) In order to homogenize the alloy in the shortest time possible, what temperature would you select? c) If the average distance between maximum compositions is 10^{-2} cm, determine the time to achieve a residual segregation ratio of 0.1 at 950°C. A diffusion coefficient can be obtained from Fig. 12.9. d) The alloy is given a "step" homogenization treatment which consists of 10 hours at 700°C, 10 hours at 800°C and 10 hours at 900°C. What is the residual segregation index after this treatment?

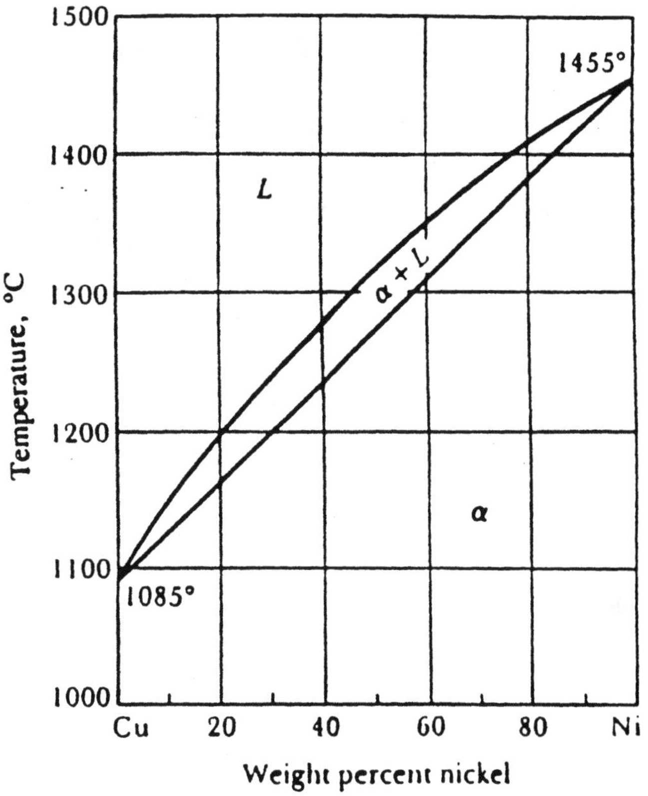

a. $S = \frac{C_M^0}{C_m^0}$ and $C_0 = \frac{1}{2}(C_M^0 + C_m^0)$

$C_M^0 = SC_m^0 = S(2C_0 - C_M^0)$; $C_M^0 = \frac{2SC_0}{1+S} = \frac{2(1.4)(10)}{1+1.4} = 11.67\%$ Ni Maximum

$C_m^0 = \frac{C_M^0}{S} = \frac{11.67}{1.4} = 8.34\%$ Ni Minimum

b. Solidus increases $\frac{1455-1085}{100} = 3.70$°C above 1085°C for each percent Ni content.

∴ homogenize just below 1116°C

c. From Fig. 12.9 $D = 3.7\times10^{-14}\ m^2 s^{-1}$; $\ell = 0.5\times10^{-4}$ m, $\delta = 0.1$

$\ln\delta = -\frac{\pi^2 D}{\ell^2}t$; $t = \frac{\ell^2}{\pi^2 D}\ln\frac{1}{\delta} = \frac{(0.5\times10^{-4})^2}{\pi^2(3.7\times10^{-14})}\ln\frac{1}{0.1} = 1.576\times10^4\ s = 4.38\ h$

d. $D_{700} = 1.9\times10^{-15}\ m^2s^{-1}$; $\left.\frac{Dt}{\ell^2}\right)_{700} = 0.0274$

$D_{800} = 5.8\times10^{-15}\ m^2s^{-1}$; $\left.\frac{Dt}{\ell^2}\right)_{800} = 0.0835$

$D_{900} = 2.1\times10^{-14}\ m^2s^{-1}$; $\left.\frac{Dt}{\ell^2}\right)_{900} = 0.3024$ Total $\frac{Dt}{\ell^2} = 0.413$

$\delta = \exp\left(-\frac{Dt}{\ell^2}\right) = \exp(-0.413) = 0.662$

13.19 A junction in silicon is made by doping with boron using predeposition followed by drive-in diffusion. a) Five minutes at 1100°C are required to deposit the dopant. At what distance from the surface is the concentration of boron raised to 3×10^{18} atoms cm^{-3}. Assume that the silicon is initially pure. b) How much boron (per cm^2 of silicon surface) will have been taken up by the silicon during the deposition step? c) To prevent loss of boron during the drive-in step, the surface is masked with silica (SiO_2). Now calculate the time required to achieve a concentration of boron equal to 3×10^{18} atoms cm^3 at a depth of 6×10^{-4} cm if the drive-in treatment is carried out at 1150°C.

a. $\frac{c - c_s}{c_o - c_s} = \mathrm{erf}\left(\frac{x}{2\sqrt{Dt}}\right)$; $t = \frac{1}{12}$ h; $c = 3 \times 10^{18}$ atoms cm^{-3}; $c_o = 0$;

$c_s = 5.1 \times 10^{20}$ atoms cm^{-3} (Table 13.3); $D = 5.8 \times 10^{-2}\ \mu m^2 h^{-1}$ (Table 13.3)

$$\frac{c - c_s}{c_o - c_s} = \frac{3 \times 10^{18} - 5.1 \times 10^{20}}{0 - 5.1 \times 10^{20}} = 0.994; \quad \frac{x}{2\sqrt{Dt}} = 1.95$$

$$x = (1.95)(2)\left[(5.8 \times 10^{-2})\left(\frac{1}{12}\right)\right]^{1/2} = 0.271\ \mu m$$

b. $$\frac{J}{A} = \frac{2}{\pi^{1/2}}(c_s - c_o)(Dt)^{1/2} = \frac{2}{\pi^{1/2}}\left(5.1 \times 10^{20}\ \frac{\text{atoms}}{cm^3}\right)\left[(5.8 \times 10^{-2})\left(\frac{1}{12}\right)\right]^{1/2} \mu m \left|\frac{cm}{10^4\ \mu m}\right.$$

$$\frac{J}{A} = 4.00 \times 10^{15}\ \text{atoms}\ cm^{-2}$$

c. $c_s = 5.1 \times 10^{20}$ (from part a); $D = 1.6 \times 10^{-1}\ \mu m^2 h^{-1}$; $c = 3 \times 10^{18}$ atoms cm^{-3}; $x = 6\ \mu m$;

$\lambda = (Dt)^{1/2} = \left[(5.8 \times 10^{-2})\left(\frac{1}{12}\right)\right]^{1/2} = 6.95 \times 10^{-2}\ \mu m$ (from part a).

$$\frac{3 \times 10^{18}}{5.1 \times 10^{20}} = \frac{2}{\pi}\left(\frac{6.95 \times 10^{-2}}{\sqrt{1.6 \times 10^{-1} t}}\right)\exp\left(-\frac{6^2}{(4)(1.6 \times 10^{-1})t}\right)$$

$$5.882 \times 10^{-3} = \frac{0.1106}{t^{1/2}}\exp\left(-\frac{56.25}{t}\right)$$ Solve by trial and error

$t = 148$ h

13.20 By ion implantation, lithium can be concentrated in a very thin surface layer (10^{-6} cm) on a nickel substrate. After implanting the surface layer, it has a lithium concentration of 10^{20} atoms cm^{-3}. Determine the time at 1000 K for reducing the surface concentration to 10^{19} atoms cm^3. At 1000 K, the interdiffusion diffusion coefficient of lithium in nickel is $5 \times 10^{-8}\ cm^2 s^{-1}$.

Eq. (13.22) $c(x,t) = \frac{B}{(\pi \tilde{D} t)^{1/2}} \exp\left(-\frac{x^2}{4\tilde{D}t}\right)$; $B = C_i \Delta x = 10^{20} 10^{-6} = 10^{14}$ atoms cm^{-2}

$\therefore 10^{19} = \frac{10^{14}}{\sqrt{(\pi)(5\times10^{-8})t}}$; $t^{1/2} = \frac{10^{14}}{\pi^{1/2}(5\times10^{-8})^{1/2}10^{19}} = 0.0252\,s$

$t = 6.37\times10^{-4}\,s$ very, very short time

13.21 A cylindrical bar of Fe (dia. of 1 cm) is suspended in a well-mixed and small melt of Mn maintained at 1300°C. Assume that there is local equilibrium at the solid-liquid interface and calculate the time required to raise the manganese composition at the center of the bar to 1 wt.%. The interdiffusion coefficient of Mn in Fe is given by

$$D = 0.49 \exp\left[-\frac{33\ 200}{T}\right]$$

with T in K and D in $cm^2\ s^{-1}$.

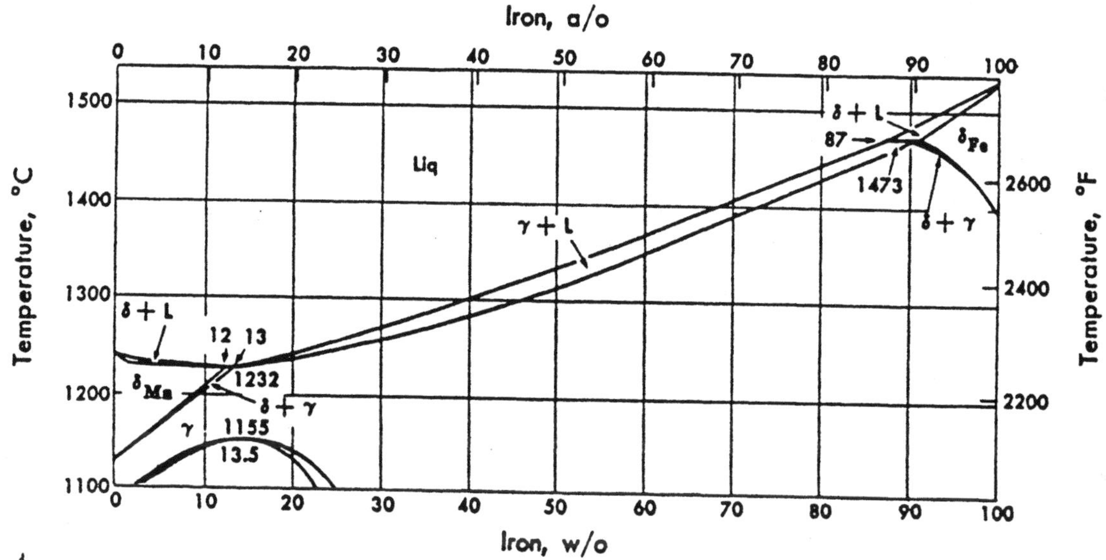

Assume that the Mn melt becomes saturated with respect to Fe and there is equilibrium at the melt/solid interface. Then C_s (surface concentration) is 54.5% Mn.

$$D = 0.49 \exp\left[-\frac{33\ 200}{1573}\right] = 3.34\times10^{-10}\,cm^2\,s^{-1}$$

$C = 1\%$ Mn ; $R = 0.5\,cm$

Apply Fig. 9.9 (a) with $Bi = 1000$

$$\frac{C - C_s}{C_i - C_s} = \frac{1 - 54.5}{0 - 54.5} = 0.982\ ;\ \frac{Dt}{R^2} \simeq 0.055$$

$$t = (0.055)\frac{(0.5)^2}{3.34\times10^{-10}} = 4.12\times10^{7}\,s = 476\ \text{days}$$

(We doubt that anyone will verify this result in the laboratory.)

13.22 A very thin sheet of Fe-0.2 atom fraction B is "sandwiched" between two large pieces of iron, and then the entire assemble is heated to 1000°C. The sheet is only 5×10^{-3} cm thick and at 1000°C diffusion bonding occurs as the boron diffuses into the iron. Assume that the boron is completely soluble and that there is only a single phase. a) Calculate the time required for the concentration of B to achieve its maximum at a distance of 1 mm from the original joint. b) What is the maximum concentration at 1 mm? c) At the time corresponding to part a), what is the atom fraction of B at the original joint?

Eq. (13.22) $c = \frac{\beta}{2(\pi D t)^{1/2}} \exp\left(\frac{-x^2}{4Dt}\right) : \beta = C_i \Delta x'$

$\therefore \quad c = \frac{C_i \Delta x'}{2(\pi D t)^{1/2}} \exp\left(-\frac{x^2}{4Dt}\right)$

a. $\frac{\partial c}{\partial t} = \frac{C_i \Delta x}{2(\pi D)^{1/2}} \exp\left(\frac{-x^2}{4Dt}\right)\left[\left(\frac{x^2}{4D}\right)\left(\frac{1}{t}\right)^{2.5} - \left(\frac{1}{2}\right)\left(\frac{1}{t}\right)^{3/2}\right]$

$\frac{\partial c}{\partial t} = 0 = \left(\frac{x^2}{4D}\right)\left(\frac{1}{t}\right)^{5/2} - \left(\frac{1}{2}\right)\left(\frac{1}{t}\right)^{3/2} ; \; t = \frac{x^2}{2D}$

$t = \frac{(1 \times 10^{-3})^2}{2(4.5 \times 10^{-11})} = 1.11 \times 10^4 \text{ s} = 3.09 \text{ h}$

or

$\frac{\partial c}{\partial t} = \frac{c}{t}\left(\frac{x^2}{4Dt} - \frac{1}{2}\right) ; \; \frac{\partial c}{\partial t} = 0 ; \; D = 4.5 \times 10^{-7} \text{ cm}^2 \text{ s}^{-1}$ (Fig. 12.10)

$\therefore \frac{x^2}{4Dt} = \frac{1}{2} ; \; t = \frac{2x^2}{4D} = \frac{x^2}{2D}$

$t = \frac{(0.1)^2}{(2)(4.5 \times 10^{-7})} = 1.11 \times 10^4 \text{ s} = 3.09 \text{ h}$

b. $c = \frac{(0.2)(5 \times 10^{-5})}{2\sqrt{\pi (4.5 \times 10^{-11})(1.11 \times 10^4)}} \exp\left[-\frac{(1 \times 10^{-3})^2}{4(4.5 \times 10^{-11})(1.11 \times 10^4)}\right] = 2.42 \times 10^{-3}$ atom fraction

Maximum concentration at 1mm = 0.00242 atom fraction

c. at $x = 0$, $c = \frac{C_i \Delta x'}{2(\pi D t)^{1/2}} = \frac{(0.2)(5 \times 10^{-5})}{2[\pi (4.5 \times 10^{-11})(1.11 \times 10^4)]^{1/2}} = 0.0040$ atom fration

13.23 A batch of steel exhibits "banding," which is a form of microsegregation in the wrought condition. The spacing between the bands is 50 μm. After 10 h of a high temperature homogenization treatment, the residual segregation index is 0.2 (determined by electron beam microanalysis). A second batch of the same type of steel has a band spacing of 100 μm. How long must this batch be maintained at the high temperature to achieve the same residual segregation index of 0.2?

$$\delta = f\left(\frac{Dt}{L^2}\right)$$

Since $\delta_1 = \delta_2$, then $\frac{D_1 t_1}{L_1^2} = \frac{D_2 t_2}{L_2^2}$, also $D_2 = D_1$

$$\therefore\ t_2 = t_1\left(\frac{L_2}{L_1}\right)^2 = (10\,h)\left(\frac{100\,\mu m}{50\,\mu m}\right)^2 = 40\,h$$

13.24 A melt with uniform concentration of solute, C_∞, is solidified with a planar interface. Chemical equilibrium is maintained at the interface, which moves with a constant velocity V. The concentration profile at steady state, with no convection in the liquid, is as depicted:

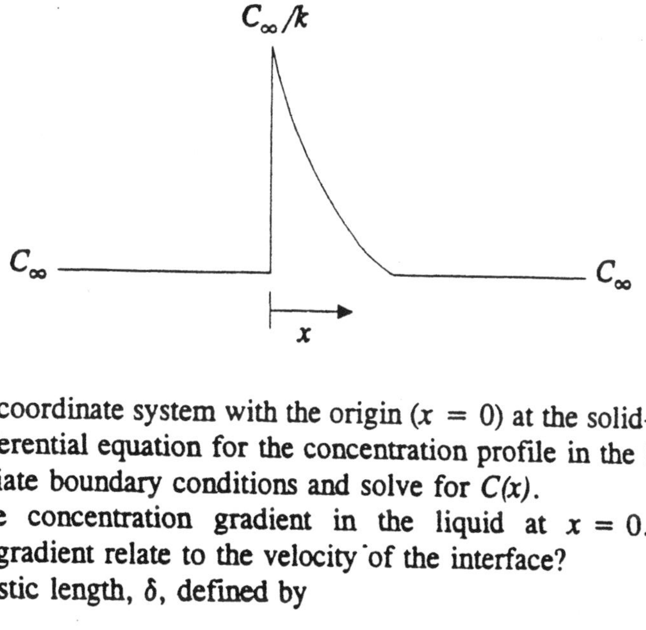

a) Use a moving coordinate system with the origin ($x = 0$) at the solid-liquid interface and derive the differential equation for the concentration profile in the liquid.
b) Write appropriate boundary conditions and solve for $C(x)$.
c) Determine the concentration gradient in the liquid at $x = 0$. How does the concentration gradient relate to the velocity of the interface?
d) The characteristic length, δ, defined by

$$j_0 = \frac{D}{\delta}\left[\frac{C_\infty}{k} - C_\infty\right],$$

where j_0 is the diffusional flux at the interface ($x = 0$). Determine δ in terms of the solidification velocity and the concentration gradient at the interface.

a. The goal is to describe the concentration profile in the liquid.

Steady-state

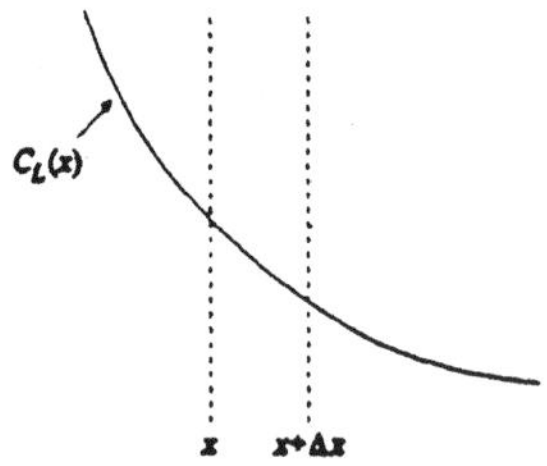

$$j|_x + RC_L|_{x+\Delta x} = j|_{x+\Delta x} + RC_L|_x$$

$$\lim_{\Delta x \to 0} \; - \frac{(j|_{x+\Delta x} - j|_x)}{\Delta x} + R \frac{(C_L|_{x+\Delta x} - C_L|_x)}{\Delta x} = 0$$

where

j = flux, kg m^{-2} s

x = distance from interface, m

R = solidification velocity, m s^{-1}

C_L = concentration, kg m^{-3}

But

$$j = -D \frac{dC_L}{dx}$$

where D = diffusion coefficient in liquid, m^2 s^{-1}

$$\frac{d^2C_L}{dx^2} + \frac{R}{D}\frac{dC_L}{dx} = 0$$

b. B.C.'s: $C_L(0) = C_\infty/k$

$C_L(\infty) = C_\infty$

Solution:

$$\frac{C_L}{C_\infty} = 1 + \left[\frac{1-k}{k}\right] \exp\left[-\frac{R}{D}x\right]$$

c.

$$\left[\frac{dC_L}{dx}\right]_{x=0} = -C_\infty\left[\frac{1-k}{k}\right]\left[\frac{R}{D}\right]$$

d.

$$\delta \approx \frac{R}{D}$$

13.25 A powder-ceramic compact is outgassed at 500°C in a chamber filled with pure argon in order to remove air before sintering. The tortuosity of the compact is 4, its porosity is 0.2, and the average pore radius is 200 Å. The compacts are 50 mm long × 25 mm diameter. Calculate the fraction of air remaining after 1 h of outgassing treatment.

We need D for gas comprising air and argon. From Table 12.10 $v_1 = v(\text{air}) = 20.1$ and $v_2 = v(\text{Ar}) = 16.1$. Then Eq. (12.54) gives:

$$D_{12} = \frac{(1\times10^{-3})(773)^{1.75}}{(1)(20.1^{1/3} + 16.1^{1/3})^2}\left[\frac{1}{28.84} + \frac{1}{39.95}\right]^{1/2} = 1.01 \text{ cm}^{-2}\text{ s}^{-1}$$

Check for Knudsen diffusion:

$$D_K = 9700\, r\sqrt{\frac{T}{M}} = (9700)(200\times10^{-8})\sqrt{\frac{773}{28.84}} = 0.100 \text{ cm}^2\text{ s}^{-1} \text{ (maximum)}$$

$\frac{D_{12}}{D_K} = 10$. Based on this we will use D_K.

$$D_K \text{ (minimum)} = 0.100\sqrt{\frac{39.95}{28.84}} = 0.118 \text{ cm}^2\text{ s}^{-1}$$

Hence, $D_{12} \simeq \bar{D}_K = 0.109 \text{ cm}^2\text{ s}^{-1}$

$$D_{eff} = \frac{D_{12}\,\omega}{\tau} = \frac{(0.109)(0.2)}{4} = 5.45\times10^{-3} \text{ cm}^2\text{ s}^{-1}$$

Now we treat diffusion in a short cylinder

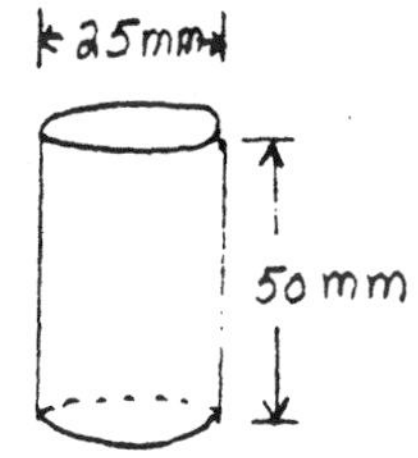

Let c = concentration of air at surface, $C_s = 0$, pure Ar.

Initial concentration, $C_i = 1$, all air.

$$\frac{Dt}{L^2} = \frac{(5.45\times10^{-3})(3600)}{(2.5)^2} = 3.14$$

$$\frac{Dt}{R^2} = \frac{(5.45\times10^{-3})(3600)}{(1.25)^2} = 12.6$$

Fig. 9.8a: $\frac{C - C_s}{C_i - C_s} \simeq 0.01$ and Fig. 9.9a: $\frac{C - C_s}{C_i - C_s} = 0$

∴ After 1 h treatment, the compact is completely outgassed.

13.26 In order to make a transformer steel with the proper hysteresis loop, a low silicon steel sheet (2 mm thick) is to be exposed on both sides to an atmosphere of $SiCl_4$ which dissociates to Si(g) and Cl_2(g). The Si(g) dissolves in the steel up to 3 wt.% at equilibrium. a) Indicate what partial differential equation and what boundary and initial conditions would apply in order to calculate the diffusion of Si into the sheet. b) Using the data in Fig. 12.11, calculate the time to achieve an average concentration of 2.85 wt.% Si at 1255 K.

a. $\frac{\partial c}{\partial t} = D \frac{\partial^2 c}{\partial x^2}$ where $D = \tilde{D}$ = chemical diffusion coefficient of Si in γ; c = concentration of Si; x = position within sheet; t = time.

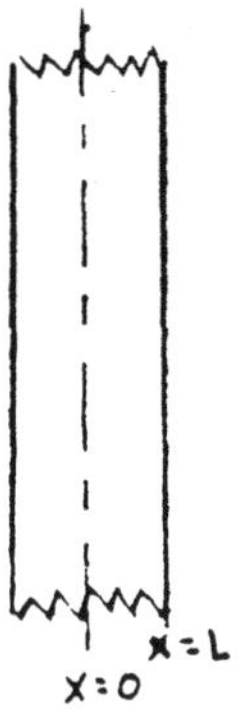

I.C.: $c(x,0) = c_i$ (initial concentration of Si) = 0% Si

B.C.: $c(L,t) = c_s$ (surface concentration of Si) = 3% Si

$\frac{\partial c}{\partial x}(0,t) = 0$

b. $\frac{1}{T} = 7.97 \times 10^{-4}\ K^{-1}$

From Fig. 12.11 $D = 8.2 \times 10^{-13}\ m^2\ s^{-1}$

The solution we seek is from Eq. (13.43) for the average concentration; or Fig. 13.6 can be used with an extrapolation of the "slab" line,

$$\frac{\bar{c} - c_s}{c_i - c_s} = \frac{2.85 - 3.0}{0 - 3.0} = 0.05\ ;\ \frac{Dt}{L^2} \simeq 1.13$$

$$\therefore t = (1.13)\frac{(1 \times 10^{-3})^2}{(8.2 \times 10^{-13})} = 1.378 \times 10^6\ s = 383\ h$$

14.1 A solute is being desorbed from the surface of a melt. The concentration of the solute varies according to

$$\frac{C - C_s}{C_\infty - C_s} = a\left[\frac{x}{\delta_c}\right] + b\left[\frac{x}{\delta_c}\right]^2,$$

where C = concentration, moles cm^{-3}; C_s = concentration at the surface; C_∞ = bulk concentration; a, b = constants; x = distance from the surface; and δ_c = thickness of the concentration boundary layer. What is the mass transfer coefficient?

$$\frac{C-C_s}{C_\infty - C_s} = a\frac{x}{\delta_c} + b\left(\frac{x}{\delta_c}\right)^2 \; ; \; k_M = \frac{j_{Ax}|_{x=0}}{|C_\infty - C_s|}$$

$$j_{Ax} = -D_A\frac{dC}{dx} = -D_A(C_\infty - C_s)\left[\frac{a}{\delta_c} + \frac{2b}{\delta_c^2}x\right]$$

$$j_{Ax}|_{x=0} = -D_A(C_\infty - C_s)\frac{a}{\delta_c}$$

$$\therefore k_M = \frac{D_A(C_\infty - C_s)a}{(C_\infty - C_s)\delta_c} = \frac{a D_A}{\delta_c}$$

14.2 a) Calculate the diffusion coefficient D_{AB} for zinc vapor diffusing through helium at 773 K. The atomic weight of Zn is 65.4 g mol^{-1}. b) Calculate the flux of Zn evaporation if the vapor pressure of Zn at 773 K is 0.1 atm. Express the flux in mol s^{-1} cm^{-2}.

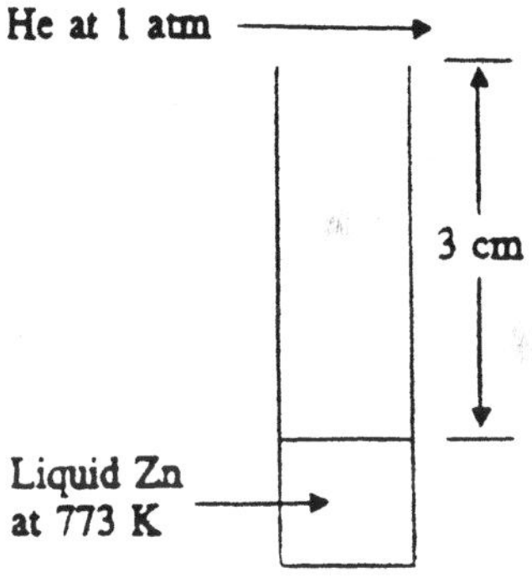

a. Parameters for Zn:

Eq. (12.52) $\frac{\varepsilon}{k_B} = 1.15\,T_b = (1.15)(1179) = 1356$ K

Eq. (12.53) $\frac{\varepsilon}{k_B} = 1.92\,T_M = (1.92)(692.5) = 1330$ K

Use $\frac{\varepsilon}{k_B}$ (avg.) $= 1343$ K ; Table 12.9 $\sigma = 0.5(2.59 + 2.51) = 2.55$ Å

Parameters for He:

Table 1.1 $\frac{\varepsilon}{k_B} = 10.22$ K ; $\sigma = 2.551$ Å

Parameters for gas mixture

$\sigma_{AB} = 0.5(2.55 + 2.551) = 2.55$ Å

$\frac{\varepsilon}{k_B} = \sqrt{(1343)(10.22)} = 117$ K ; $T^* = \frac{k_B T}{\varepsilon} = \frac{773}{117} = 6.61$

$$\Omega_{D,AB} = \frac{1.06036}{(6.61)^{0.15610}} + \frac{0.19300}{\exp[(0.47635)(6.61)]} + \frac{1.03587}{\exp[(1.52996)(6.61)]} + \frac{1.76474}{\exp[(3.89411)(6.61)]}$$

$$= 0.798$$

Eq. (12.50) $$D_{Zn-He} = \frac{0.0018583\,T^{3/2}}{P(\sigma_{AB})^2\,\Omega_{D,AB}}\left(\frac{1}{M_A} + \frac{1}{M_B}\right)^{1/2} = \frac{(0.0018583)(773)^{3/2}}{(1)(2.55)^2(0.798)}\left(\frac{1}{65.4} + \frac{1}{4.0}\right)^{1/2}$$

$$D_{Zn-He} = 3.96\ \text{cm}^2\,\text{s}^{-1} = 3.96\times10^{-4}\ \text{m}^2\,\text{s}^{-1}$$

b. Let A ≙ Zn vapor and B ≙ He: Eq. (14.16) $$(x_B)_{ln} = \frac{x_B^{\ell} - x_B^{o}}{\ln(x_B^{\ell}/x_B^{o})} = \frac{1-0.9}{\ln\left(\frac{1}{0.9}\right)} = 0.949$$

$$C = \frac{P}{RT} = \frac{1.0132\times10^5\ \text{N}}{\text{m}^2}\left|\frac{\text{mol K}}{8.314\ \text{J}}\right|\frac{1}{773\ \text{K}}\left|\frac{1\ \text{J}}{1\ \text{N m}}\right. = 15.76\ \text{mol m}^{-3}$$

Eq. (14.15) $$N_{Ax|x=0} = \frac{CD_A}{(x_B)_{ln}}\left(\frac{x_A^{o} - x_A^{\ell}}{\ell}\right) = \frac{(15.76)(3.96\times10^{-4})(0.1-0)}{(0.949)(0.03)} = 2.19\times10^{-2}\ \text{moles m}^{-2}\,\text{s}^{-1}$$

$$N_{Ax|x=0} = 2.19\times10^{-2}\ \text{moles m}^{-2}\,\text{s}^{-1}$$

14.3 Silicon can be grown by the chemical vapor deposition of silane (SiH_4). Given a system in which silicon grows from the bottom of an inert crucible as shown adjacent, predict the growth rate of silicon. Assume that at the gas/silicon interface equilibrium is established according to the reaction

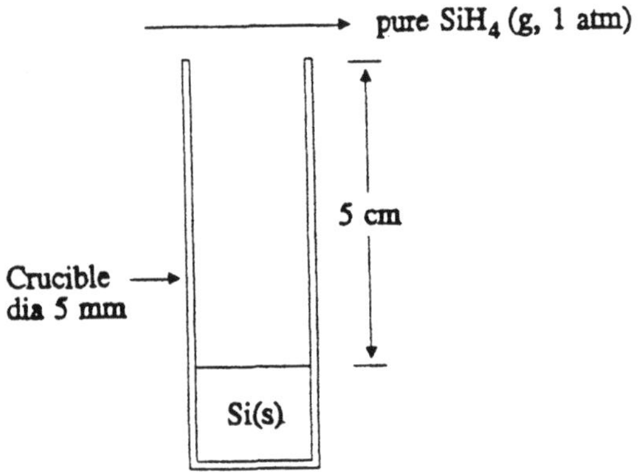

$$SiH_4(g) = Si(s) + 2H_2(g),$$

with the equilibrium constant given by

$$K = \frac{P_{H_2}^2}{P_{SiH_4}} = 10^{-4}.$$

It is also known that $D_{H_2\text{-}SiH_4} = 1\ \text{cm}^2\,\text{s}^{-1}$, and the temperature is 700 K.

Assume equilibrium at the vapor-liquid interface

$$K = \frac{P_{H_2}^2}{P_{SiH_4}} = 10^{-4}$$; also $P_{H_2} + P_{SiH_4} = 1$ atm.

Obviously the gas is very dilute in H_2, so $P_{H_2}^{o} \simeq 10^{-2}$ atm

$x_{H_2}^{o} = 0.01$ and $x_{SiH_4}^{o} = 0.99$

Because the gas is so dilute, we can write the flux of $H_2^{(A)}$ as

$$j_{H_2} = \left|-D\frac{C_A^{\ell} - C_A^{o}}{\ell}\right| = \left|-DC\frac{x_A^{\ell} - x_A^{o}}{\ell}\right| ;\ C = \frac{P}{RT} = \frac{1.0132\times10^5}{(8.314)(700)} = 17.41\ \text{mol m}^{-3}$$

$$j_{H_2} = (1\times10^{-4})(17.41)\left(\frac{0.01-0}{0.05}\right) = 3.48\times10^{-4} \text{ mol s}^{-1}$$

The stoichiometry of the reaction requires $j_{Si} = \frac{j_{H_2}}{2} = 1.74\times10^{-4} \text{ mol s}^{-1}$

The growth rate:

$$A\frac{d\delta}{dt} = \frac{1.74\times10^{-4}\text{ mol}}{\text{s}}\left|\frac{28.09\text{ g}}{1\text{ mol}}\right|\frac{1\text{ kg}}{1000\text{ g}}\left|\frac{1\text{ m}^3}{2300\text{ kg}}\right. = 2.12\times10^{-9}\text{ m}^3\text{ s}^{-1}$$

$$\frac{d\delta}{dt} = \frac{(4)(2.12\times10^{-9})}{(\pi)(5\times10^{-3})^2} = 1.08\times10^{-4}\text{ m s}^{-1} = 108\ \mu\text{m s}^{-1}$$

This hypothetical problem gives a rate which is 3 to 4 times greater than found in practice.

14.4 Consider diffusion through a stagnant gas as in Section 14.1; however, in this case the gas is not isothermal, but rather temperature varies according to

$$\frac{T - T_0}{T_\ell - T_0} = \frac{x}{\ell},$$

where x and ℓ are indicated in Fig. 14.1, T is temperature, T_0 is the temperature at $x = 0$, and T_ℓ is the temperature at $x = \ell$. Assume that D_{AB} varies with temperature according to Eq. (12.49); i.e., $D_{AB} = AT^{3/2}$ and A is constant. Obtain the concentration profile which gives the mole fraction of A (X_A) as a function of distance up the tube (x).

Start with Eq. 14.8 $\frac{d}{dx}\left[\frac{C D_{AB}}{1-X_A}\frac{dX_A}{dx}\right] = 0$

$C = \frac{P}{RT}$ but T varies with x, as does D_A

$$\frac{d}{dx}\left[\frac{D_{AB}}{T}\frac{1}{1-X_A}\frac{dX_A}{dx}\right] = 0$$

Integrate and substitute $\frac{D_{AB}}{T} = AT^{1/2}$

$$T^{1/2}\left(\frac{1}{1-X_A}\right)\frac{dX_A}{dx} = C_1$$

but $T = (T_\ell - T_0)\frac{x}{\ell} + T_0$; $T^{1/2} = \left[(T_\ell - T_0)\frac{x}{\ell} + T_0\right]^{1/2}$

$$\frac{dX_A}{(1-X_A)} = C_1\frac{dx}{\left[(T_\ell - T_0)\left(\frac{x}{\ell}\right) + T_0\right]^{1/2}}$$

Integrate again

$$-\ln(1-X_A) = \frac{2C_1}{T_\ell - T_0}\left[(T_\ell - T_0)\frac{x}{\ell} + T_0\right]^{1/2} + C_2$$

or

$$\ln(1-X_A) = c_1'\left[(T_\ell - T_0)\frac{x}{\ell} + T_0\right]^{1/2} + c_2'$$

at $\frac{x}{\ell} = 0$, $X_A = X_A^0$; $\frac{x}{\ell} = 1$, $X_A = X_A^\ell$

$$\ln(1-X_A^0) = c_1' T_0^{1/2} + c_2' \; ; \; \ln(1-X_A^\ell) = c_1' T_\ell^{1/2} + c_2'$$

$$c_1' = \frac{\ln(1-X_A^0) - \ln(1-X_A^\ell)}{T_0^{1/2} - T_\ell^{1/2}} = \frac{1}{T_0^{1/2} - T_\ell^{1/2}} \ln\left(\frac{1-X_A^0}{1-X_A^\ell}\right)$$

and

$$c_2' = \frac{\ln(1-X_A^0)}{c_1' T_0^{1/2}} = \frac{\ln(1-X_A^\ell)}{c_1' T_\ell^{1/2}}$$

14.5 Derive expressions for diffusion through a spherical shell that are analogous to Eq. (14.12) (concentration profile) and Eq. (14.14) (molar flux).

a. B.C.: at $r = R_0$, $X_A = X_A^0$; $r = R_\ell$, $X_A = X_A^\ell$

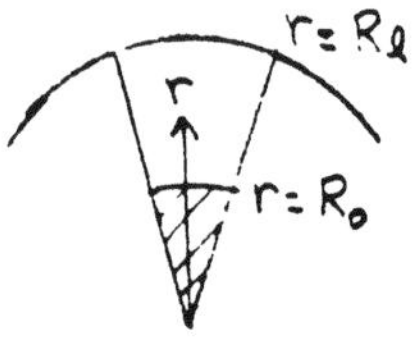

Flux of A is $N_{Ar} = X_A(N_{Ar} + N_{Br}) - cD_A \frac{dX_A}{dr}$

At steady state $N_{Br} = 0$ and $\frac{dX_A}{dr} = \frac{dX_A}{dr}$

Therefore, $N_{Ar} = -\frac{cD_A}{1-X_A}\frac{dX_A}{dr}$ (1)

Mass balance on A: $4\pi r^2 N_{Ar}\big|_r = 4\pi r^2 N_{Ar}\big|_{r+\Delta r}$ or $\frac{d}{dr}(r^2 N_{Ar}) = 0$

$\therefore r^2 N_{Ar} = C_1$ or, by substituting (1), $\frac{r^2}{1-X_A}\frac{dX_A}{dr} = C_1$

$$\frac{dX_A}{1-X_A} = C_1 \frac{dr}{r^2}$$

$\ln(1-X_A) = C_1 \frac{1}{r} + C_2$ From the boundary conditions:

$$\ln(1-X_A^0) = \frac{C_1}{R_0} + C_2$$

$$\ln(1-X_A^\ell) = \frac{C_1}{R_\ell} + C_2$$

$$\ln\frac{1-X_A^0}{1-X_A^\ell} = C_1\left(\frac{1}{R_0} + \frac{1}{R_\ell}\right)$$

$$C_1 = \left(\frac{1}{R_0} + \frac{1}{R_\ell}\right)^{-1} \ln\frac{1-X_A^0}{1-X_A^\ell} \; ; \; C_2 = \ln(1-X_A^0) - \frac{C_1}{R_0}$$

$$\ln(1-X_A) = C_1\frac{1}{r} + \ln(1-X_A^0) - C_1\frac{1}{R_0}$$

$$\ln(1-X_A) = C_1\left(\frac{1}{r} - \frac{1}{R_0}\right) + \ln(1-X_A^0)$$

$$\ln \frac{1-X_A}{1-X_A^0} = \left(\frac{1}{R_0} + \frac{1}{R_\ell}\right)^{-1} \ln \frac{1-X_A^0}{1-X_A^\ell} \left(\frac{1}{r} - \frac{1}{R_0}\right)$$

To obtain the flux at $r = R_0$

$$\frac{dX_A}{dr} = C_1 (1-X_A) \frac{1}{r^2} \quad ; \quad \left.\frac{dX_A}{dr}\right|_{r=R_0} = C_1 \frac{(1-X_A^0)}{R_0^2}$$

Then $\left.N_{Ar}\right|_{r=R_0} = -\frac{C D_A}{1-X_A^0} C_1 \frac{(1-X_A^0)}{R_0^2}$ or $\left.N_{Ar}\right|_{r=R_0} = \frac{-C_1 C D_A}{R_0^2}$

$$\therefore \left.N_{Ar}\right|_{r=R_0} = -\frac{C D_A}{R_0^2 \left(\frac{1}{R_0} + \frac{1}{R_\ell}\right)} \ln \frac{1-X_A^0}{1-X_A^\ell}$$

14.6 Liquid flows in one direction, from $x = 0$ to $x = L$, with a velocity V_M. Because of the "no-slip" condition, the velocity of the gas along the surface $y = 0$ is also V_M. When the gas crosses the plane at $x = 0$, its concentration of A is uniform and constant, $C_{A\infty}$. However, at $y = 0$ and $0 \leq x \leq L$, its concentration is C_A^0.

Assume steady state. a) Write a partial differential equation for $C_A(x,y)$ in the gas. b) Write boundary conditions for C_A that are needed to solve part a). c) Solve for $C_A(x,y)$.

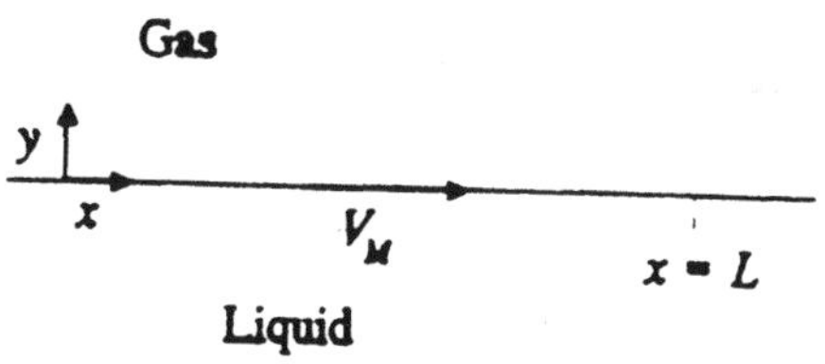

a. Assume fully developed flow and $v_y = 0$. Also diffusion in y-direction dominates, so that $\frac{\partial^2 C_A}{\partial Y^2} \gg \frac{\partial^2 C_A}{\partial x^2}$ So we neglect $\frac{\partial^2 C_A}{\partial x^2}$.

$$V_x \frac{\partial C_A}{\partial x} = D_A \frac{\partial^2 C_A}{\partial Y^2}$$

b. $C_A(0,Y) = C_{A\infty}$

$C_A(x,0) = C_A^0$

$C_A(x,\infty) = C_{A\infty}$

velocity varies from V_M at $Y=0$ to $V_x=0$ at $Y=\infty$

c. For short contact time, we approximate $V_x = V_M$ (constant). Then the solution is Eq. (14.38).

$$\frac{C_A - C_A^0}{C_A^\infty - C_A^0} = erf \frac{Y}{2\left(\frac{D_A x}{V_M}\right)}$$

14.7 At 1000°C metal A is soluble in liquid B but B is not soluble in solid A as shown in the pertinent part of the phase diagram. A 5 cm diameter cylinder of A is rotated at 1000 revolutions per minute (rpm) in a large melt of 0.5 atom fraction B at 1000°C, and it is noted that after 15 minutes the bar diameter is 4.8 cm. For the same temperature, estimate the bar diameter after 15 minutes if another 5 cm diameter cylinder of A is rotated in a large melt of the same composition, but now it is rotated at 4000 rpm. By analogy with Eq. (14.78) assume that

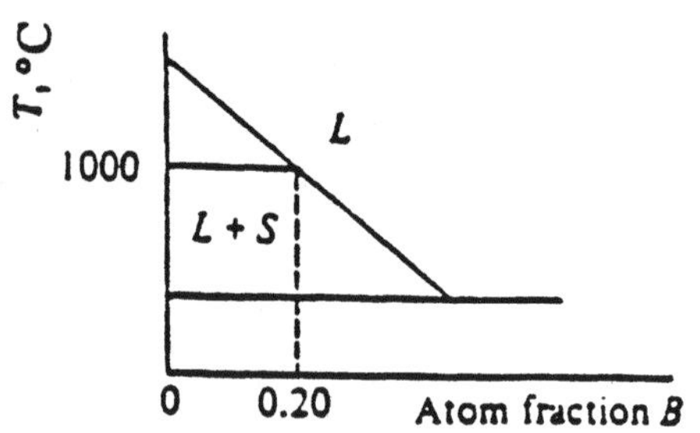

$$Sh_R = (\text{constant})\ Sc^{0.343}\ Re_R^{1/2},$$

where R is the radius of the bar, and the velocity in the Reynolds number is the tangential velocity of the rotating bar.

The melt of B is large enough that as the solid A dissolves, the composition of the melt does not change.

$j_A = k_{M,A}(C_A^* - C_A^\infty)$ where C_A^* = concentration of A in the melt at the S/L interface and $C_A^\infty \simeq 0$ because the melt is large.

For the two cases, $C_A^* - C_A^\infty$ remains the same.

$$\frac{j_{A1}}{j_{A2}} = \frac{k_{M,A1}}{k_{M,A2}} \quad \text{but} \quad \frac{k_{M,A1}}{k_{M,A2}} = \frac{Sh_1}{Sh_2} = \left(\frac{Re_{R1}}{Re_{R2}}\right)^{1/2} = \left(\frac{1000}{4000}\right)^{1/2} = \frac{1}{2}$$

For small changes in diameters and with the same times, the changes in diameter are proportional to the fluxes:

Hence, $\frac{\Delta D_1}{\Delta D_2} = \frac{1}{2}$ or $\Delta D_2 = 2\Delta D_1 = 2(0.2) = 0.4$ cm

$\therefore D_2$ (after 15 min.) = 4.6 cm

14.8 Use dimensional analysis to show:

a) Sh = f(Re, Sc) for forced convection;

b) Sh = f(Gr, Sc) for natural convection.

a. The functional relationship is $k_m = f(\bar{v}, D, \rho, \eta, x)$ where x is the geometrical term. The dimensions of each are:

$\bar{v}: L\theta^{-1}$, $\rho: ML^{-3}$, $k_m: L\theta^{-1}$, $\eta: ML^{-1}\theta^{-1}$, $D: L^2\theta^{-1}$, $x: L$

There are 3 fundamental units ∴ 3 primary variables.

Choosing ρ, D, x as primary variables (arbitrary), then $k, \eta, \bar{\nu}$ are the secondary variables.

Forming the π's, $\pi_1 = \frac{k}{\rho^a D^b x^c} \Rightarrow \frac{L\theta^{-1}}{(ML^{-3})^a (L^2\theta^{-1})^b (L)^c}$

L: $1 = -3a + 2b + c$

M: $0 = a$

θ: $-1 = -b$

Thus, $a = 0$, $b = 1$, $c = -1$ and $\pi_1 = \frac{kx}{D}$ Sherwood No.

Similarly, $\pi_2 = \frac{\eta}{\rho D}$ Schmidt No.

Then $\pi_3 = \frac{Vx}{D}$. This is not recognizable as the Reynolds No., but inspection shows that the dimensions of D and ν are both $L^2\theta^{-1}$, and therefore, substitution of ν for D in π_3 will result in another dimensionless number. $\pi_3' = \frac{Vx}{\nu}$ Reynolds No.

b. The fundamental relationship is $k_m = f(g, D, \rho, \eta, \xi, \Delta c, x)$ – $\xi = \left(\frac{\partial \rho}{\partial c}\right)\frac{1}{\rho}$

$\Delta c: ML^{-3}$, $g: L\theta^{-2}$, $x: L$, $k_m: L\theta^{-1}$, $D: L^2\theta^{-1}$, $\rho: ML^{-3}$, $\eta: ML^{-1}\theta^{-1}$, $\xi: M^{-1}L^3$

There are 3 fundamental units ∴ 3 primary variables.

Choosing ρ, η, x as primary variables, then $k_m, \xi, D, g, \Delta c$ are the secondary variables.

Forming the π's, $\pi_1 = \frac{k_m}{\rho^a \eta^b x^c} \Rightarrow \frac{L\theta^{-1}}{(ML^{-3})^a (ML^{-1}\theta^{-1})^b L^c}$

L: $1 = -3a - b + c$

M: $0 = a + b$

θ: $-1 = -b$

Thus, $a = -1$, $b = 1$, $c = -1$

∴ $\pi_1 = \frac{k\rho x}{\eta} = \frac{kx}{\nu}$; simarly $\pi_2 = \xi\rho$; $\pi_3 = \frac{D\rho}{\eta}$; $\pi_4 = \frac{gx^3\rho^2}{\eta^2} = \frac{gx^3}{\nu^2}$;

$\pi_5 = \frac{\Delta c}{\rho}$

By inspection, π_3 = Schmidt No., and, as in (a.), substitution of D for ν in π_1 will result in π_1 = Sherwood No.

The remaining three dimensionless Π's may be multiplied together to obtain another dimensionless number:

$$\Pi_3 \cdot \Pi_4 \cdot \Pi_5 = \Pi_6 = (\xi\rho)\left(\frac{gx^3}{\nu^2}\right)\left(\frac{\Delta c}{\rho}\right) = \frac{g\xi\Delta c x^3}{\nu^2}$$ = Grashoff No. for mass transfer.

14.9 Levitation melting is a means of supporting a metallic melt by an electromagnetic field. No impurities are added in melting and operation under an inert atmosphere removes dissolved gases. At 1920 K and 1 atm hydrogen pressure, the solubility of hydrogen in iron is 31 cm^3 per 100 g of iron. Estimate the rate at which hydrogen can be removed from a levitated drop of iron that initially contains 10 ppm in the set-up shown to the right. Assume that no convection occurs within the iron drop, and that the gas temperature is 1920 K so that Eq. (8.11), in mass-transfer form, applies.

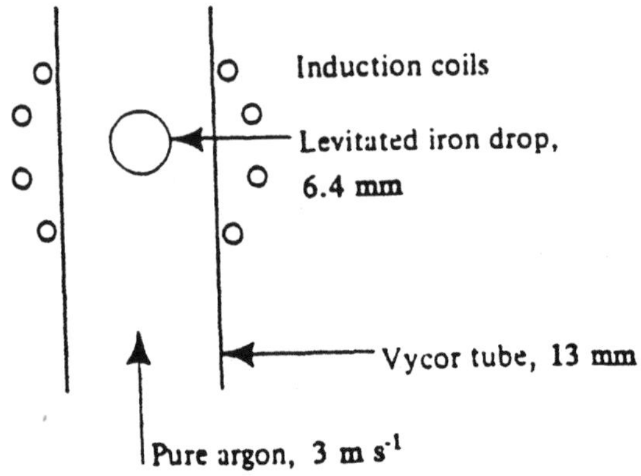

$$\frac{2Rk_m}{D_{AB}} = 2.0 + 0.60\left(\frac{2RV_\infty\rho}{\eta}\right)^{1/2} Sc^{1/3}$$ This is Eq. (8.11) in mass transfer terms, with the diameter written as $2R$ to avoid confusion with the diffusion coefficient, D_{AB}.

Solubility of $\underline{H}$ in Fe with $P_{H_2} = 1$ atm.

$$C_{\underline{H}} = \frac{31\ cm^3\ H_2(STP)}{100\ g\ Fe}\left|\frac{6.89\ g\ Fe}{1\ cm^3\ Fe}\right|\frac{1\ mol\ H_2}{22{,}414\ cm^3\ H_2(STP)}\left|\frac{2\ mol\ \underline{H}}{1\ mol\ H_2}\right. = 1.906\times10^{-4}\ mol\ \underline{H}\ cm^{-3}\ Fe$$

From this we obtain equilibrium constant, K, for the reaction $\frac{1}{2}H_2(g) = \underline{H}$ where $\underline{H}$ represents dissolved hydrogen. $K = \frac{C_{\underline{H}}}{P_{H_2}^{1/2}} = 1.906\times10^{-4}\ mol\ \underline{H}\ cm^{-3}\ (27.66\ ppm\ \underline{H})$

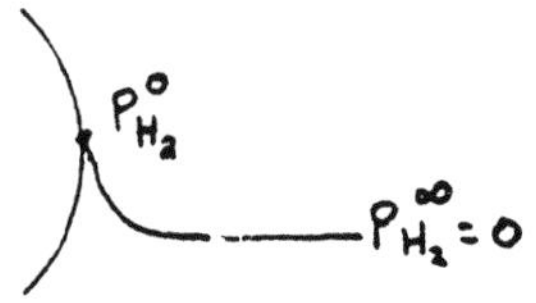

$$\frac{\sqrt{P^\circ_{H_2}}}{10\ ppm} = \frac{\sqrt{1\ atm}}{27.66\ ppm}$$ or $P^\circ_{H_2} = 0.131$ atm

Mass transfer in the gas phase

$J\left(\frac{moles}{s}\right) = k_M A (c^o_{H_2} - c^\infty_{H_2}) = \left(\frac{k_M A}{RT}\right)(p^o_{H_2} - p^\infty_{H_2})$

Diffusion Coefficient, Eq. (12.54)

$$D_{AB} = \frac{(1\times10^{-3})(1920)^{1.75}}{(1)(7.07^{1/3} + 16.1^{1/3})^2}\left(\frac{1}{39.95} + \frac{1}{2}\right)^{1/2} = 20.4\ cm^2 s^{-1} = 2.04\times10^{-3}\ m^2 s^{-1}$$

Viscosity (depends on composition of gas)

$p_{H_2}(avg.) = \frac{1}{2}(0.131 + 0) = 6.54\times10^{-2}$ atm.

$\therefore X_{H_2} = 0.0654$ and $X_{Ar} = 0.9346 : \eta = \eta_{MIX} \simeq X_{H_2}\eta_{H_2} + X_{Ar}\eta_{Ar}$

Fig. 1.7 can be used for estimating the viscosities of H_2 and Ar.

Then $\eta = (0.0654)(3\times10^{-4}) + (0.9346)(7.2\times10^{-4}) = 6.91\times10^{-4} P = 6.91\times10^{-5}\ N\ s\ m^{-2}$

Density

$\bar{M}$ (avg. mol. wgt.) $= (0.0654)(2) + (0.9346)(39.95) = 37.47\ g\ mol^{-1} = 3.747\times10^{-2}\ kg\ mol^{-1}$

$$\rho = \frac{P}{RT} = \frac{1.0132\times10^5\ N}{m^2}\left|\frac{mol\ K}{8.314\ J}\right|\frac{1}{1920\ K}\left|\frac{0.03747\ kg}{1\ mol}\right| = \frac{0.2378\ N\ kg}{m^2\ J}\left|\frac{1\ J}{1\ Nm}\right.$$

$$= 0.2378\ kg\ m^{-3} = 2.378\times10^{-4}\ g\ cm^{-3}$$

Schmidt No.

$$Sc = \frac{\nu}{D_{AB}} = \frac{6.91\times10^{-5}}{(0.2378)(2.06\times10^{-3})} = 0.141$$

Now we have enough data to evaluate the Sherwood No. and the mass transfer coefficient.

$$Sh = \frac{2k_M R}{D_{AB}} = 2.0 + 0.60\left[\frac{(6.4\times10^{-3})(3)(0.2378)}{(6.91\times10^{-5})}\right]^{1/2}(0.141)^{1/3} = 4.54$$

$$\therefore k_M = \frac{(4.54)(20.6)}{0.64} = 146\ cm\ s^{-1} = 1.46\ m\ s^{-1}$$

Initial rate of loss of hydrogen

$-V\frac{dc_H}{dt} = \frac{k_M A}{RT}(p^o_{H_2} - p^\infty_{H_2})$ where V is volume of droplet.

Then $\frac{dc_H}{dt} = \left(\frac{k_M}{RT}\right)\left(\frac{A}{V}\right)(p^o_{H_2} - p^\infty_{H_2}) : \frac{A}{V} = \frac{3}{R} = \frac{3}{3.2\ mm} = 9.375\ cm^{-1} = 937.5\ m^{-1}$

$$\frac{dC_H}{dt} = -\frac{146\ m}{s}\left|\frac{mol\ K}{8.205\times10^{-5}\ m^3\ atm}\right|\frac{}{1920\ K}\left|\frac{937.5}{m}\right|(0.131-0)\ atm$$

$$= -\frac{1.14\times10^3\ mol}{m^3\ s}\left|\frac{1\ m^3}{6890\ kg}\right|\frac{1\ g\ H}{1\ mol}\left|\frac{1\ kg}{1000\ g}\right|\frac{10^6\ g\ \ 1\ ppm}{1\ g\ H} \simeq 165\ ppm\ s^{-1}$$

$$\frac{dC_H}{dt} \simeq -165\ ppm\ s^{-1}$$

Notice that this is an extremely rapid rate. This tells us that the mass transfer in the gas phase is very high, and the rate limiting step would be the mass transfer in the liquid. In this problem, we assume no convection in the drop, so that loss of hydrogen is by diffusion only.

Examine the mass transfer Bi number. Estimate $D_H = 10^{-5}$ to 10^{-4} $cm^2\ s^{-1}$ in the liquid.

$$Bi_M(MAX) = \frac{k_M R}{D_H} = \frac{(146)(0.32)}{10^{-5}} = 4.67\times10^6$$

$$Bi_M(Min) = 4.67\times10^5$$

For certain, the concentration in the liquid at the interface drops almost immediately to zero, corresponding to $p_{H_2}^{\circ} = p_{H_2}^{\infty} = 0$ atm.

Therefore, Fig. 13.6 applies.

To drop the average concentration of $\underline{H}$ in the drop to 1 ppm, we have

$$\frac{\bar{C}-C_s}{C_i-C_s} = \frac{1-0}{10-0} = 0.1:\ \text{Thus}\ \frac{D_H t}{R^2} \simeq 0.24$$

$$t = \frac{(0.24)(0.32)^2}{10^{-5}}\ \text{or}\ \frac{(0.24)(0.32)^2}{10^{-4}}$$

So $246 < t < 2460$ s depending on $D_{\underline{H}}$

14.10 Hydrogen gas is being absorbed from a gas in an experimental set-up shown in the figure to the right. The absorbing liquid is aluminum at 1030 K which is falling in laminar flow with an average velocity of 2.5 mm s^{-1}.

What is the hydrogen content of the aluminum leaving the tube if it enters with no hydrogen? At T = 1030 K and 1 atm hydrogen pressure, the solubility of hydrogen is 1 cm^3 per 100 g of aluminum, the density of Al = 2.5 g cm^{-3}, and $D_H = 1 \times 10^{-9}$ m^2 s^{-1}.

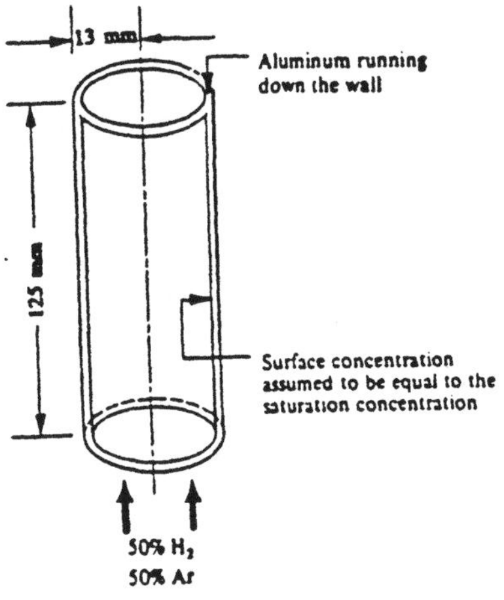

At 1030 K, $\eta = 3.7 \times 10^{-3}$ N s m^{-2} (50% H_2 − 50% Ar, Fig. 1.7)

Film thickness, from Eq. (2.14)

$$\delta^2 = \frac{3\eta\bar{V}}{\rho g} = \frac{3 \mid 3.7\times10^{-3}\,\mathrm{N\,s} \mid s^2 \mid m^3 \mid 2.5\times10^{-3}\,m}{\mid m^2 \mid 9.81\,m \mid 2500\,kg \mid s} = 1.13\times10^{-9}\,m^2$$

$$\text{Now evaluate } \frac{D_A L}{\delta^2 V_{MAX}} = \frac{2}{3}\frac{D_A L}{\delta^2 \bar{V}} = \frac{2 \mid 1\times10^{-9}\,m^2 \mid 0.125\,m \mid \quad \mid s}{3 \mid s \mid \quad \mid (1.13\times10^{-9})^2\,m^2 \mid 2.5\times10^{-3}\,m}$$

$= 26.1 \times 10^9$ Long time solution is in order.

$$\text{Eq. (14.44)} \quad \frac{\bar{C}_A^L - C_A^o}{C_A^i - C_A^o} = 0.7857 \exp\left[(-5.1213)(26.1\times10^9)\right] = 0$$

$$\bar{C}_A^L = C_A^o$$

For $H_2(g) = 2\underline{H}$; K = 1 cm^3 per 100g of Al at 1atm.

At $P_{H_2} = 0.5$ atm; $\bar{C}_A^L = C_A^o = (K P_{H_2})^{1/2} = \left[(1)(0.5)\right]^{1/2} = 0.707$ cm^3 per 100g of Al

$\bar{C}_A^L \simeq 7 \times 10^{-5}$ wt. Pct.

14.11 When ceramic oxides are used to contain molten metals, they can dissolve and add undesirable impurities to the melt. This is especially true when melting is done under vacuum. For example, magnesium oxide decomposes (slowly) according to

$$MgO(s) \rightarrow \underline{Mg} + \underline{O},$$

where the underlines indicate that the elements are dissolved in the melt. The equilibrium constant for the reaction is

$$K = C_{Mg}\, C_O = 10^{-6},$$

where C_{Mg} and C_O are the concentrations (moles per cm^3 of melt) of $\underline{Mg}$ and $\underline{O}$, respectively.

a) For flow parallel to a plate of MgO, calculate the average mass transfer coefficient for Mg dissolving in the melt. Assume that $C_{Mg} = C_O$. For $x < 0$, $C_{Mg} = 0$ in the melt and $V_\infty = 3 \text{ m s}^{-1}$.

b) What is the average flux of MgO dissolving in the melt? *Melt properties*: $\rho = 8000 \text{ kg m}^{-3}$; $k = 50 \text{ W m}^{-1} \text{ K}^{-1}$; $C_p = 840 \text{ J kg}^{-1} \text{ K}^{-1}$; $\eta = 1.24 \times 10^{-3} \text{ N s m}^{-2}$; and $D_O = D_{Mg} = 5 \times 10^{9} \text{ m}^2 \text{ s}^{-1}$.

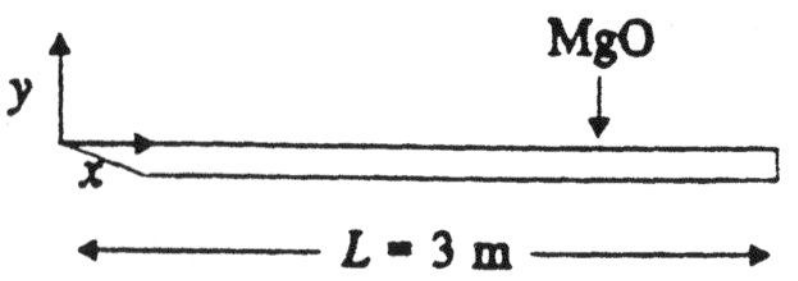

a. $Sc = \frac{\nu}{D} = \frac{\eta}{\rho D} = \frac{1.24 \times 10^{-3}}{(8000)(5 \times 10^{-9})} = 31$

$Re_L = \frac{L V_\infty \rho}{\eta} = \frac{(3)(3)(8000)}{1.24 \times 10^{-3}} = 5.81 \times 10^{7}$ The flow is turbulent so we should use the mass transfer equivalent of Eq. (8.16).

$Sh_L = 0.037\, Re_L^{0.8}\, Sc^{1/3} = (0.037)(5.81 \times 10^{7})^{0.8}(31)^{1/3} = 1.890 \times 10^{5}$

$k_M = (1.890 \times 10^{5}) \frac{D}{L} = \frac{(1.890 \times 10^{5})(5 \times 10^{-9})}{3} = 3.15 \times 10^{-4} \text{ m s}^{-1}$

b. $C_{Mg} = C_O$ and $D_O = D_{Mg}$ so we can calculate either flux.

$j_{Mg} = k_M (C^{o}_{Mg} - C^{\infty}_{Mg})$ but $C^{o}_{Mg} C^{o}_{O} = C^{o\,2}_{Mg} = 10^{-6}$ $\therefore$ $C^{o}_{Mg} = 10^{-3} \text{ mol Mg m}^{-3}$

$j_{Mg} = \frac{3.15 \times 10^{-4} \text{ m}}{\text{s}} \Big| \frac{(10^{-3} - 0) \text{ mol Mg}}{\text{m}^3} = 3.15 \times 10^{-7} \text{ mol Mg m}^{-2} \text{s}^{-1}$

14.12 A common procedure for decreasing the hydrogen content of a melt is to allow bubbles of an inert gas (*viz.*, He) to rise through the melt. Assume that Eq. (8.11), in a form suitable for mass transfer, applies. a) Estimate the mass transfer coefficient from the melt to a single bubble for the removal of dissolved hydrogen. b) Will the overall kinetics of hydrogen removal from the melt depend on mass transfer in the liquid, in the bubble, or both? Justify your answer. c) Derive an equation that can be used to calculate the concentration of dissolved hydrogen as a function of time. *Data*: bubble diameter, 2 mm; melt density, 7000 kg m^{-3}; melt viscosity, 2×10^{-3} N s m^{-2}; diffusivity (hydrogen in melt), 5×10^{-9} m^2 s^{-1}; and diffusivity (hydrogen in gas bubble), 10^{-4} m^2 s^{-1}.

Eq. (8.11) in mass transfer form is $Sh_D = 2.0 + 0.60\, Re_D^{1/2}\, Sc^{1/3}$

a. $Sc = \frac{\nu}{D_H} = \frac{\eta}{\rho D_H} = \frac{2 \times 10^{-3} \text{ N s}}{\text{m}^2} \Big| \frac{\text{m}^3}{7000 \text{ kg}} \Big| \frac{\text{s}}{5 \times 10^{-9} \text{ m}^2} \Big| \frac{\text{kg m}}{\text{N s}^2} = 57.1$

For V_∞ we can not use Stokes' law $\therefore$ Use Fig. 3.8. Assume $f = 0.4$

Then $V_\infty = \left[\frac{8Rg(\rho - \rho_{gas})}{3f\rho}\right]^{1/2} = \left[\frac{(8)(0.001)(9.81)(7000)}{(3)(0.4)(7000)}\right]^{1/2} = 0.256 \text{ m s}^{-1}$

$$Re_D = \frac{D V_\infty \rho}{\eta} = \frac{(0.002)(0.256)(7000)}{2 \times 10^{-3}} = 1.79 \times 10^{-3}$$

Hence, assumption that $f = 0.4$ is satisfactory.

$$Sh_D = 2.0 + (0.60)(1.79 \times 10^{-3})^{1/2}(57.1)^{1/3} = 99.8$$

$$\therefore \frac{k_M D}{D_H} = 99.8 \Rightarrow k_M = \frac{(99.8)(5 \times 10^{-9})\, m^2}{s \; 2 \times 10^{-3}\, m} = 2.49 \times 10^{-4} \ m\ s^{-1}$$

$$k_M = 2.49 \times 10^{-4} \ m\ s^{-1}$$

b. $$Bi_M = \frac{k_M D}{D_H} = \frac{2.49 \times 10^{-4}\, m \; 2 \times 10^{-3}\, m \; s^{-1}}{s \; 10^{-4}\, m^2} = 4.99 \times 10^{-3}$$

(D_H ← in gas)

This indicates that even with no convection in the bubble (unlikely), that the rate of diffusion in the bubble is very high. Therefore, mass transfer is not limited within the bubble itself. The overall kinetics are controlled by mass transfer in the liquid.

c. $$\left\{\begin{matrix}\text{Mass transfer of hydrogen} \\ \text{from melt to the bubble}\end{matrix}\right\} = \left\{\begin{matrix}\text{Decrease of hydrogen} \\ \text{in the melt}\end{matrix}\right\}$$

$$k_M A (C_H^\infty - C_H^*) = -V \frac{dC_H^\infty}{dt}$$

Where k_M = Mass transfer coefficient in the liquid

A = Surface area of the bubble

V = Volume of the melt

C_H^∞ = Concentration of $\underline{H}$ in the melt, mol m^{-3}

C_H^* = Concentration of $\underline{H}$ at the bubble/melt interface, mol m^{-3}

C_H^* can be determined from the thermodynamics involving the pressure of hydrogen in the bubble and the appropriate equilibrium constant. In turn the pressure in the bubble as it rises through the melt will be controlled by the mass transfer of hydrogen in the melt. Hence, C_H^* will change as the bubble rises. A numerical solution is required, in order to estimate the hydrogen removed from the melt when the bubble leaves the melt.

14.13 Consider an electrodeposition process that is controlled by mass transfer at the anode. Calculate the average mass transfer coefficient for the dissolution of the copper anode assuming that the process is controlled by the flux of cupric ions. Assume natural convection and an isothermal solution. *Data*: Mole fraction of Cu^{2+} ions at surface is 0.05; in the bulk it is 0.01. Properties of solution: $\rho = 1200 \text{ kg m}^{-3}$, $\eta = 2 \times 10^{-3} \text{ N s m}^{-2}$, $D = 5 \times 10^{-9} \text{ m}^2 \text{ s}^{-1}$, and $\xi = 4$.

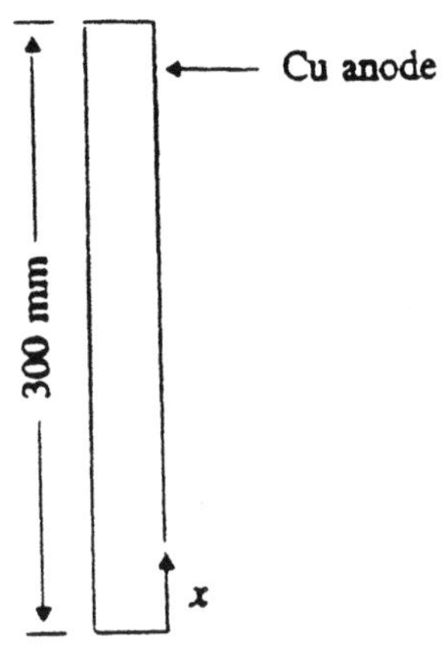

$$Gr_M = g\xi(x_A^{\circ} - x_{A\infty})\frac{L^3}{\nu^2} \; ; \; \nu = \frac{\eta}{\rho} = \frac{2\times10^{-3}}{1200} = 1.667\times10^{-6}\ m^2 s^{-1}$$

$$Gr_M = \frac{9.81\ m}{s^2}\Big|\,4\,\Big|\,0.04\,\Big|\frac{(0.3)^3\ m^3}{}\,\frac{s^2}{(1.667\times10^{-6})^2\ m^4} = 1.525\times10^{10}$$

$$Sc = \frac{\nu}{D} = \frac{1.667\times10^{-6}}{5\times10^{-9}} = 333$$

Assume low mass transfer so that heat transfer correlation is applicable to mass transfer. Use section 8.3 with appropriate change of dimensionless variables.

$$Gr_M \cdot Sc = 5.078\times10^{12} \Rightarrow \log_{10} Gr_M \cdot Sc = 12.7$$

Use Fig. 8.8 $\log_{10} Sh \simeq 2.90 \Rightarrow Sh = 790$

$$\therefore \frac{k_M L}{D} = 790 \; ; \; k_M = \frac{(790)(5\times10^{-9})}{(0.3)} = 1.3\times10^{-5}\ m\ s^{-1}$$

14.14 A schematic diagram of a vapor-phase epitaxial growth system is shown.

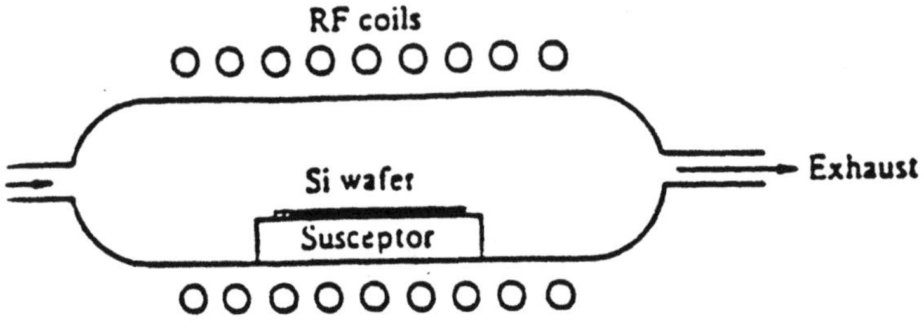

Assume that the reaction at the surface is

$$SiCl_4(g) + 2H_2(g) = Si(s) + 4HCl(g)$$

a) Sketch the concentration of each gas species as a function of the vertical distance from the silicon surface. b) Assume two dimensional flow (rectangular coordinates) and constant ρ and D for all species in the gas phase. Give an appropriate equation of diffusion for HCl(g) in an isothermal system. Also give the momentum equation. c) Select an origin and write boundary conditions that are appropriate for the equation of diffusion, with the intent of predicting the growth rate of the epitaxial silicon.

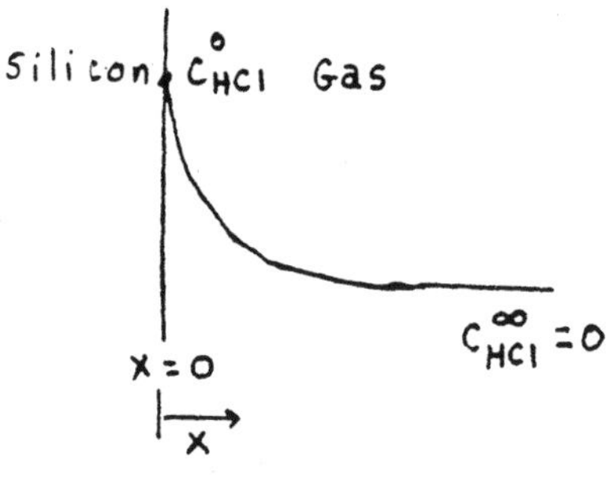

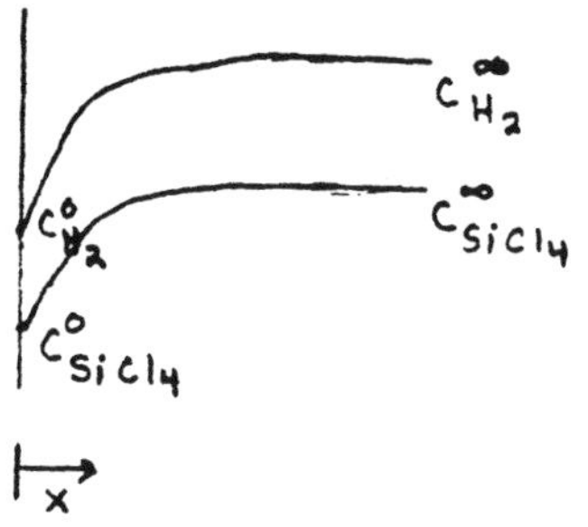

Product must be transported away from the interface

Reactants must be transported to the interface.

Flow past flat substrate. Let C_A = concentration

$$V_x \frac{\partial C_A}{\partial x} + V_Y \frac{\partial C_A}{\partial Y} = D \frac{\partial^2 C_A}{\partial Y^2}$$

: Assumptions → • Steady state
• ~~Constant~~ properties
• Ignore diffusion in direction of flow
• Ignore $\frac{\partial^2 V_x}{\partial x^2}$

The momentum equation

$$V_x \frac{\partial V_x}{\partial x} + V_Y \frac{\partial V_x}{\partial Y} = \nu \frac{\partial^2 V_x}{\partial Y^2}$$

∴ $C_A(0,Y) = C_A^{\infty}$ (upstream concentration of HCl)

$C_A(x,0) = C_A^{0}$ (equil. conc. of HCl at the solid/gas interface)

$C_A(x,\infty) = C_A^{\infty}$

14.15 Boron fibers for composites can be made by running a very fine tungsten wire (12.5 μm dia.) through a reactor tube that is continuously fed with a gas comprising 20% BCl_3 and 80% H_2. The substrate wire is heated electrically and causes the gases to react and deposit boron. Assume that the reaction is

$$BCl_3(g) + {}^3/_2\, H_2(g) \rightarrow B(s) + 3HCl(g),$$

and is controlled solely by mass transfer. The standard free energies of formation are

$$\Delta G^\circ\ (HCl) = -22\ 240 + 1.53\ T,\ \text{cal gmol}^{-1};$$

$$\Delta G^\circ\ (BCl_3) = -96\ 560 - 12.35\ T,\ \text{cal gmol}^{-1}.$$

a) Pick a substrate temperature for depositing the boron. b) Estimate the mass transfer coefficients for the reactants and products. Which species will control the overall deposition rate? (For a discussion of the technology and applications of these fibers see M. E. Buck, *Advanced Materials and Progress In Metal Processes*, 9/87, pp. 61-65.)

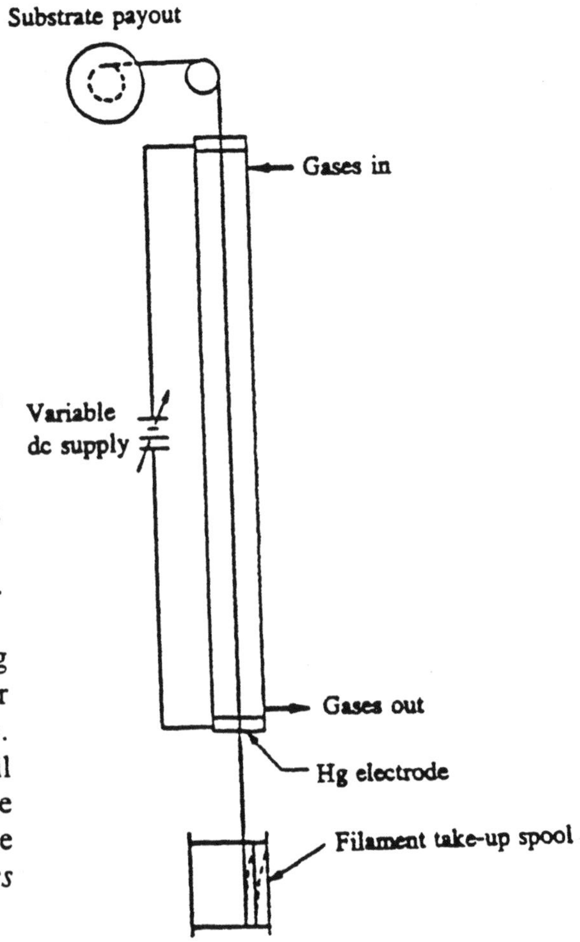

a. $\frac{3}{2} H_2(g) + \frac{3}{2} Cl_2(g) \longrightarrow 3HCl(g) \qquad \Delta G_1^\circ = 3(-22{,}240 + 1.53T) \qquad (1)$

$B(s) + \frac{3}{2} Cl_2(g) \longrightarrow BCl_3(g) \qquad \Delta G_2^\circ = -95{,}560 - 12.35T \qquad (2)$

Subtract (1)-(2): $\frac{3}{2} H_2(g) + BCl_3(g) \longrightarrow 3HCl(g) + B(s) \quad \Delta G^\circ = 28{,}840 + 16.94T$

Imagine a tiny reactor at the interface

Basis: 1 mol feed gas

0.8 mol H_2 → Reactor → $(0.8 - \frac{3}{2}n)$ mol H_2

0.2 mol BCl_3 → $(0.2 - n)$ mol BCl_3

$3n$ mol HCl

Before reaction $\qquad 1 + \frac{1}{2}n$ mol total $\qquad$ After reaction

Let n = mol BCl_3 reacted

$\frac{3}{2}n$ = mol H_2 reacted

$3n$ = mol HCl produced

Composition after reaction

$$x_{BCl_3} = \frac{0.2 - n}{1 + n/2}$$

$$x_{H_2} = \frac{0.8 - 3n/2}{1 + n/2}$$

$$x_{HCl} = \frac{3n}{1 + n/2}$$

To decide on a temperature, we use $\Delta G^\circ = -RT \ln K$, cal mol^{-1}

With $R = 1.987$ cal mol^{-1}K^{-1}; $\ln K = -\left(\frac{28,840 + 16.94T}{RT}\right) = -\frac{14,514}{T} - 8.525$

$\therefore T = -\frac{14,514}{\ln K + 8.525}$; To get a positive temperature $\ln K < -8.525$

ln K	T, K
-8.600	1.94×10^5
-10	9840
-12	4180
-14	2651
-16	1942
-18	1532

The melting point of boron is 2303 K, so we need T < 2303 K. Notice that as T decreases, K decreases rapidly. Hence, the value of P_{HCl} at the interface also decreases with decreasing T. As P_{HCl} decreases the mass transfer also decrease, so we should select as high a temperature as possible without melting the boron. Use 1950 K.

b. First we should calculate the composition of the gas at the interface, which is assumed to be the equilibrium composition.

$$\ln K = -\frac{14,514}{1950} - 8.525 \Rightarrow K = 1.162 \times 10^{-7}$$

$$K = \frac{\left(\frac{3n}{1+n/2}\right)^3}{\left(\frac{0.2-n}{1+n/2}\right)\left(\frac{0.8-3n/2}{1+n/2}\right)^{3/2}} \cong \frac{(3n)^3}{(0.2)(0.8)^{3/2}}; \quad n \simeq 8.51 \times 10^{-4} \text{ mol}$$

$\therefore$ At the interface: $x_{BCl_3} = 0.1991$

$x_{H_2} = 0.7984$

$x_{HCl} = 0.00255$

Look at the fluxes:

$$j_{BCl_3} \cong \frac{DC}{\delta}(0.2000 - 0.1991) = 9 \times 10^{-4}\frac{DC}{\delta}$$

$$j_{H_2} \cong \frac{DC}{\delta}(0.8000 - 0.7984) = 16 \times 10^{-4}\frac{DC}{\delta}$$

$$-j_{HCl} \cong \frac{DC}{\delta}(0.00255 - 0) = 26 \times 10^{-4}\frac{DC}{\delta}$$

The controlling species is BCl_3

15.1 Argon containing 2 volume percent of hydrogen is bubbled through a melt of aluminum at 970 K. The bubbling is continued to equilibrium with this gas at 1 atm. Initially the melt contains 5×10^{-6} m^3 (STP) of hydrogen per kg of aluminum. It is known that the solubility of hydrogen is 10×10^{-6} m^3 (STP) kg^{-1}. Determine whether the rate of mass transfer is gas-phase or liquid-phase controlled, a) at the beginning of degassing and b) near the end of degassing.

Initial concentration of $\underline{H}$ in Al.

$$X_{H\infty} = \frac{5\times10^{-6}\,m^3(STP)}{kg\,Al}\left|\frac{1\,mol\,H_2}{2.24\times10^{-2}\,m^3(STP)}\right|\frac{2\,mol\,\underline{H}}{mol\,H_2}\left|\frac{kg\,Al}{1\times10^{3}\,g\,Al}\right|\frac{26.98\,g\,Al}{mol\,Al}$$

$$= 1.204\times10^{-5}\ mol\,\underline{H}\ mol^{-1}Al$$

At the gas-liquid interface, the maximum X_H^* is found by first determining the equilibrium constant for the reaction $\frac{1}{2}H_2(g) = \underline{H}$.

with $P_{H_2} = 1\,atm$, $C_{\underline{H}} = 10^{-5}\,m^3(STP)\,kg^{-1}$; therefore $K = \frac{C_{\underline{H}}}{P_{H_2}^{1/2}} = 10^{-5}$

with $P_{H_2} = 0.02$ atm, $C_{\underline{H}} = K\,P_{H_2}^{1/2} = 10^{-5}(0.02)^{1/2} = 1.414\times10^{-6}\,m^3(STP)\,kg^{-1}$

a. $\therefore X_H^*(MAX) = \frac{(1.414\times10^{-6})}{(5\times10^{-6})}(1.204\times10^{-5}) = 3.405\times10^{-6}$

Then there is mass transfer of $\underline{H}$ from the liquid to the gas.

The driving force in the liquid is only: $X_{H\infty} - X_H^* = 8.6\times10^{-6}$.

Since $Y_{H\infty} \simeq 0.02$ we can expect $Y_{H\infty} - Y_H^* \ggg X_{H\infty} - X_H^*$

so that mass transfer is liquid-phase controlled.

b. Near the end of degassing the concentration driving forces in both phases will be small, so it becomes a competition between the mass transfer coefficients. Since $D_{AB}(gas) \gg \tilde{D}(liquid)$, the mass transfer will be controlled by the mass transfer in the liquid.

15.2 Iron wire (4 mm dia.) is boronized at 1200°C in a gas that establishes an equilibrium concentration of 15×10^{-3} wt.pct. boron. The transfer of boron is partly controlled by the reaction kinetics between the gas and the iron, as given by Eq. (15.11) with $r = 5 \times 10^{-8}$ m s^{-1} at the surface. a) Make a plot of the concentration of boron (in wt.pct.) at the surface. b) When the concentration at the surface is 90% of the equilibrium value, what is the concentration in the center of the wire?

a. $r = 5\times10^{-8}\ \mathrm{m\,s^{-1}}$ at the surface.

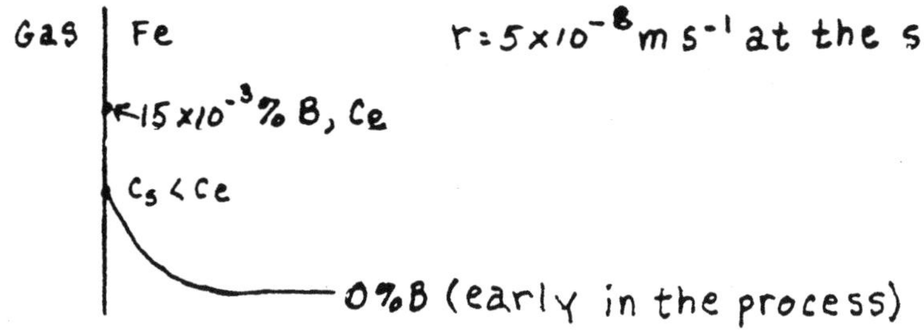

b. $C = C_s = 0.90\,Ce = 1.35\times10^{-2}\,\%\,B$

$$\frac{C - Ce}{C_i - Ce} = \frac{1.35\times10^{-2} - 1.50\times10^{-2}}{0 - 1.50\times10^{-2}} = 0.10\ ; \quad D(B \text{ in Fe}) = 1.4\times10^{-10}\ \mathrm{m^2\,s^{-1}}\ (\text{Fig. } 12.10)$$

$$\frac{rR}{D} = \frac{5\times10^{-8}\ \mathrm{m}}{\mathrm{s}}\ \frac{2\times10^{-3}\ \mathrm{m}}{}\ \frac{\mathrm{s}}{1.4\times10^{-10}\ \mathrm{m^2}} = 0.714$$

For a cylinder, we use Fig. 9.9 a (center) and get: $\frac{Dt}{R^2} \simeq 2.2$

Then from Fig. 9.9 b (surface): $\frac{C - Ce}{C_i - Ce} = 0.07$

$$C = (0.07)(0 - 15\times10^{-3}) + 15\times10^{-3} = 1.40\times10^{-2}\,\%\,B$$

15.3 Refer to Example 15.2. a) Plot the distance x that satisfies the specification in Example 15.2 as a function of the gas velocity up to 1 m s^{-1}. b) Repeat if the separation between the sheets is reduced to 5 mm.

a. From Example 15.2: $x = 0.450\sqrt{\frac{t}{D}\ \frac{\dot{n}}{2b}\ \frac{P}{K_{eq}}}$

Keeping the same specifications, except for changing $\dot{n}$, we have $x = A\dot{n}$, but $\dot{n}$ is proportional to velocity. Hence $x = B\bar{V}$ where A and B are constants.

From Example 15.2: $x = 0.15$ m when $\bar{V} = 0.1$ m s^{-1}. So when $\bar{V} = 1$ m s^{-1}, $x = 1.5$ m.

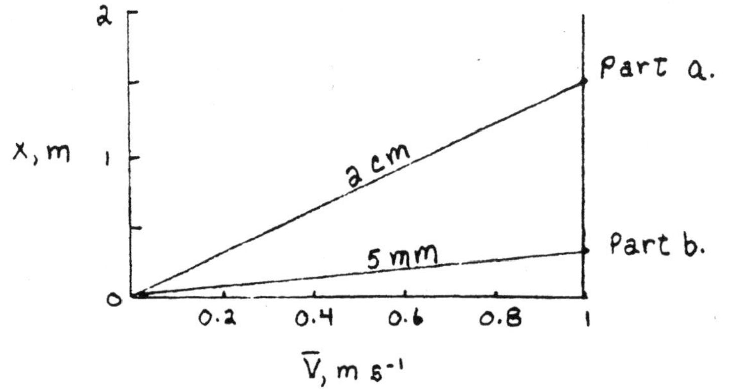

b. $\dot{n}$ is proportional to the spacing between the sheets. So if we reduce the spacing from 2 cm to 5 mm, then when $\bar{v} = 1\,ms^{-1}$,

$$x = \frac{1.5}{4}\,m = 0.375\,m$$

15.4 Assume that the initial thickness of an oxide layer on silicon is 10 nm. Determine the thickness of the oxide after oxidation in dry oxygen at 1350 K for 7000 s.

$$\delta_i = 10\,nm = 10 \times 10^{-9}\,m$$

Eq. (15.34) where $A = 1.18 \times 10^{-10} \exp\left(\frac{8840}{1350}\right) = 8.236 \times 10^{-8}\,m$

$$B = 1.81 \times 10^{-13} \exp\left(\frac{-14050}{1350}\right) = 5.468 \times 10^{-18}\,m$$

$$\tau = \frac{(10 \times 10^{-9})^2 + (8.236 \times 10^{-8})(10 \times 10^{-9})}{5.468 \times 10^{-18}} = 168.9\,s$$

$$\delta = \frac{8.236 \times 10^{-8}}{2}\left[\left(1 + \frac{(4)(5.468 \times 10^{-18})(7000 + 168.9)}{(8.236 \times 10^{-8})^2}\right)^{1/2} - 1\right] = 1.61 \times 10^{-7}\,m = 161\,nm$$

15.5 The standard free energy of formation of $SiO_2(s)$ is

$$\Delta G° = -215\,600 + 41.5\,T,\ \text{cal mol}^{-1},$$

for $700 \le T \le 1700$ K. Suppose the pure oxygen in Problem 15.4 is replaced with 90% Ar-10% O_2. Would this appreciably change the oxidation kinetics?

$$\text{Eq. (15.30a)}: \quad j = \frac{p_{O_2}^{\infty}}{\frac{k_B T}{K_M} + \frac{1}{k_r K} + \frac{\delta}{DK}}$$

$p_{O_2}^{\infty}$ is 1 atm in Prob. 15.4 whereas it is 0.1 atm now. Hence, we can expect the flux to be only 10% of that in Example 15.4.

15.6 The vapor pressure of Zn(s) is

$$\log_{10} P(\text{mm Hg}) = -\frac{6850}{T} - 0.755 \log_{10} T + 11.24.$$

(Note: 760 mm Hg = 1 standard atm = 1.0133×10^5 N m^{-2}.) Thermodynamic data at 1200 K give

$$\frac{\ln \gamma_{Zn}}{(1 - X_{Zn})} = 3.875\,X_{Zn} - 3.425.$$

For other temperatures, assume that the regular solution model applies so that $T \ln \gamma_{Zn}$ = constant. a) Calculate the initial flux of Zn from 70-30 brass by sublimation into a vacuum as a function of temperature. b) Calculate the surface concentration as a function of time with sublimation into the vacuum at 1200 K.

a. Eq. (15.38) can be written $j = \dfrac{A_e W_e \gamma_A P_A^o C_A^s}{A_s \rho (2\pi MRT)^{1/2}}$

Assume $\dfrac{A_e}{A_s} = 1$ and $W_e = 1$

P_A^o can be calculated

C_A^s = 30 wt % Zn, convert to $\dfrac{\text{mol Zn}}{\text{cm}^3 \text{ alloy}}$ density is in the Appendix

ρ see the Appendix

Since we must compute $j = f(T)$, then it is convenient to set up a computer program.

b. Apply Eq. (15.41), with $x = 0$. Use Fig. 12.9 for D.

The following program solves a. and b.

```
10 'Problem 15.6
20  AE = 1: AS = 1 : WE = 1 : PI = 3.1416        'areas and Clausing factor
30  MZN = 65.38 : MCU = 63.54: R = 8315     'atom. masses of Zn and Cu, kg/kmol
                                  and gas constant, J/( kmol K )
40  CZN = 30 : MASSZN = CZN : MASSCU = 100 - CZN        'CZN is wgt.% Zn
50  MOLZN = MASSZN/MZN : MOLCU = MASSCU/MCU : MOL = MOLZN + MOLCU
60  XZN = MOLZN/MOL : XCU = 1 - XZN               'atom fractions of Zn and Cu
70 'calculate activity coefficient at 1200 K to get the constant
80  ZNLN = ( 1 - XZN )*( 3.875*XZN - 3.425 )
90  GAMMAR = EXP(ZNLN)                      'activity coefficient of Zn at 1200 K
100 CON = 1200*ZNLN                        'CON = T*ln(GAMMA)
110 'part a
120 LPRINT "                 Evap. flux"
130 LPRINT "     T, K        kmol/m^2 s"
140 LPRINT "    ******       **********"
150  FOR T = 500 TO 1200 STEP 100
160       TLOG = LOG(T)/2.303        'log(base 10) of T
170       PLOG = -6850/T - .755*TLOG + 11.24
180       P = 10^PLOG/760               'vapor pressure of Zn in atm
190       P = 101330! * P             'vapor pressure of Zn in N/m^2
200       IF T = 1200 THEN PR = P         'PR is used in part b as v.p. at 1200 K
210       GAMMA = EXP(CON/T)          'activity coeff. of Zn at T
220       NUMER = AE*WE*GAMMA*P*XZN : DENOM = AS*SQR(2*PI*MZN*R*T)
230       FLUX = NUMER/DENOM         'evaporation flux, kmol Zn/(m^2 s)
240       LPRINT USING "    ####       ##.##^^^^";T,FLUX
250 NEXT T
252 LPRINT : LPRINT
260 '
```

```
270 'part b       at T = 1200 K
280 T = 1200
290 RHO600 = 8240 : RHO900 = 8080   'density of brass at 600 and 900 K, kg/m^3
300 VOLEXP = RHO900*( 1/RHO900 - 1/RHO600 )/300   'vol. expansion coeff., 1/K
310 VOL1200 = (1/RHO900)*( 1 + VOLEXP*(T - 900) )
320 RHO1200 = 1/VOL1200                       'density at 1200 K in kg/m^3
330 MAVG = XZN*MZN + XCU*MCU                  'avg. atomic mass of brass, kg/kmol
340 RHO = RHO1200/MAVG                        'kmol/m^3
350 D = 3E-12    'diff. coeff. of Zn in Cu at 1200 K, extrap. in Fig. 12.9
360 NUMER = AE*ME*GAMMAR*PR : DENOM = AS*D*RHO*SQR(2*PI*MZN*R*T)
370 Y = NUMER/DENOM                           'under Eq. (15.40c)
380 CZNO = (CZN/100)*RHO1200/MZN          'initial concen. of Zn, kmol/m^3
390 LPRINT "   Time,s      % Zn at surface "
400 LPRINT "   ******      *************** "
410 FOR TIME = 0 TO .000001 STEP .0000001
420     TERM1 = EXP(Y*Y*D*TIME)
430     ARG = Y*SQR(D*TIME) : PRINT " The arguement is";ARG".              " :
 PRINT " Now use Table 9.3 on page 302 and input the error function here, or hit
 control break to end. " : INPUT ERF
440     ERFC = 1 -ERF : RATIO = TERM1 * ERFC
450     CZNS = CZNO* RATIO                'kmol/m^3
460     CZNS = 100*CZNS*MZN/RHO1200       'wt pct Zn
470     LPRINT USING "   ##.##^^^^        ##.##       ";TIME,CZNS
480 NEXT TIME
490 END
```

a.

T, K	Evap. flux kmol/m^2 s
******	**********
500	6.28E-10
600	1.83E-07
700	1.02E-05
800	2.04E-04
900	2.06E-03
1000	1.29E-02
1100	5.73E-02
1200	1.97E-01

b.

Time,s	% Zn at surface
******	***************
0.00E+00	30.00
1.00E-07	13.03
2.00E-07	10.10
3.00E-07	8.91
4.00E-07	8.11
5.00E-07	0.00
6.00E-07	0.00
7.00E-07	0.00

At 1200K, the evaporative flux at the surface is so great that the concentration of Zn at the surface is zero in only 0.5 μs.

15.7 Ferritic stainless steel parts are vacuum heat-treated in order to maintain a shiny surface finish. If AISI 410 (12% Cr) parts are heat treated in a vacuum of 10 μm of Hg at 1140 K for 2 hours, what will the concentration of chromium on the surface of the parts be? At 1140 K, the vapor pressure of pure Cr is 1.33×10^{-5} Pa, and D_{Cr} in iron is approximately 10^{-16} m^2 s^{-1}.

Eq. (15.41) $\frac{C_A}{C_A^o} = \text{erf}\frac{x}{2\sqrt{Dt}} + \exp(Yx + Y^2Dt)\,\text{erfc}\left(\frac{x}{2\sqrt{Dt}} + Y\sqrt{Dt}\right)$

at the surface $x = 0$; $\frac{C_A^s}{C_A^o} = \exp(Y^2Dt)\,\text{erfc}(Y\sqrt{Dt})$

Where $Y = \frac{A_e W_e \gamma_A P_A^o}{A_s D \rho (2\pi MRT)^{1/2}}$

Assume $\gamma_A = 1$, $\frac{A_e}{A_s} = 1$, $W_e = 1$, $M(Cr) = 52$ kg kmol^{-1}, $\bar{M}(\text{alloy}) \simeq 55.4$ kg kmol^{-1}

$\rho = \frac{7920 \text{ kg}}{m^3} \cdot \frac{\text{kmol}}{55.4 \text{ kg}} = 143 \text{ kmol m}^{-3}$

$Y = \frac{(1.33 \times 10^{-5})\,1}{(10^{-16})(143)[2\pi(52)(8315)(1140)]^{1/2}} = 16.71 \text{ m}^{-2}$

$\text{erfc}(Y\sqrt{Dt}) = 1 - \text{erf}(16.71\sqrt{10^{-16} \times 7200}) \simeq 1$

$\frac{C_A^s}{C_A^o} = \exp(Y^2Dt) = \exp[(16.71)^2(10^{-16})(7200)] \simeq 1$

Hence, $C_A^s = C_A^o$ and we would not be concerned that there will be a depletion of Cr at the surface.

15.8 A model for the reduction of a spherical oxide is illustrated to the right. Assume mixed control, with mass transfer in the gas and diffusion through the porous product layer controlling the overall kinetics. A reducing gas flows past the sphere with a concentration C_∞, and at the product-oxide interface its concentration is C^*. Derive an equation that gives the rate at which the oxide is reduced in kmol s^{-1}. The overall reaction is

$$MO(s) + CO(g) \rightarrow M(s) + CO_2(g),$$

where MO is the metal oxide.

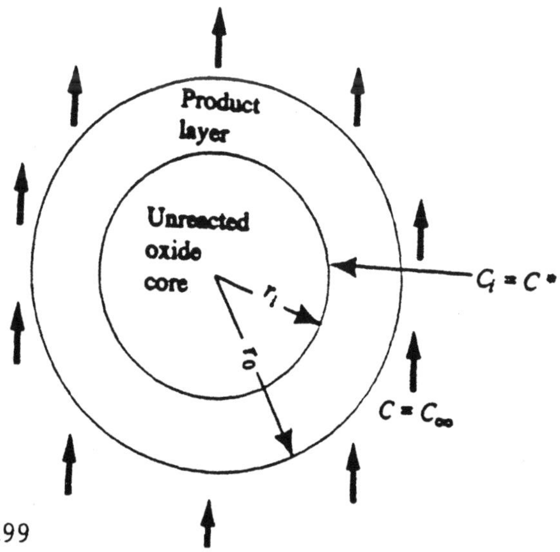

Mass transfer of the reducing gas (co) from the bulk stream to the surface at r_o is: $J = 4\pi r_o k_M (C_\infty - C_o)$, kmol s^{-1}

Mass transfer through the spherical shell, $r_i \le r \le r_o$, is:

$$J = 4\pi D_{eff}(C_o - C^*)/\left(\frac{1}{r_o} - \frac{1}{r_i}\right)$$

where C_o is the concentration at $r = r_o$

C^* is the concentration at $r = r_i$

D_{eff} is the diffusion coefficient of co through the porous product layer.

This is the steady state mass transfer through a spherical shell (analogous to heat transfer considered in Problem 7.12). D_{eff} can be determined by the method described in section 12.7. The total resistance for the mass transfer is: $R_M = \frac{1}{r_o^2 k_M} + \frac{1}{D_{eff}}\left(\frac{r_i - r_o}{r_i r_o}\right)$ so that the mass flow can be written in terms of $(C_\infty - C^*)$ as $J = \frac{4\pi(C_\infty - C^*)}{R_M}$. C^* can be determined from the equilibrium constant for the given reaction. The result for the mass flow can be used to calculate the rate at which the oxide core shrinks. Such a model is referred to as the "shrinking core model".

15.9 Using data in Table 15.2, plot the change in concentration with time (up to 3600 s) of a melt of iron containing 1% Mn, 1% Cr, and 0.05% S. The melt is contained in a nonreacting crucible, which has the dimensions 0.5 m dia. by 1 m height.

Use Equation (15.47): $\ln \frac{C_A^\circ}{C_{A\infty}} = \frac{A_s}{V} kt$; $\frac{A_s}{V} = \frac{\frac{\pi D^2}{4}}{\frac{\pi D^2}{4} H} = \frac{1}{H} = 1\ m^{-1}$

$k_{Mn} = 8.4 \times 10^{-5}\ m\ s^{-1}$

$k_{Cr} = 2.1 \times 10^{-6}\ m\ s^{-1}$

$k_S = 7.0 \times 10^{-6}\ m\ s^{-1}$

Hence, $C_{Mn} = 1 \exp(-8.4 \times 10^{-5} t)$; C_{Mn} in % and t in s

$C_{Cr} = 1 \exp(-2.1 \times 10^{-6} t)$

$C_S = 0.05 \exp(-7 \times 10^{-6} t)$

Time, s	C(Mn), %	C(Cr), %	C(S), %
0	1.000	1.000	0.0500
300	0.975	0.999	0.0499
600	0.951	0.999	0.0498
900	0.927	0.998	0.0497
1200	0.904	0.997	0.0496
1500	0.882	0.997	0.0495
1800	0.860	0.996	0.0494
2100	0.838	0.996	0.0493
2100	0.817	0.995	0.0492
2700	0.797	0.994	0.0491
3000	0.777	0.994	0.0490
3300	0.758	0.993	0.0489
3600	0.739	0.992	0.0488

In vacuum refining of steel there is little loss of Cr and S but the Mn loss can be substantial.

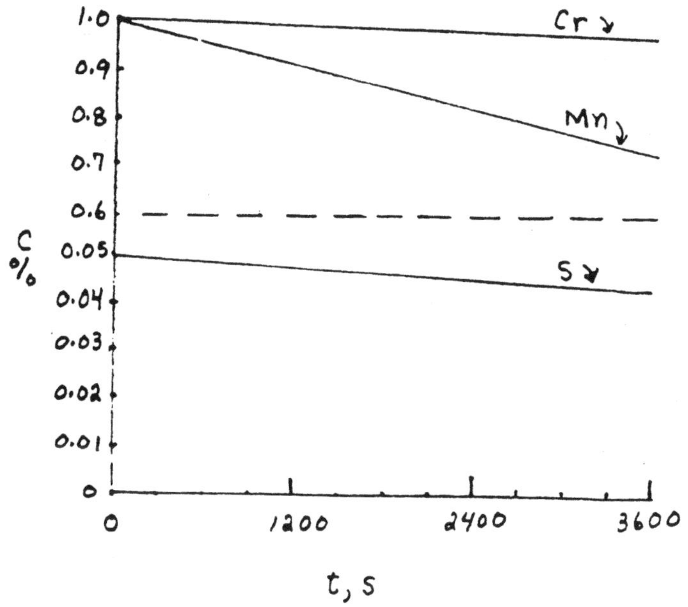

15.10 The melt of Problem 15.9 also contains carbon, which reacts with the magnesia crucible according to the overall reaction:

$$MgO(s) + \underline{C} \rightarrow CO(g) + Mg(g).$$

Assume that evaporation can only occur at the surface of the well-mixed melt. a) Hypothesize the reduction reaction at the melt-crucible interface. b) Hypothesize vaporization reactions at the melt surface. c) What are the mass transfer steps for oxygen? for magnesium? for carbon? For each mass transfer step, make a schematic concentration profile.

Carbon, magnesium and oxygen are dissolved in the melt as $\underline{C}$, $\underline{Mg}$ and $\underline{O}$.

a.

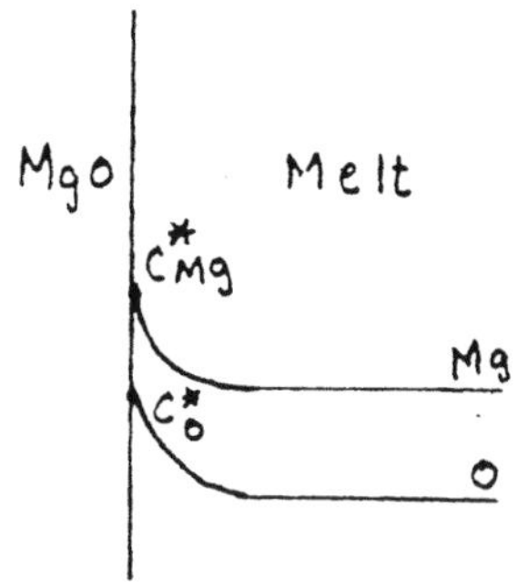

$MgO(s) \longrightarrow \underline{Mg} + \underline{O}$ (1)

(Better to describe this as the decomposition of the MgO.)

C^*_{Mg} and C^*_O would be at equilibrium with MgO. Hence, $C^*_{Mg}\,C^*_O = K$

b.

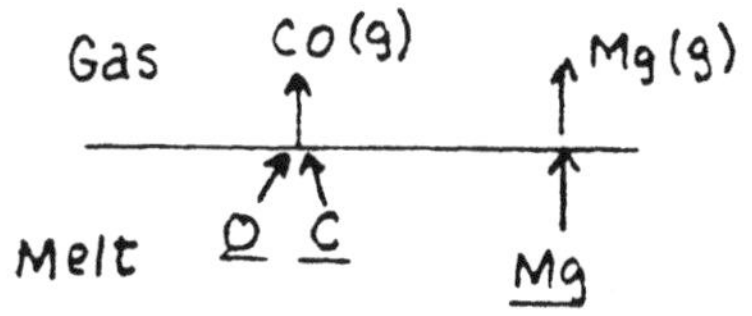

$\underline{O} + \underline{C} \longrightarrow CO(g)$ (2)

$\underline{Mg}$ — $Mg(g)$ (3)

If we add reactions (1), (2) and (3), we get $MgO(s) + \underline{C} \longrightarrow CO(g) + Mg(g)$

c. Mass transfer steps for oxygen:

1. MgO/melt interface to the bulk melt.
2. Bulk melt to the melt/gas interface.
3. Reaction with $\underline{C}$ to form CO(g).
4. Melt/gas interface to the chamber as CO(g).

Mass transfer steps for magnesium:

1. MgO/melt interface to the bulk melt.
2. Bulk melt to the melt/gas interface.
3. Reaction to form Mg(g).
4. Melt/gas interface to the chamber.

Mass transfer steps for carbon:

1. Bulk melt to melt/gas interface.
2. Reaction with $\underline{O}$ to form CO(g)
3. Melt/gas interface to chamber as CO(g)

15.11 In vacuum laser welding, it is difficult to measure the temperature of the molten pool. One estimate of the temperature can be made from collecting a sample of vapor evaporated from the weld pool. The ratio of elements in the sample is the ratio of vaporization rates of the elements. If 201 stainless steel is welded and the ratio of Cr/Mn in the condensed vapor sample is *0.056* estimate the pool temperature, assuming that the pool is an ideal thermodynamic solution.

We need the vapor pressure of chromium and manganese. They are each given by $\log_{10} P = \frac{A}{T} + B + C \log_{10} T$, *mm Hg*

	A	B	C
Cr	20680	14.56	-1.31
Mn	13900	17.27	-2.52

201 SS contains 17% Cr and 6.5% Mn, as well as 4.5% Ni and balance Fe. We can solve Eq. (15.38) to get the evaporation rates of both Cr and Mn and the temperature that gives the correct ratio is the temperature of the melt.

The following program determines the ratios.

```
10 'Problem 15.11
20 PI = 3.1416 : R = 8314          'pi and gas constant, J/(kmol K)
30 AE = 1: AS = 1 : WE = 1 : PI = 3.1416       'areas and Clausing factor
40 MMN = 54.94 : MCR = 52: MFE = 55.85 : MNI = 58.71  'atomic masses, kg/kmol
50 CMN = 6.5 : CCR = 17 : CFE = 72 : CNI = 4.5          'wt pcts
60 MASSMN = CMN : MASSCR = CCR : MASSFE = CFE : MASSNI = CNI
70 MOLMN = MASSMN/MMN : MOLCR = MASSCR/MCR : MOLFE  = MASSFE/MFE
80 MOLNI = MASSNI/MNI : MOL = MOLMN + MOLCR + MOLFE + MOLNI
90 XMN = MOLMN/MOL : XCR = MOLCR/MOL              'atom fractions of Mn and Cr
100 XFE = MOLFE/MOL : XNI = MOLNI/MOL             'atom fractions of Fe and Ni
110 LPRINT "                Evap. flux (Mn)   Evap. flux (Cr)    Mass ratio"
120 LPRINT "    T, K         kmol/(m^2 s)        kmol/(m^2 s)       Cr/Mn "
130 LPRINT "   ******    ***************    ***************    **********"
140  FOR T = 2200 TO 2400 STEP 20
150      TLOG = LOG(T)/2.303        'log(base 10) of T
160      PCR = -20680/T + 14.56 -1.31*TLOG : PCR = (10^PCR/760)*101330:  'N/m^2
170      PMN = -13900/T + 17.27 -2.52*TLOG : PMN = (10^PMN/760)*101330!  'N/m^2
180      NUMER = AE*WE*PCR*XCR : DENOM = AS*SQR(2*PI*MCR*R*T)
190      FLUXCR = NUMER/DENOM        'evaporation flux, kmol Cr/(m^2 s)
200      NUMER = AE*WE*PMN*XMN : DENOM = AS*SQR(2*PI*MMN*R*T)
210      FLUXMN = NUMER/DENOM        'evaporation flux, kmol Mn/(m^2 s)
220      RATIO = FLUXCR*MCR/(FLUXMN*MMN)     'ratio put on mass basis
230      LPRINT USING "    ####        ##.##^^^^             ##.##^^^^            ##.#
#^^^^";T,FLUXMN,FLUXCR,RATIO
240  NEXT T
250  END
```

T, K	Evap. flux (Mn) kmol/(m^2 s)	Evap. flux (Cr) kmol/(m^2 s)	Mass ratio Cr/Mn
******	***************	***************	**********
2200	3.72E-02	1.89E-03	4.80E-02
2220	4.12E-02	2.25E-03	5.17E-02
2240	4.57E-02	2.69E-03	5.57E-02
2260	5.04E-02	3.19E-03	5.99E-02
2280	5.56E-02	3.78E-03	6.43E-02
2300	6.12E-02	4.46E-03	6.90E-02
2320	6.72E-02	5.25E-03	7.39E-02
2340	7.37E-02	6.16E-03	7.91E-02
2360	8.06E-02	7.20E-03	8.45E-02
2380	8.81E-02	8.41E-03	9.03E-02
2400	9.61E-02	9.78E-03	9.63E-02

By interpolation, T = 2241 K

16.1 Derive an alternative to Eq. (16.7) for an insulated surface, in a way which is different than that in the text. The insulated surface is at node N, and there is an imaginary node at $N+1$ with the same temperature as T_{N-1}.

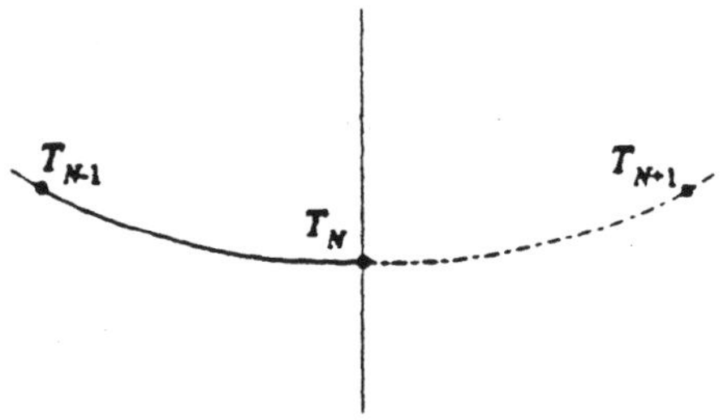

Start with Eq. (16.6), with $m = N$.

$$\frac{\partial T_N}{\partial t} = \frac{\alpha}{(\Delta x)^2}\left[T_{N-1} - 2T_N + T_{N+1}\right]$$

But $T_{N+1} = T_{N-1}$

Hence

$$\frac{\partial T_N}{\partial t} = \frac{2\alpha}{(\Delta x)^2}\left[T_{N-1} - T_N\right]$$

16.2 Use finite difference approximations and set up the one-dimensional heat conduction problem that is specified as follows:

$$\frac{\partial T}{\partial t} = \alpha\,\frac{\partial^2 T}{\partial x^2},$$

with

$$T(L,t) = T_L \qquad t > 0,$$

$$T(0,t) = T_0 \qquad t > 0,$$

and

$$T(x,0) = T_i.$$

Divide the domain $0 \le x \le L$ into four segments so that $\Delta x = L/4$. a) Write the set of algebraic equations in matrix form using the Euler method. b) Repeat using the Crank-Nicolson method.

$$u = \frac{T - T_i}{T_0 - T_i}\ ; \quad x' = \frac{x}{L}\ ; \quad \theta = \frac{\alpha t}{L^2}$$

$$\frac{\partial u}{\partial \theta} = \frac{\partial^2 u}{\partial x'^2}$$

$$u(1,\theta) = 0$$

$$u(0,\theta) = 1$$

$$u(x',0) = 0$$

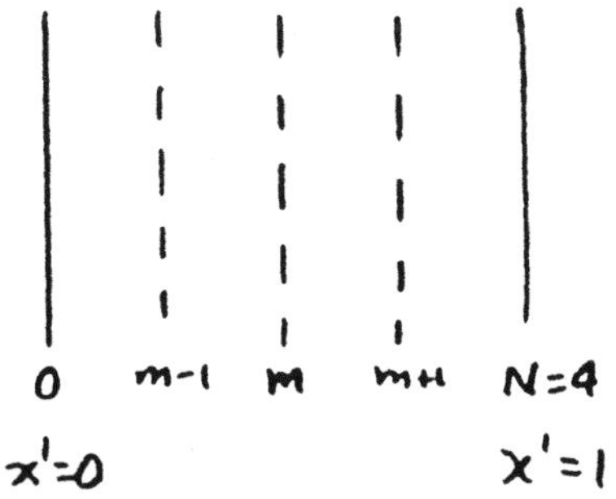

a. Initial temperatures: $u_0^0 = u_1^0 = u_2^0 = u_3^0 = u_4^0 = 0.$

From Eqs. (16a) and (16b), we deduce:

$$u_0^{\nu+1} = 1 \qquad (\text{except } u_0^0 = 0)$$

$$u_1^{\nu+1} = P u_0^{\nu} + (1-2P) u_1^{\nu} + P u_2^{\nu}$$

$$u_2^{\nu+1} = P u_1^{\nu} + (1-2P) u_2^{\nu} + P u_3^{\nu}$$

$$u_3^{\nu+1} = P u_2^{\nu} + (1-2P) u_3^{\nu} + P u_4^{\nu}$$

$$u_4^{\nu+1} = 0$$

b.

Set up for Crank-Nicholson method:

$$u_m^{\nu+1} - u_m^{\nu} = \frac{P}{2}\Big[u_{m+1}^{\nu} - 2u_m^{\nu} + u_{m-1}^{\nu} + u_{m+1}^{\nu+1} - 2u_m^{\nu+1} + u_{m-1}^{\nu+1} \Big]$$

$$\begin{bmatrix} 1 & 0 & 0 & 0 & 0 \\ -\frac{P}{2} & (1+P) & -\frac{P}{2} & 0 & 0 \\ 0 & -\frac{P}{2} & (1+P) & -\frac{P}{2} & 0 \\ 0 & 0 & -\frac{P}{2} & (1+P) & -\frac{P}{2} \\ 0 & 0 & 0 & 0 & 1 \end{bmatrix} \begin{bmatrix} u_0 \\ u_1 \\ u_2 \\ u_3 \\ u_4 \end{bmatrix}^{\nu+1} = \begin{bmatrix} 1 & 0 & 0 & 0 & 0 \\ \frac{P}{2} & (1-P) & \frac{P}{2} & 0 & 0 \\ 0 & \frac{P}{2} & (1-P) & \frac{P}{2} & 0 \\ 0 & 0 & \frac{P}{2} & (1-P) & \frac{P}{2} \\ 0 & 0 & 0 & 0 & 0 \end{bmatrix} \begin{bmatrix} u_0 \\ u_1 \\ u_2 \\ u_3 \\ u_4 \end{bmatrix}^{\nu}$$

16.3 Write a computer program that can be used to solve for $T(x,t)$ in Problem 16.2a.

```
10 'Problem 16.3      Euler method (also known as explicit method)
20 P = .25 : L = 1 : DX = L/4 : DTHETA = P*DX*DX      'P is the modulus, L is
     nondimensional length, x is nondimensional coordinate, theta is nondimen-
     sional time, DX = delta x, and DTHETA = delta theta.
30 LPRINT "               ------------ T E M P E R A T U R E S -------------"
40 LPRINT "    Time    x/L = 0   x/L = 1/4   x/L = 1/2   x/L = 3/4   x/L = 1"
50 LPRINT "    *****   *******   *********   *********   *********   *******"
60 UL = 0 : U0 = 1                        'set end temperatures
70 THETA = 0
80 FOR I = 0 TO 4                         'set initial temperatures
90     U(I) = 0
100 NEXT I
110 LPRINT USING "    #.###    ##.###    ##.###    ##.###    ##.###    ##
.###";THETA, U(0), U(1), U(2), U(3), U(4)
```

```
120  INDEX = 0                   'counting index for printing
130  FOR J = 1 TO 40             'this sets the number of time steps
140       INDEX = INDEX + 1
150       THETA = THETA + DTHETA
160       U(0) = U0
170       FOR I = 1 TO 3          'internal nodes
180           U(I) = P*U(I-1) + (1-2*P)*U(I) + P*U(I+1)
190       NEXT I
200       U(4) = UL : IF INDEX < 5 THEN 220
210       LPRINT USING "   #.###     ##.###     ##.###     ##.###     ##.###
   ##.###";THETA, U(0), U(1), U(2), U(3), U(4) : INDEX = 0
220  NEXT J
230  END
```

------------- T E M P E R A T U R E S -------------

Time	x/L = 0	x/L = 1/4	x/L = 1/2	x/L = 3/4	x/L = 1
*****	*******	*********	*********	*********	*******
0.000	0.000	0.000	0.000	0.000	0.000
0.078	1.000	0.582	0.295	0.122	0.000
0.156	1.000	0.690	0.424	0.201	0.000
0.234	1.000	0.728	0.472	0.232	0.000
0.313	1.000	0.742	0.489	0.243	0.000
0.391	1.000	0.747	0.496	0.248	0.000
0.469	1.000	0.749	0.499	0.249	0.000
0.547	1.000	0.750	0.499	0.250	0.000
0.625	1.000	0.750	0.500	0.250	0.000

16.4 Write a computer program that can be used to solve for $T(x,t)$ in Problem 16.2b.

Use the program on pages 581 and 582, with the following modifications:

```
120  T = I * 0.03125
155  IF I > 1 THEN 170
160  E(1,1) = E(1,1) + 1
```

($\Delta x'$ changed to 0.25, so $\Delta\theta$ also changed appropriately.)

Change A Matrix (DATA), P = 0.5

```
1,      0,     0,     0,     0
-0.25,  1.5,   -.25,  0,     0
0,      -0.25, 1.5,   -.25,  0
0,      0,     -.25,  1.5,   -.25
0,      0,     0,     0,     1
```

Change C Matrix (DATA), P = 0.5

```
1,    0,    0,    0,    0
.25,  .5,   .25,  0,    0
0,    .25,  .5,   .25,  0
0,    0,    .25,  .5,   .25
0,    0,    0,    0,    0
```

θ	u_0	u_1	u_2	u_3	u_4
.0938	1	.525	.209	.0619	0
.1875	1	.664	.380	.166	0
.2812	1	.715	.450	.215	0
.4688	1	.744	.492	.244	0
1.5000	1	.750	.500	.250	0

16.5 Consider the problem in Fig. 16.1 and the VAX-BASIC program for solving Eq. (16.31). Modify and use the program to solve for the time required for the center of the core to heat to 873 K, when molten aluminum encapsulates a plaster core. Thermal data can be found with the problems of Chapter 10.

Thermal properties of plaster: $k = 0.35\ \text{W m}^{-1}\text{K}^{-1}$, $\rho = 1120\ \text{kg m}^{-3}$, $C_p = 840\ \text{J kg}^{-1}\text{K}^{-1}$, $\alpha = 3.72 \times 10^{-7}\ \text{m}^2\text{s}^{-1}$.

The program on pages 581 and 582 is in terms of nondimensional variables. Of course, we can extract the dimensional variables from the nondimensional variables.

$$u = \frac{T - T_i}{T_M - T_i} = \frac{T - 300}{933 - 300} \Rightarrow T = (T_M - T_i)\,u + T_i$$

$$u(x', 0) = u_i \Rightarrow T(x, 0) = T_i$$

$$\frac{\partial u}{\partial x'}(1, \theta) = 0 \Rightarrow \frac{\partial T}{\partial x}(L, t) = 0$$

$$u(0, \theta) = 1 \Rightarrow T(0, t) = T_M$$

Also, $x = Lx'$ (assume $L = 10\ \text{mm} = 0.01\ \text{m}$) and

$$t = (L^2/\alpha)\,t.$$

We want t when $T_4 = T\big|_{x=L} = 873\ \text{K}$.

Insert lines 15-18:

```
15 ALPHA = 3.72 E-7
16 L = 0.01
17 TM = 933
18 TI = 300
```

Insert lines 212–214:

```
212  TEMP=(TM-TI)*B(5,1)+TI
214  IF TEMP > 873 THEN 251
```

Omit lines 220–230 (optional).

Insert lines 251 – 253

```
251  REM --- ANSWER IN DIMENSIONAL VARIABLE
252  TT = L*L*T/ALPHA
253  PRINT TT, TEMP
```

Output

Time = 285 s when temperature at core center = 873.3 K

16.6 Consider one-dimensional homogenization as in Section 13.5. The following microsegregation data in an Fe-Ni alloy are reported.

x, μm	% Ni	x, μm	% Ni	x, μm	% Ni
0	12	30	16	60	29
10	13	40	18.5	70	37
20	14	50	23	80	45

a) Plot the data and determine the average concentration of nickel.

b) Set up the set of algebraic equations that result from making finite difference approximations. For simplicity, use Δx = 20 μm and the Euler method.

a.

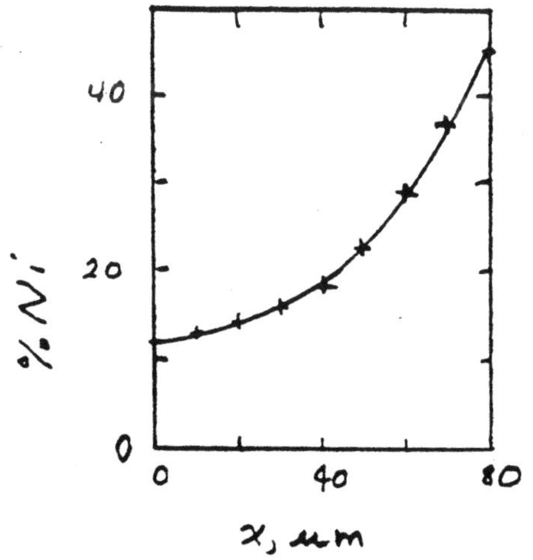

$\bar{C}_{Ni}$ = 22.38 % Ni

(In determing the average, the end points should be weighted by one-half.)

b. $$\frac{\partial C}{\partial t} = D\frac{\partial^2 C}{\partial x^2}$$

$C(x,0) = f(x)$, as given in part a)

$$\frac{\partial C}{\partial x}(0,t) = \frac{\partial C}{\partial x}(L,t) = 0$$

With $\Delta x = 20$ um, the nodes will be from $m = 0$ to $m = N = 4$.

Euler method:

$$D\frac{\partial^2 C}{\partial x^2} \cong \frac{D}{(\Delta x)^2}\left(C_{m-1}^{\nu} - 2C_m^{\nu} + C_{m+1}^{\nu}\right)$$

$$\frac{\partial C}{\partial t} \cong \frac{C_m^{\nu+1}}{\Delta t} - \frac{C_m^{\nu}}{\Delta t}$$

$$C_m^{\nu+1} = \frac{D\Delta t}{(\Delta x^2)}\left(C_{m-1}^{\nu} - 2C_m^{\nu} + C_{m+1}^{\nu}\right) + C_m^{\nu}$$

Let $\Delta\theta = D\Delta t/(\Delta x^2)$

$$C_m^{\nu+1} = \Delta\theta\, C_{m-1}^{\nu} + (1-2\Delta\theta)C_m^{\nu} + \Delta\theta\, C_{m+1}^{\nu}$$

at node 0 : $C_{m-1}^{\nu} = C_{m+1}^{\nu}$ (no flux)

at node N : $C_{N+1}^{\nu} = C_{N-1}^{\nu}$ (no flux)

Hence, the following set of equations are obtained:

<u>Node</u>

Node	Equation
0	$C_0^{\nu+1} = (1-2\Delta\theta)C_0^{\nu} + 2\Delta\theta\, C_1^{\nu}$
1	$C_1^{\nu+1} = \Delta\theta\, C_0^{\nu} + (1-2\Delta\theta)C_1^{\nu} + \Delta\theta\, C_2^{\nu}$
2	$C_2^{\nu+1} = \Delta\theta\, C_1^{\nu} + (1-2\Delta\theta)C_2^{\nu} + \Delta\theta\, C_3^{\nu}$
3	$C_3^{\nu+1} = \Delta\theta\, C_2^{\nu} + (1-2\Delta\theta)C_3^{\nu} + \Delta\theta\, C_4^{\nu}$
4	$C_4^{\nu+1} = 2\Delta\theta\, C_3^{\nu} + (1-2\Delta\theta)C_4^{\nu}$

16.7 Extend the one-dimensional homogenization model of Problem 16.6 to the two-dimensional model considered in Section 16.1.3. For $L = \ell = 80\ \mu m$, the following microsegregation data are given (see Fig. 16.7 for identification of the nodes).

Nodes	% Ni	Nodes	% Ni
0,15	14	5,10	16
1,11	23	6	29
2,7	37	8,13	13
3	45	9	14
4,14	13	12	12
		Average	20.25

a) The diffusion coefficient for nickel in γ-iron at $1273 \leq T \leq 1560$ K is

$$D = \exp\,(0.0519\ X_{Ni} + 1.51)\ \exp\left[-\frac{38\ 380 + 5.85\ X_{Ni}}{T}\right],$$

where D = diffusion coefficient, $cm^2\ s^{-1}$; X_{Ni} = atom percent Ni; T = temperature, K. Calculate an appropriate diffusion coefficient for homogenizing the alloy at 1350 K.

b) Determine the residual segregation index as a function of time for homogenization at 1350 K.

We can use the program on pp. 588–590 and merely replace the values of concentrations in Matrix D (lines 800–960). In the program, however, the time steps are 900s. The time steps for this problem depend on the diffusion coefficient.

a. $T = 1350$ K

$$X_{Ni} = \frac{\left(\frac{20.25}{58.71}\right)(100)}{\frac{20.25}{58.71} + \frac{(100-20.25)}{55.85}} = 19.46\ \%$$

$$D = \exp(0.0519 \times 19.46 + 1.51)\exp\left[-\frac{38\,380 + 5.85 \times 19.46}{1350}\right]$$

$$D = 5.14 \times 10^{-12}\ cm^2\ s^{-1} = 5.14 \times 10^{-16}\ m^2\ s^{-1}$$

b. Next we calculate the time steps ($\Delta\theta = 0.01$).

$$\Delta x = \frac{80}{3}\ \mu m = \frac{80}{3} \times 10^{-6}\ m$$

$$\Delta t = 0.01\,(\Delta x)^2/D$$

$$= (0.01)\left(\frac{80\times10^{-6}}{3}\right)^2 / 5.14\times10^{-16}$$

$\Delta t = 13.835\times10^3$ s $= 3.843$ h (line 120).

The residual segragation index;

$$\delta = \frac{C_M - C_m}{C_M^0 - C_m^0} = \frac{C_M - C_m}{45-12}$$

From the output, we can select C_M and C_m as a function of time and calculate δ.

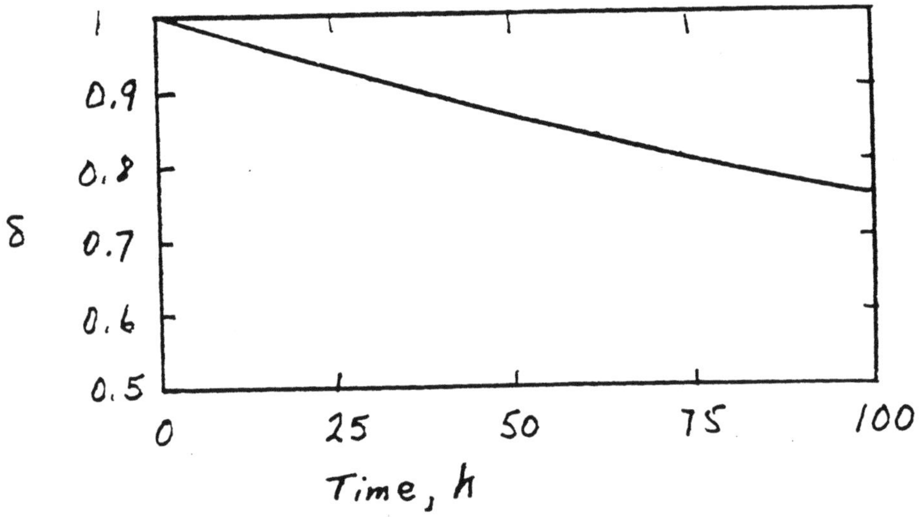

Even after four days at 1350 K, δ is reduced to only 0.75. To effect more homogenization in a reasonable period, a higher temperature would be required.

16.8 Consider radiant heating of a poorly conducting slab as described in Section 16.1.4. Using the same thermal properties and furnace temperature (2020 K), determine the time to heat the center to 2000 K as a function of the thickness of the slab, in the range of 30 mm to 90 mm.

We use the program on pp. 596–597. Line 90 gives Δx for semithickness of 60 mm. We run the following cases:

L (semithickness), m	$\Delta x = L/4$, m	Time, minutes
15×10^{-3}	3.75×10^{-3}	13.28
30×10^{-3}	7.50×10^{-3}	48.36
45×10^{-3}	11.25×10^{-3}	105.40

The time steps, Δt, are not equal in this program. They are calculated in lines 220 – 250.

16.9 Gas is injected into the bottom of a melt contained in a large cylindrical vessel; the recirculating flow is turbulent. Assume axisymmetric flow and write an appropriate momentum equation for the liquid.

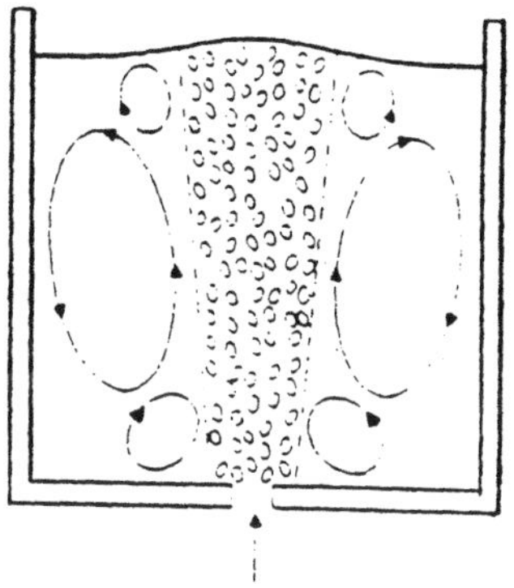

$\bar{v}_\theta = 0 \qquad \partial \bar{v}_r / \partial t = \partial \bar{v}_z / \partial t = 0$

r component: $\tau_{rr} \simeq 0$, $g_r = 0$

$$\bar{v}_r \frac{\partial \bar{v}_r}{\partial r} + \bar{v}_z \frac{\partial \bar{v}_r}{\partial z} = -\frac{1}{\rho}\frac{\partial P}{\partial r} + \frac{\partial}{\partial z}(\nu + \epsilon_M)\frac{\partial \bar{v}_r}{\partial z}$$

z component: $\tau_{zz} \simeq 0$

$$\bar{v}_r \frac{\partial \bar{v}_z}{\partial r} + \bar{v}_z \frac{\partial \bar{v}_z}{\partial z} = -\frac{1}{\rho}\frac{\partial P}{\partial z} + \frac{1}{r}\frac{\partial}{\partial r}\left(r(\nu + \epsilon_M)\frac{\partial \bar{v}_z}{\partial r}\right)$$

These equations were obtained by combining Eqs. (A) and (C), Table 2.3 and Eq. 16.80 (with appropriate change of notation).

9 780470 9237